NATO ASI Series

Advanced Science Institutes Series

A series presenting the results of activities sponsored by the NATO Science Committee, which aims at the dissemination of advanced scientific and technological knowledge, with a view to strengthening links between scientific communities.

The Series is published by an international board of publishers in conjunction with the NATO Scientific Affairs Division

A Life Sciences **B Physics**	Plenum Publishing Corporation London and New York
C Mathematical and Physical Sciences **D Behavioural and Social Sciences** **E Applied Sciences**	Kluwer Academic Publishers Dordrecht, Boston and London
F Computer and Systems Sciences **G Ecological Sciences** **H Cell Biology** **I Global Environmental Change**	Springer-Verlag Berlin Heidelberg New York London Paris Tokyo Hong Kong Barcelona Budapest

PARTNERSHIP SUB-SERIES

1. Disarmament Technologies	Kluwer Academic Publishers
2. Environment	Springer-Verlag/Kluwer Academic Publishers
3. High Technology	Kluwer Academic Publishers
4. Science and Technology Policy	Kluwer Academic Publishers
5. Computer Networking	Kluwer Academic Publishers

The Partnership Sub-Series incorporates activities undertaken in collaboration with NATO's Cooperation Partners, the countries of the CIS and Central and Eastern Europe, in Priority Areas of concern to those countries.

NATO-PCO DATABASE

The electronic index to the NATO ASI Series provides full bibliographical references (with keywords and/or abstracts) to about 50 000 contributions from international scientists published in all sections of the NATO ASI Series. Access to the NATO-PCO DATABASE compiled by the NATO Publication Coordination Office is possible in two ways:

- via online FILE 128 (NATO-PCO DATABASE) hosted by ESRIN, Via Galileo Galilei, I-00044 Frascati, Italy.
- via CD-ROM "NATO Science & Technology Disk" with user-friendly retrieval software in English, French and German (© WTV GmbH and DATAWARE Technologies Inc. 1992).

The CD-ROM can be ordered through any member of the Board of Publishers or through NATO-PCO, Overijse, Belgium.

Series H: Cell Biology, Vol. 99

Springer
Berlin
Heidelberg
New York
Barcelona
Budapest
Hong Kong
London
Milan
Paris
Santa Clara
Singapore
Tokyo

Tumor Biology

Regulation of Cell Growth, Differentiation and Genetics in Cancer

Edited by

Asterios S. Tsiftsoglou

Laboratory of Pharmacology
Department of Pharmaceutical Sciences
Aristotle University of Thessaloniki
Thessaloniki 54006, Greece

Alan C. Sartorelli

Department of Pharmacology
Yale University School of Medicine
333 Cedar Street
New Haven, CT 06510, USA

David E. Housman

Department of Biology
Center for Cancer Research
Massachusetts Institute of Technology
77 Massachusetts Avenue, E17-536
Cambridge, MA 02139, USA

T. Michael Dexter

Department of Experimental Hematology
Christie Hospital, NHS Trust
Paterson Institute for Cancer Research
Wilmslow Road
Manchester M20 9BX, UK

Springer
Published in cooperation with NATO Scientific Affairs Division

Proceedings of the NATO Science Institute "Regulation of Cell Growth, Differentiation and Genetics in Cancer", held at Porto Carras, Halkidiki, Greece, September 1–12, 1995

Library of Congress Cataloging-in-Publication Data

Tumor biology : regulation of cell growth, differentiation, and genetics in cancer / edited by Asterios S. Tsiftsoglou ... [et al.].
p. cm. -- (NATO ASI series. Series H, Cell biology ; vol. 99)
"Proceedings of the NATO Science Institute "Regulation of Cell Growth, Differentiation, and Genetics in Cancer," held at Porto Carras, Halkidiki, Greece, September 1-12, 1995"--T.p. verso.
"Published in cooperation with NATO Scientific Affairs Division."
Includes bibliographical references and index.

1. Cancer cells--Congresses. 2. Cancer--Genetic aspects--Congresses. 3. Cancer--Molecular aspects--Congresses. I. Tsiftsoglou, Asterios S., 1948- . II. NATO Scientific Affairs Organization. Scientific Affairs Division. III. NATO Science Institute "Regulation of Cell Growth, Differentiation, and Genetics in Cancer" (1995 : Chalkidikē, Greece) IV. Series.
[DNLM: 1. Neoplasms--genetics--congresses. 2. Cell Cycle--congresses. 3. Cell Differentiation--congresses. 4. Apoptosis--congresses. QZ 200 T911 1996]
RC254.6.T82 1996
616.99'407--dc20
DNLM/DLC
for Library of Congress 96-32244
CIP

ISBN-13: 978-3-642-64735-2 e-ISBN-13: 978-3-642-61180-3
DOI: 10.1007/978-3-642-61180-3

Softcover reprint of the hardcover 1st edition 1996

Typesetting: Camera ready by authors/editors
Printed on acid-free paper
SPIN 10477178 31/3137 - 5 4 3 2 1 0

FOREWORD

With the aim of providing an international forum for the communication of both the basic and clinical aspects of molecular and cellular biology of cancer, a NATO ASI was held in Porto Carras, Halkidiki, Greece, September 1-12, 1995. The principles as well as recent developments in tumor biology were discussed in depth, with emphasis on the regulation of the cell cycle, differentiation, programmed cell death (apoptosis) and genetics of cancer. This book constitutes the proceedings of that meeting.

Specifically, the following areas were addressed: (a) enzymes and proteins (cyclins) that control the cell cycle, as well as the role of *mos* gene in meiosis and transformation; (b) the structural basis for specificity in protein-tyrosine kinase reactions; (c) the differentiation of normal as well as neoplastic cells with respect to molecular mechanism(s) by which chemical agents or growth factors trigger maturation; (d) phenotypic and genetic aspects of apoptosis; (e) the role of growth factors, like IGF-1, FGF, TN, IL-6, etc., in cell cycle regulation, apoptosis (cell death) and senescence; (f) molecular mechanisms of transcriptional activation of globin genes and stability of mRNAs related to growth proteins and iron metabolism; (g) the cellular and molecular biology of bone marrow hemopoiesis; and (h) neurotrophic factors and the generation of cellular diversity in the central nervous system.

It was obvious from the studies presented that neoplastic cell growth, differentiation and apoptosis in many cell types are regulated at several levels. These include (a) the membrane-mediated events which promote growth via signal transduction pathways and activation of various genes, including oncogenes and mutated tumor suppressor genes (e.g., p53, Rb); (b) transcription factors or inducer-binding proteins that promote expression of a differentiated phenotype and maturation of neoplastic cells; (c) growth factors which activate growth, differentiation and/or cell death in hemopoietic, squamous epithelial cells and neurons; and (d) structural regulatory cis-elements that regulate transcription of globin genes and stability of RNA transcripts.

Some of the lecturers and contributors presented the most recently developed methodology to resolve questions related to signal transduction and transcriptional activation. In addition, emphasis was given to transgenic and gene knockout animal models bearing deletions in certain oncogenes and other genes as valuable systems for examining the role of some genes in tumor development. The role of the p53 gene in the process of apoptosis induced by anticancer agents was discussed. Finally, differentiation therapy with retinoids was presented as an alternative way of treating patients with acute promyelocytic leukemia (APL).

The meeting was attended by selected participants from Belgium, Bulgaria, Canada, Finland, France, Germany, Greece, Israel, Italy, Poland, Romania, Russia, Switzerland, Turkey, United Kingdom and USA. Presentations by the guest lecturers were followed by intense and insightful discussions, which continued following the afternoon sessions (tutorials). Moreover, several interesting contributions were made by participants who presented their work in such areas as differentiation of epithelial cells and megakaryocytes, as well as apoptosis of colon carcinoma cells.

In summary, the participants in the Porto Carras meeting were exposed to both fundamental and more advanced aspects of cancer research, which may have potential application in their work. The Institute also provoked many new questions regarding the regulation of cancer cells. Equally significant was the opportunity for scientific interchange on a personal level, especially between participants from East Europe and those from the NATO countries. We wish to express our appreciation to the NATO Scientific Affairs Division for sponsoring and financing the ASI, as well as the Greek Secretariat of Science and Technology and the Greek National Tourist Organization (Central Macedonia Branch) for their cooperation. We would also like to acknowledge Ms. Brenda Fasnacht for her dedicated effort and efficiency in coordinating the conference.

The Editors

TABLE OF CONTENTS

LAP (NF-IL6) MODULATES HEPATOMA CELL PROLIFERATION

Martina Buck and Mario Chojkier
Department of Medicine and
Center for Molecular Genetics (9111-D)
University of California
San Diego, CA 92161

LAP, a liver-enriched activator protein, in addition to its role as a transactivator of liver-specific genes (Descombes et al., 1990; Poli et al., 1990; Cao et al., 1991; Williams et al., 1991), consistently arrests the progression of the cell cycle from G_1 to S-phase in HepG2 hepatoma cells (Buck et al., 1994). Alternatively, LAP may induce entry of hepatoma cells into G_0. In addition, a relatively low LAP/LIP ratio, as well as the disruption of the LAP leucine zipper, abolishes the inhibitory effect of LAP on hepatoma cell proliferation. By contrast, C/EBPα did not affect HepG2 cell proliferation.

Unlike LAP, C/EBPα did not block HepG2 cell proliferation either in transient, or in stable expression experiments. Both LAP and C/EBPα transactivate, as expected, an albumin promoter-CAT reporter gene. Although LAP and C/EBPα show an indistinguishable DNA-binding specificity *in vitro* and readily heterodimerize, we suggested that by virtue of their divergent N-terminal domains LAP and C/EBPα would be likely to affect different target genes (Descombes et al., 1990). Although it is of great interest that LAP and C/EBPα may have different roles on hepatic cell proliferation, other examples of their dissimilar biological functions have been found (Wegner et al., 1992; Poli et al., 1990; Houglum et al., 1994).

In a series of experiments, we assessed the effect of various LAP constructs on the cell cycle in order to characterize the structure/functional relationship. As expected, a deletion of 21 amino acids from the N-terminal domain, resulted in a protein expressed in normal liver (Descombes et al., 1990; Descombes and Schibler, 1991), and did not affect the inhibitory activity of LAP upon the cell cycle. Similarly, mutation of 4 amino acids on the basic DNA binding domain ($R^{238} \rightarrow G$, $K^{239} \rightarrow T$, $S^{240} \rightarrow P$ and $R^{241} \rightarrow G$), which inhibits binding to the cognate D-site of the albumin promoter did not prevent the inhibition of cell proliferation by LAP. By contrast, an intact DNA binding domain is required for C/EBPα to arrest adipocyte proliferation (Umek et al., 1991), suggesting that different mechanisms may be responsible for the inhibition of hepatoma cell and adipocyte proliferation by LAP and C/EBPα, respectively.

NATO ASI Series, Vol. H 99
Tumor Biology
Regulation of Cell Growth, Differentiation and
Genetics in Cancer
Edited by Asterios S. Tsiftsoglou, Alan C. Sartorelli,
David E. Housman and T. Michael Dexter

On the other hand, LAP_{LZ}, a leucine zipper mutant, which also exhibits impaired binding to the D-site, did not alter the rates of cell entry into S-phase, suggesting that integrity of the leucine zipper may be crucial for the role of LAP upon cell proliferation (Buck et al., 1994). The interaction of LAP with C/EBPα and c-Jun, has been previously documented (Descombes et al., 1990; Williams et al., 1991; Cao et al., 1991; Hsu et al., 1994).

Given that LAP arrests proliferation of HepG2 cells, one would predict that it might inhibit the expression of c-*jun*. Indeed, we found that LAP markedly decreases the transcription from the c-*jun* promoter. These preliminary experiments suggest that LAP inhibits c-*jun* expression indirectly, and at least in part, through the formation of LAP/c-Jun dimers. Whether this inhibition of c-*jun* expression occurs *in vivo* remains to be elucidated. However, we (Leffert et al., 1990) and others (Alcorn et al., 1990) have shown that c-*jun* expression is negligible in differentiated quiescent hepatocytes of adult rats, when LAP is maximally expressed (Descombes and Schibler, 1991). Conversely, after partial hepatectomy (Alcorn et al, 1990), or treatment with transforming growth factor-α (Leffert et al., 1990), c-*jun* expression is increased and hepatocyte proliferation is induced. These experiments suggest that the relative expression of LAP and c-Jun may determine whether the hepatocyte is in a state of quiescence or proliferation. Similarly, Rb reduces the AP-1 transcriptional activity through negative regulation of *c-fos* expression (Robbins et al., 1990).

The striking inhibitory competition of LIP on LAP suppression of cell proliferation is in agreement with the reported interaction of LAP/LIP on the albumin promoter site D (Descombes and Schibler, 1991). The inhibitory competitive domain seems circumscribed essentially to the DNA-binding and leucine zipper region. On the other hand, although the leucine zipper domain is necessary to exert the suppressive effects of LAP on the cell cycle, it is not sufficient since additional domains within the activation domain region are also needed to achieve this effect. It would be of interest to assess whether the LAP/LIP ratio modulates hepatocyte proliferation during prenatal development or following partial hepatectomy.

Alternately, as it has been shown for Rb and p53 (Marshall, 1991), mutations of indispensable domains may explain why LAP fails to exert an effect on the cell cycle in some tumor cell lines. Possible relevant mutations may include phosphoacceptor domains which may be important to modulate post-transcriptionally the function of LAP as an inhibitor of cell proliferation. For example, it has been shown that mitogenic stimuli induce Rb phosphorylation while differentiation signals activate Rb dephosphorylation (DeCaprio et al., 1988; Chen et al., 1989). Furthermore, only the underphosphorylated, active form of Rb is present in the nucleus. In late G_1 or early S-phase, dephosphorylation and nuclear dissociation occurs (Mittnacht and Weinberg,

1991). Likewise, activation of signal transduction pathways may result in the phosphorylation of LAP, as we suggested previously (Descombes et al., 1990). In rat pheochromocytoma PC12 cells, cAMP-mediated phosphorylation of LAP is associated with nuclear translocation and induction of c-*fos* transcription (Metz and Ziff, 1991). In pituitary G/C cells, calcium-regulated phosphorylation within the leucine zipper of LAP stimulates transcription from a Ca^{2+}-calmodulin-dependent protein kinase II responsive element (Wegner et al., 1992). We have demonstrated that activation of protein kinase C pathway stimulates phosphorylation of LAP on Ser^{105}, which enhances its transcriptional activity (Trautwein et al., 1993). It remains to be determined whether the phosphorylation state of the activation domain of LAP may modulate progression of the hepatic cell cycle, in response to signal transduction pathways activated during hepatic regeneration.

Although under normal conditions the liver is quiescent, it has the extraordinary ability to proliferate following hepatocellular necrosis or partial hepatectomy (Michalopoulos, 1990). Our results suggesting that LAP is a modulator of hepatic proliferation may provide insights into this biological enigma.

References

Alcorn, J., Feitleberg, S.P., and Brenner, D.A. (1990). *Hepatology*. **11**, 909-915.
Buck M, Turler H, and Chojkier M. (1994). *EMBO J.* **13**, 851-860.
Cao, Z., Umek, R.M., and McKnight, S.L. (1991). *Genes Dev*. **5**, 1538-1552.
Chen, P.-L., Scully, P., Shew, J.-Y., Wang, J.Y.J., and Lee, W.-H. (1989). *Cell*. **58**, 1193-1198.
DeCaprio, J.A., Ludlow, J.W., Figge, J., Shew, J.-Y., Huang, C.-M., Lee, W.-H., Marsilio, E., Paucha, E., and Livingston, D.W. (1988). *Cell*. **54**, 275-283.
Descombes, P., Chojkier, M., Lichtsteiner, S., Falvey, E., and Schibler, U. (1990). *Genes Dev*. **4**, 1541- 1551.
Descombes, P., and Schibler, U. (1991). *Cell*. **67**, 569-579.
Houglum, K., M. Buck, V. Adir, and M. Chojkier. (1994). *J Clin Invest*. **94**, 808-814.
Hsu, W., Kerppola, T.K., Chen, P.L., Curran, T., Chen-Kiang, S. (1994). *Mol. Cell Biol.* **14**, 268-276.
Leffert, H.L., Koch, K.S., Lu, X.P., Brenner, D.A., Karin, M., Skelly, H.F., Rippe, R.A., Fey, G., and Chojkier, M. (1990). The Biology of Medicine and Signal Transduction (Ed. Nishizuka et al.) Raven Press, New York.
Marshall, C.J. (1991). *Cell*. **64**, 313-326.
Metz, R., and Ziff, E. (1991). *Genes and Dev*. **5**, 1754-1766.
Michalopoulos, G.K. (1990). *FASEB J.* **4**, 176-187.
Mittnacht, S., and Weinberg, R.A. (1991). *Cell*. **65**, 381-393.
Poli, V., Mancini, F.P., and Cortese, R. (1990). *Cell*. **63**, 643-653.
Robbins, P.D., Horowitz, J.M., Mulligan, R.C. (1990). *Nature*. **346**, 668-671.
Trautwein, C., Caelles, C., van der Geer P., Hunter T., Karin, M. and Chojkier, M. (1993). *Nature*. **364**,544-547.
Umek, R.M., Friedman, A.D., and McKnight, S.L. (1991). *Science*. **251**, 288-291.
Wegner, M., Cao, Z., Rosenfeld, M.G. (1992). *Science*. **256**, 370-373.
Williams, S.C., Cantwell, C.A., and Johnson, P.F. (1991). *Genes Dev*. **5**, 1553-1567.

The Structural Basis for Specificity in Protein-Tyrosine Kinase Signaling

Lewis C. Cantley and Zhou Songyang
Department of Cell Biology
Harvard Medical School and
Division of Signal Transduction
Beth Israel Hospital
200 Longwood Ave.
Boston, MA 02115

Introduction

Mammalian cells make decisions to divide, differentiate, adhere, chemotax, secrete, quiesce or apoptose on the basis of their developmental history, extracellular matrix composition, and the presence or absence of growth factors/differentiatiation agents and other cellular activators. From extensive research over the past ten years it has become clear that these decisions are mediated in most cases by the activation of specific protein kinases in the cytosol of the cell. A typical mammalian cell expresses hundreds of different protein-Ser/Thr kinases and protein-Tyr kinases. Thus, in order for a cellular activator to elicit a specific cellular response it is necessary that a specific subset of these kinases be activated and that these kinases find specific targets to phosphorylate. This manuscript focuses on a new approach to determining the structural basis for how protein kinases find their specific targets in the cell interior. The primary focus is on protein-Tyr kinases.

Research over the past 10 years has led to the conclusion that growth factor/cytokine-stimulation of cell division is mediated by protein-tyrosine kinases (Hunter and Cooper, 1985; Cantley et al., 1991; Pawson, 1995). In many cases the receptors for growth factors have been shown to have endogenous protein-tyrosine kinase activity. In other cases the growth factor receptors associate with cytosolic protein-tyrosine kinases. Binding of growth factors to their receptors usually results in the dimerization of the receptors such that the

NATO ASI Series, Vol. H 99
Tumor Biology
Regulation of Cell Growth, Differentiation and Genetics in Cancer
Edited by Asterios S. Tsiftsoglou, Alan C. Sartorelli, David E. Housman and T. Michael Dexter

kinase domains of the homodimer transphosphorylate and activate each other. As a consequence, tyrosine-phosphorylation of the the growth factor receptors occurs very quickly following growth factor binding. In addition, the activated growth factor receptor protein-tyrosine kinases phosphorylate other cytosolic and plasma membrane proteins that ultimately transmit signals to the cell interior (Pawson and Schlessinger, 1993).

SH2 Domains

A major advance in understanding how receptor protein-Tyr kinases find specific targets to activate was provided by the discovery of src-homology 2 (SH2) domains. The SH2 domain was first elucidated by Pawson's laboratory as a ~100 amino acid regulatory region of the fps oncogene product that had significant homology to a non-catalytic region of $pp60^{c\text{-}src}$ (Pawson, 1988). Mutations in this domain were shown to affect the activity of the fps protein-Tyr kinase. Later, Hanafusa's laboratory discovered a retrovirus-encoded oncogene, v-crk) that encoded an SH2 domain but lacked a protein-Tyr kinase domain (Mayer et al., 1988). The function of this domain became clear when it was shown by both Pawson's laboratory and by Hanafusa's laboratory that this domain is capable of directly binding to tyrosine-phosphorylated proteins and that the binding involved the phosphoTyr moiety (Anderson et al., 1990; Mayer et al., 1991). These results led to a model in which phosphorylation of receptors on Tyr residues leads to the recruitment of SH2-containing proteins to the receptors (Cantley et al., 1991). Indeed, most of the proteins that have been found to bind to activated growth factor receptors contain SH2 domains or bind to adaptor proteins with SH2 domains. To date more that 50 SH2-containing proteins have been discovered in mammalian cells and many of these proteins have been shown to mediate downstream signaling cascades such as hydrolysis of PtdIns-4,5-P_2, synthesis of PtdIns-3,4,5-P_3, and activation of ras.

The discovery of SH2-phosphoTyr complexes provided a biochemical explanation for why certain cytosolic proteins are recruited to receptors, but left unanswered the question of why only a subgroup of SH2-containing proteins is recruited to a specific receptor. A possible explanation for this specificity came from comparing the primary sequences around a phosphoTyr site in polyoma middle t required for binding to the SH2 domains of Phosphoinositide 3-kinase (PI3K) and phosphoTyr sites on the PDGF receptor also required for binding PI3K (Cantley et al., 1991). These sites had the consensus sequence phosphoTyr-Met/Val-Xxx-Met.

A peptide library for studying SH2 Domain specificity

In order to test the idea that individual SH2 domains recognize phosphoTyr in a specific sequence context, we developed a new "oriented peptide library" approach (Songyang et al., 1993). We synthesized a mixture of soluble peptides with the sequence Gly-Asp-Gly-phosphoTyr-Xxx-Xxx-Xxx-Ser-Pro-Leu-Leu-Leu where Xxx indicates all amino acids but Cys and Trp. The total degeneracy of this library is $18^3 = 5832$. The strategy we chose was to make an affinity column out of the SH2 domain of interest and use the column to purify the subgroup of phosphopeptides that have highest affinity. The peptides are then specifically eluted with a high concentration of phenylphosphate (which competes with the phosphoTyr moiety of the peptide for binding to the SH2 domain). Rather than isolate individual phosphopeptides and sequence them one at a time, we then sequence the entire batch of affinity-purified peptides. The reason this approach can work is that all peptides in the library are of the same length and have the same sequence except for the three residues C-terminal of the phosphoTyr. Thus, the fifth residue in the sequence must be bound to the SH2 domain at a location immediately adjacent to the phosphoTyr binding pocket (pY+1 position). If the SH2 domain has a binding pocket at this position with specificity for a particular amino acid then those peptides that have the preferred amino acid at position 5 in the sequence will be preferentially retained. Thus, a comparison of the amino acids at cycle 5 in the sequence of the affinity-purified peptides to the amino acids at cycle 5 of the starting mixture indicates the SH2 domain preference for acids at cycle 5 of the starting mixture indicates the SH2 domain preference for the pY+1 position of the phosphopeptide.

When the N-terminal SH2 domain of the 85 kd subunit of PI3K was used as the affinity column, there was a preferential retention of peptides that contain Met or Val at the pY+1 position and for peptides that contain Met at the pY+3 position. Thus, the technique predicted that the optimal peptide for binding to this domain has the sequence pY-Met/Val-Xxx-Met, in good agreement with the sequences at the sites in polyoma middle t and the PDGF receptor where PI3K binds in vivo (see above).

This same approach was then used to investigate numerous SH2 domains and show that different SH2 domains prefer different optimal motifs (Songyang et al., 1993; Songyang et al., 1994). For example, the src SH2 domain (and other members of the src subfamily such as lck and fyn) preferentially retained peptides with the motif phosphoTyr-Glu-Glu-Ile. The optimal phosphopeptides were then synthesized and shown to bind to the SH2 domains with dissociation constants in the range of 0.1 to 1 micromolar.

The Structural Basis for SH2 domain Specificity

While this work was in progress, John Kuriyan's laboratory was crystallizing the SH2 domain of src and Stephen Harrison's laboratory was crystallizing the SH2 domain of lck. Both laboratories obtained crystals of these SH2 domains bound to the optimal phosphoTyr-Glu-Glu-Ile peptide that was determined by our library approach (Eck et al., 1993; Waksman et al., 1993). The crystal structures were quite similar and provided a structural explanation for the specificity observed. For example, the Ile residue at the pY+3 position was shown to fit into a hydrophobic pocket in the SH2 domain. The Glu at the pY+1 position interacted with a Tyr residue (Y-βD5) and a Lys residue (K-βD3) on the surface. The Glu at the pY+2 position also lay on the surface of the SH2 domain and interacted through a water molecule with an Arg residue (R-βD'1). Significantly, the residues that made up the contacts with the side chains of the associated phosphopeptide were conserved between src and lck, explaining why both SH2 domains selected for the same optimal phosphopeptide motif.

When the primary sequences of other SH2 domains were compared to those of src and lck, it was apparent that the residues that make up the contact sites for the side chains of associated phosphopeptides are not generally conserved. Table I below provides the identities of the residues at the phosphopeptide contact site for some of the SH2 domains we have studied. We have divided these SH2 domains into 4 major groups on the basis of the identity of the residue at the βD5 position. This residue is of particular importance in determining specificity of the src SH2 domain since on one side it contacts the aliphatic part of the pY+1 Glu and on the other side it forms a wall of the pY+3 pocket. The relatively large size of the Tyr side chain is essential for this function and all SH2 domains with Tyr or Phe at the βD5 position are in group I. We have noticed that most of these SH2 domains select for peptides with the general motif phosphoTyr-hydrophilic-hydrophilic-hydrophobic. The selection for hydrophilic amino acids at the pY+1 and pY+2 positions can be explained by the fact that the bulky aromatic side chain at the βD5 position forces this region of the associated peptide to sit on the surface of the SH2 domain (Songyang et al., 1995b).

SH2 domains that lack aromatic amino acids at the βD5 position select the general motif pY-hydrophobic-Xxx-hydrophobic (where Xxx indicates little selectivity). To date, vav is the only SH2 domain found with Thr at the βD5 position and it is placed in a group by itself (group II). A large number of SH2 domains have Ile, Val, Cys or Leu at the βD5 position and these domains have some common features and are placed in group III. The crystal or solution structures of some of these domains have now been determined (Pascal et al., 1994;Lee et al., 1994). They differ from the src and lck SH2 domains in that the βD5 residue is buried in the SH2 domain. As a consequence the pY+1 residue of the associated phosphopeptide fits deeper into a cavity, explaining the selection for hydrophobic amino acids at that position. The side chain of the pY+2 residue extends into the aqueous medium, explaining why there is little selectivity at this

TABLE I PHOSPHOPEPTIDE MOTIFS FOR SH2 DOMAINS: RESIDUES PREDICTED TO INTERACT WITH THE SIDE CHAINS OF THE ASSOCIATED PHOSPHOPEPTIDES

SH2 DOMAIN	+1	SRC-SH2 200 βD3	202 βD5	+2	SRC-SH2 205 βD'1	+3	SRC-SH2 202 βD5	214 βE4	215 EF1	230 αB9	237 BG4	
SRC	E	K	Y	E	R	**I**	Y	I	T	Y	L	
FYN	**E**	K	Y	**E**	R	**I**	Y	I	T	Y	L	
LCK	**E**	K	Y	E	R	**I**	Y	I	S	Y	L	GROUP 1A
FGR	**E**	K	Y	**E**	R	IV	Y	I	T	Y	L	
SYK N	TI	H	Y	-	E	**ILM**	Y	I	S	H	L	
SYK C	QIE	L	Y	eqt	D	**L**	Y	I	P	Y	L	
ITK	E	K	Y	NY	K	vyi	Y	V	A	H	L	
ABL	E	Y	Y	**N**	N	P	Y	V	S	H	L	GROUP 1B
ARG		Y	Y		N		Y	V	T	H	L	
CSK	**T**	E	Y	K	M	MIVr	Y	I	D	Y	?	
CRK	D	S	Y	H	N	**P**	Y	A	G	Y	T?	
NCK	**D**	K	F	**E**	Q	**P**	F	I	G	Y	T?	
fes/fps	**E**	R	F	-	Q	vi	F	R	L	L	G	
SEM5	LV	Q	F	**N**	L	vp	F	L	W	H	R?	
DGBR2	y	Q	F	**N**	L	**-**	F	L	W	H	R?	
GRB2	qy	Q	F	**N**	L	y	F	L	W	H	R?	
TENSIN	**E**	R	F	N	T	ifv	F	?	?	H	?	
3BP2	**E**	R	Y	**N**	F	-	Y	E	G	Y	P?	
VAV	**M**	K	T	E	I	P	T	I	T	Y	?	GROUP 2
p85aN	MIVE	K	I	-	F	**M**	I	F	S	Y	A?	
p85aC	mli	K	C	-	N	**M**	C	F	A	Y	V?	
PLC g1C	VI	K	C	IL	N	**PIV**	C	L	G	Y	Y?	
PLC g1N	**LIV**	Q	C	Ei	H	LIV	C	K	F	Y	L?	GROUP 3
HCP N	F	T	I	-	Q	**Flyp**	I	D	L	Y	V	
SHPTP2 N	IV	T	I	-	Q	VI	I	D	L	Y	L	
SHC	EI	K	L	-	V	**ILM**	L	T	K	H	P?	
ShB	T	M	M		A	**L**	M	L	G	Y	?	

Columns +1, +2 and +3 comprise the 1st, 2nd and 3rd residue C-terminal to P-tyrosine of the optimal phosphopeptide selected by each SH2 domain (eg. P-YEEI for src SH2). SRC-SH2 200 and 202 indicate the residues of src (and residues at analogous positions of other SH2 domains) predicted to contact the +1 residue side chain of the associate peptide. SRC-SH2 205 is predicted to be near the +2 side chain and SRC 202, 214, 215 and 237 are predicted to form a hydrophobic pocket to bind the +3 residue side chain. The alignments were made on the basis of Waksman et al. (1992)

Bold letters indicate strong selection. Upper case without bold indicates medium selection. Lower case indiates weak selection. A hyphen indiates no selection. Motifs not yet determined or not submitted for publication are left blank.

position. In addition, the group III SH2 domains show selectivity at the pY+5 position due to extension of the open cavity beyond the binding site of at the pY+3 position.

In summary, the peptide library results combined with the SH2 domain structures have provided an explanation for why different receptors specifically bind and activate different SH2-containing protein. These studies have led to the realization that the primary sequence surrounding the Tyr phosphorylation sites on receptors determine which downstream signals will be activated.

Substrate Specificities of the Catalytic Sites of Protein-Tyrosine kinases

Although the identification of SH2 domain specificities has provided a partial explanation for how protein kinases maintain fidelity in downstream signaling, it is clear that additional factors are involved. In many cases, elimination of autophosphorylation sites (or transphosphorylation sites) on receptors does not eliminate cellular responses. These findings suggest that catalytic site specificity is also important. In order to investigate this idea, we modified our "oriented peptide library" approach to investigate the substrate specificity of protein kinases (Songyang et al., 1994).

A peptide library was constructed with the sequence Met-Ala-Xxx-Xxx-Xxx-Xxx-Tyr-Xxx-Xxx-Xxx-Xxx-Ala-Lys-Lys-Lys where Xxx indicates all amino acids but Ser, Thr, Tyr, Cys and Trp (Songyang et al., 1995a). The total degeneracy of this library is 15^8 = ~2.5 billion. The protein kinase of interest is allowed to phosphorylate about 0.5% of the total peptide mixture (assayed by carrying out the reaction with a trace of [γ-^{32}P]-ATP and spotting the mixture on phosphocellulose). The peptide mixture is then added to a ferric-IDA column (Pierce) that binds phosphopeptides with high affinity (Fig. 1). The column is extensively washed to elute non-phosphorylated peptides and then the phosphopeptides are eluted at high pH and high ionic strength. This procedure provides a collection of phosphopeptide products that were preferentially

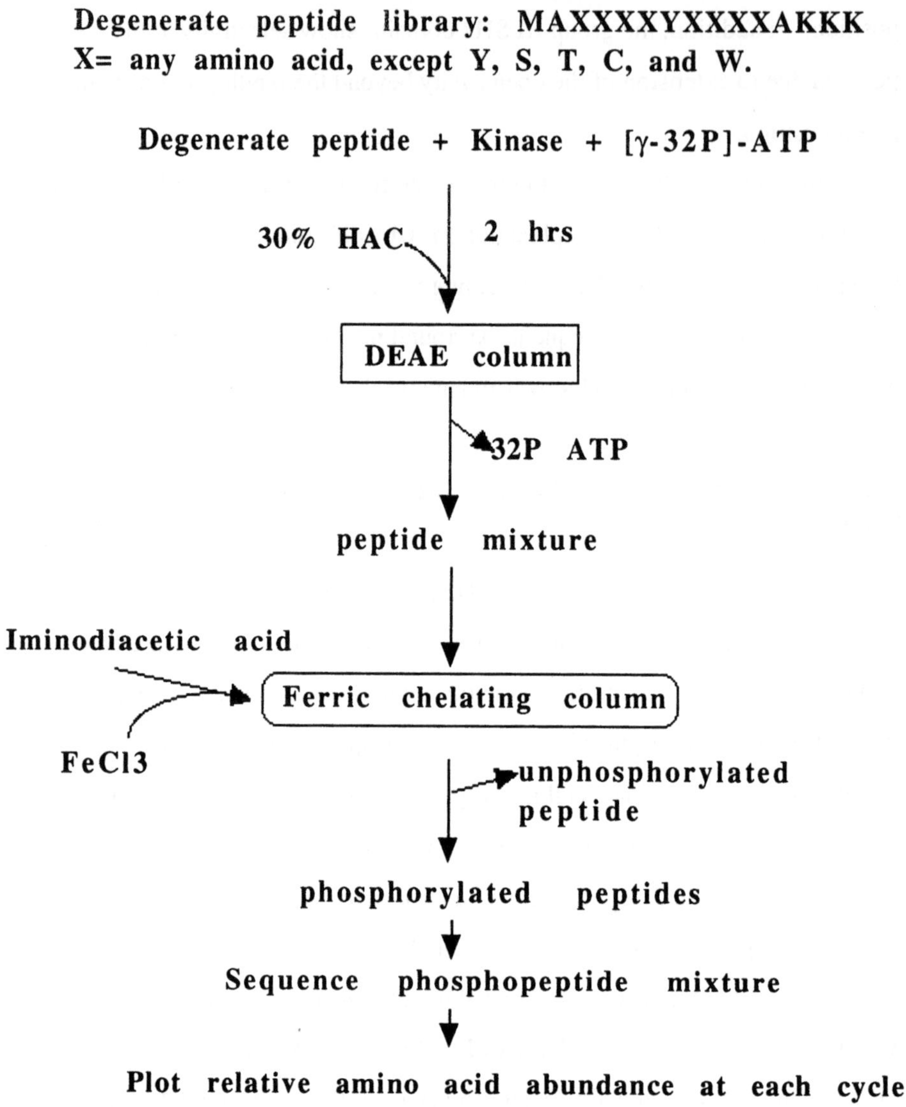

Figure 1: An oriented peptide library technique for determining protein kinase specificity

Table II Optimal Substrate Sequences Recognized by Different Protein-tyrosine Kinases:

Tyrosine Kinases	Position -4	-3	-2	-1	0	+1	+2	+3	+4
c-fps/fes	**E**	**E**	**E**	**I**	**Y**	**E**	**E**	**I**	**E**
MT/c-src	D	**E**	**E**	**I**	**Y**	**G/E**	E	**F**	F
v-src	E	**E**	**E**	**I**	**Y**	**G/E**	**E**	**F**	**D**
lck	X	E	X	**I**	**Y**	**G**	V	**L**	F
c-abl	A	X	V	**I**	**Y**	**A**	A	**P**	F
EGF Receptor	E	**E**	**E**	**E**	**Y**	**F**	E	L	V
PDGF receptor	E	**E**	**E**	**E**	**Y**	**V**	F	I	X
FGF recptor	**A**	E	E	E	**Y**	F	**F**	L	F
Insulin receptor	X	E	E	E	**Y**	**M**	**M**	**M**	M

phosphorylated by the kinase compared to the bulk of non-phosphorylated peptides. This entire mixture is then sequenced and the amino acid abundance at each of the positions of degeneracy is compared to the abundance of amino acids in the starting mixture. Since there is only a single Tyr in each peptide at position 7, the amino acid selected at cycle 8 will indicate the amino acid preference of the kinase for the Y+1 position.

We have now used this technique to study the specificity of a number of protein-Tyr kinases and found that each kinase has a different optimal motif (Songyang et al., 1995a). Some of the results are summarized in Table II. We have also used a variation on this technique (with Ser rather than Tyr at position 7) to study the specificity of protein-Ser/Thr kinases (Songyang and Cantley, unpublished). These studies have led us to conclude that protein-Tyr kinases (and protein-Ser/Thr kinases) have much greater specifities for short linear sequences that was previously appreciated.

In considering the protein-Tyr kinase specifities, one feature that emerged is that src-like protein-Tyr kinases prefer substrates with the general motif Glu-Glu-Ile-Tyr-hydrophilic-hydrophilic-hydrophobic while receptor-type protein-Tyr kinase preferentially phosphorylate substrates with the general motif Glu-Glu-Glu-Tyr-hydrophobic-Xxx-hydrophobic. These motifs are reminiscent of the optimal motifs for group I and group III SH2 domains respectively. Thus, these studies suggest that group I SH2 domains and the catalytic cleft of src-like protein-Tyr kinases have converged to recognize similar optimal motifs. In contrast, group III SH2 domains and catalytic clefts of receptor-protein-Tyr kinases have converged to recognize common motifs.

The fact that src-like protein-Tyr kinases prefer to phosphorylate substrates favorably recognized by Group I SH2 domains (including their own SH2 domains), implied that these protein-Tyr kinases may processively phosphorylate their substrates (Mayer et al., 1995; Songyang and Cantley, 1995). When multiple tyrosine residues of a protein substrate are to be phosphorylated, the src-like protein-Tyr kinase will first phosphorylate the optimal site. The phosphorylated sequence then anchors the SH2 domain of the kinase. As a consequence, other less optimal sites on the substrate are sequentially phosphorylated.

In summary, the peptide library technique has enabled us to systematically study the substrate specificities of protein-Tyr kinases. These studies indicated a close functional relationship between the specificities of both the SH2 and the kinase domains. Thus generation of specific signals for growth and differentiation may well be a result of double selection by the SH2 and the kinase domains. Since the primary sequence specificities of these domains can be defined, it is possible now to predict their in vivo targets. Therefore, the peptide library technique will also provide a shortcut to ping-pong signal transduction networks.

Reference

Anderson, D., Koch, C. A., Grey, L., Ellis, C., Moran, M. F., and Pawson, T. (1990). Binding of SH2 domains of phospholipase C gamma 1, GAP, and Src to activated growth factor receptors. Science *250*, 979-82.

Cantley, L. C., Auger, K. R., Carpenter, C. L., Duckworth, B., Graziani, A., Kapeller, R., and Soltoff, S. (1991). Oncogenes and Signal Transduction. Cell *64*, 281-302.

Eck, M. J., Shoelson, S. E., and Harrison, S. C. (1993). Recognition of a high-affinity phosphotyrosyl peptide by the Src homology-2 domain of p56lck. Nature *362*, 87-91.

Hunter, T., and Cooper, J. A. (1985). Protein-tyrosine kinases. Ann. Rev. Biochem. *54*, 897-930.

Lee, C. H., Kominos, D., Jacques, S., Margolis, B., Schlessinger, J., Shoelson, S. E., and Kuriyan, a. J. (1994). Crystal structures of peptide complexes of the amino-terminal SH2 domain of the Syp tyrosine phosphatase. Structure *2*, 423-38.

Mayer, B. J., Hamaguchi, M., and Hanafusa, H. (1988). Characterization of p47gag-crk, a novel oncogene product with sequence similarity to a putative modulatory domain of protein-tyrosine kinases and phospholipase C. Cold Spring Harb Symp Quant Biol *2*, 907-14.

Mayer, B. J., Hirai, H., and Sakai, R. (1995). SH2 domains and processive phosphorylation. Current Biology *5*, 296-305.

Mayer, B. J., Jackson, P. K., and Baltimore, D. (1991). The noncatalytic src homology region 2 segment of *abl* tyrosine kinase binds to tyrosine-phosphorylated cellular proteins with high affinity. Proc. Natl. Acad. Sci. *88*, 627-631.

Pascal, S. M., Singer, A. U., Gish, G., Yamazaki, T., Shoelson, S. E., Pawson, T., Kay, L. E., and Forman Kay, J. D. (1994). Nuclear magnetic resonance structure of an SH2 domain of phospholipase C-gamma 1 complexed with a high affinity binding peptide. Cell *77*, 461-72.

Pawson, T. (1988). Non-catalytic domains of cytoplasmic protein-tyrosine kinases: regulatory elements in signal transduction. Oncogene *3*, 491-5.

Pawson, T. (1995). Protein modules and signalling networks. [Review]. Nature *373*, 573-80.

Pawson, T., and Schlessinger, J. (1993). SH2 and SH3 domains. Current Biology *3*, 434-442.

Songyang, Z., Shoelson, S. E., Chaudhuri, M., Gish, G., Pawson, T., Haser, W. G., King, F., Roberts, T., Ratnofsky, S., Lechleider, R. J., and B. G. Neel, R. B. B., J. E. Fajardo, M. M. Chou, H. Hanafusa, B. Schaffhausen and L. C. Cantley. (1993). SH2 domains recognize specific phosphopeptide sequences. Cell *72*, 767-78.

Songyang, Z., Shoelson, S. E., McGlade, J., Olivier, P., Pawson, T., Bustelo, R. X., Barbacid, M., Sabe, H., Hanafusa, H., Yi, T., Ren, R., Baltimore, D., Ratnofsky, S., Feldman, R. A., and Cantley, L. C. (1994). Specific Motifs Recognized by the SH2 Domains of Csk, 3BP2, fps/fes, GRB-2, HCP, SHC, Syk and Vav. Mol. Cell. Biol. *14*, 2777-85.

Songyang, Z., Blechner, S., Hoagland, N., Hoekstra, M. F., Piwnica, W. H., and Cantley, L. C. (1994). Use of an oriented peptide library to determine the optimal substrates of protein kinases. Current Biology *4*, 973-82.

Songyang, Z., Carraway III, K. L., Eck, M. J., Harrison, S. C., Feldman, R. A., Mohammadi, M., Schlessinger, J., Hubbard, S. R., Smith, D. P., Eng, E., Lorenzo, M. J., Ponder, B. A. J., Mayer, B. J., and Cantley, L. C. (1995a). Catalytic specificity of protein-tyrosine kinases is critical for selective signaling. Nature *373*, 536-539.

Songyang, Z., Gish, G., Mbamalu, G., Pawson, T., and Cantley, L. C. (1995b). A single point mutation switches the specificity of Groop III SH2 domains to that of Group I SH2 domains. J. Biol. Chem. *270*, 26029-32.

Songyang Z. and L. C. Cantley. (1995) Recognition and specificity in protein tyrosine kinase-mediated signalling. TIBS (in press).

Waksman, G., Shoelson, S. E., Pant, N., Cowburn, D., and Kuriyan, J. (1993). Binding of a high affinity phosphotyrosyl peptide to the Src SH2 domain: crystal structures of the complexed and peptide-free forms. Cell *72*, 779-90.

PML Is a Primary Target Gene of Interferon and Could Mediate Some of Its Biological Activities

Mounira. K. Chelbi-Alix, Luis Pelicano, Frederique Quignon, Marcel Koken and Hugue de Thé
UPR 43 CNRS Hôpital Saint-Louis
1 avenue Claude Vellefaux
75475 Paris CEDEX 10
France

Introduction

Interferons (IFNs) are a family of secreted proteins with antiviral, antiproliferative and immunomodulatory activities. On the basis of their antigenic properties, they are classified as three major groups α, β, and γ. IFNs α and β (type I) are produced by many cell types upon viral infection, double stranded RNA treatment, or other stimuli, whereas IFN γ (type II) is induced in T lymphocytes or natural killer cells in response to antigens or mitogens (Chelbi-Alix *et al.*, 1994; Chelbi-Alix and Chousterman, 1992; Pestka *et al.*, 1987; Sen and Ransohoff, 1993). After interacting with their cognate cell surface receptors, IFNs α/β and IFN γ activate different signal transduction cascades leading to the transcription of distinct sets of genes which mediate the biological effects of these cytokines. The IFN α/β- and the IFN γ-inducible gene promoters are characterized by the presence of consensus elements the interferon-stimulated response element (ISRE) and the gamma activated sequence (GAS) respectively. Binding of both IFN α/β and IFN γ results in the differential activation of latent cytoplasmic transcription factors termed STATS (Signal Transducer and Activator of Transcription) that act at different cis-acting DNA elements. Tyk2 and JAK1 kinases play a central roles in the induction of IFN α/β responsive genes and JAK1 and JAK2 in that of IFN γ (Pelligrini and Schindler, 1993). In response to IFN α/β, all the three proteins viz, Stat-113, Stat-91, and Stat-84 (the last two are generated from the same gene by alternative splicing) are activated by phosphorylation to form the transcriptionally active IFN-stimulated gene factor 3 (ISGF3) by association with a 48-kDa protein. ISGF3 then translocates to the nucleus and binds to the ISRE. In response to IFN γ, one of the ISGF3 components, Stat-91, is activated which then forms homodimers, translocates to nucleus and binds

NATO ASI Series, Vol. H 99
Tumor Biology
Regulation of Cell Growth, Differentiation and Genetics in Cancer
Edited by Asterios S. Tsiftsoglou, Alan C. Sartorelli, David E. Housman and T. Michael Dexter

to the GAS. These events lead to transcriptional activation (Schindler and Darnell Jr., 1995). More than 40 human proteins are known to be induced by IFN, but the physiological role of the majority of them has not yet been recognized. A few of the IFN-induced proteins were shown to have intrinsic antiviral and /or anti proliferative activities namely the p68 protein kinase, the 2'5' oligoadenylate (2'5'A) synthetase and certain Mx family proteins (Sen and Ransohoff, 1993; Staeheli, 1990; Staeheli *et al.*, 1993). The mechanism of action of the p68 kinase and that of 2'5'A synthetase are known but that of Mx protein has to be elucidated(Staeheli *et al.*, 1993). The p68 kinase is a serine-threonine-specific kinase which is activated by autophosphorylation and able to phosphorylate the α subunit of the eucaryotic translation initiation factor eIF2, a modification that causes inhibition of protein synthesis (Hovanessian, 1993; Sen and Ransohoff, 1993). The 2'5'A synthetase, after being activated by dsRNA, catalyses the synthesis, from ATP, of the unusual 2'5' linked oligoadenylates (2'5'A), with a general formula pppA (2'p 5'A) n, where n is 1 to 4. The 2'5'A in turn activate a latent cellular endonuclease which degrades mRNAs. The 2'5'A is therefore a potent inhibitor of translation, and the 2'5'A system, i.e. 2'5'A synthetase, 2'5'A and the 2'5'A-dependent endonuclease, thus forms part of the cascade of events implicated in the antiproliferative and antiviral actions of IFN (Sen and Ransohoff, 1993).

Structure of PML in normal cells and Acute Promyelocytic Leukaemia

The PML (ProMyelocytic Leukaemia) protein shares a C3HC4 (RING finger) zinc binding motif with a large group of polypeptides which perform heterogeneous functions ranging from transactivation of viral genes to DNA repair or peroxisome assembly (reviewed in Barlow *et al.*, 1994; Freemont, 1993). PML belongs to a subfamily of nine proteins defined by the additional presence of one or two other cysteine-rich motifs as well as a coiled-coil region (Reddy *et al.*, 1992) which is implicated in homo- and hetero- dimerisation (Dyck *et al.*, 1994; Kastner *et al.*, 1992; Perez *et al.*, 1993). The functional role of PML remains relatively unknown. It is interesting to note, first, that three members of the sub-family are involved in rearrangements which lead to the formation of transforming chimaeric proteins (T18/B-raf, RFP/ret and PML/RARα) where all three fusions occur just downstream of the coiled-coil motif (Kastner *et al.*, 1992; Mahé *et al.*, 1995; Miki *et al.*, 1991) and, second, that two members, PML (Chelbi-Alix *et al.*, 1995 ; Stadler *et al.*, 1995 ; this paper) and Staf-50 (Tissot and Mechti, 1995) are induced by interferon. In acute promyelocytic leukaemia (APL), PML has been identified by its fusion to the RARα gene by the t(15;17) translocation (de Thé *et al.*, 1990; de Thé *et al.*, 1991; Goddard *et al.*, 1992; Kastner *et al.*, 1992), reviewed in (Warrell *et al.*, 1993). This translocation results in formation of the PML/RARα and RARα/PML fusion genes and proteins. PML/RARα fusion transcripts and proteins are found in all APL patients whereas RARα/PML fusion transcripts, which are produced as a result of the t(17;15) reciprocal translocation, are detected in the majority but not all APL patients (Alcalay *et al.*, 1992). Therefore, the PML/RARα, whose gene transcription is under the control of PML promotor, rather than the RARα/PML proteins most probably play a crucial role in APL pathogenesis. The PML/RARα fusion protein contains the functional domains of both PML and RARα and is the likely molecular basis of APL leukaemogenesis probably through alteration of PML and/or RARα functions.

Localisation of PML in normal and APL cells

PML has, in normal cells, a speckled nuclear pattern of expression and is localized on poorly characterized nuclear organelles named nuclear bodies (NBs) (Daniel *et al.*, 1993; Dyck *et al.*, 1994; Koken *et al.*, 1994; Weis *et al.*, 1994). NBs represent a heterogeneous group of intranuclear structures which can be distinguished by morphological or antigenic criteria ; the functions of NBs are unknown, except for the coiled bodies which are considered to be storage compartments for splicing factors (Visa *et al.*, 1993). PML colocalises on NBs with primary biliary cirrhosis autoantigens such as Sp100 (Szostecki *et al.*, 1987) and with Nuclear Dot protein 52 (NDP52)(Korioth *et al.*, 1995). In APL, both the PML and Sp100 patterns are dominantly delocalised by the PML/RARα fusion protein towards distinct microspeckles which are smaller and much more numerous (Dyck *et al.*, 1994; Koken *et al.*, 1994; Weis *et al.*, 1994). The PML-RARα fusion protein may form heterodimers with PML and RXR and could be a dominant negative inhibitor of PML and RXR (Kastner *et al.*, 1992). The sequestration of RXR by PML-RARα may impair the heterodimerization of RXR with some hormone receptors like RARs, thyroid hormone receptor or vitamin D3 receptor (Perez *et al.*, 1993; Testa *et al.*, 1994) and therefore could result in a differentiation arrest. Furthermore, (Kastn PML-RARα heterodimer formation could alter PML and/or RARα/RXR er *et al.*, 1992) intracellular localisation as well as functions of these proteins.

Effects of IFN on PML expression in normal and APL cells

PML is induced by IFNs in a variety of human cell lines. All IFNs (α, β, and γ) sharply induce PML expression leading to a marked swelling of the NBs (Chelbi-Alix *et al.*, 1995). The major PML protein induced had an apparent M.W. of 70 kD, but others minor PML proteins ranging in size from 47 to 98 Kd were also induced, as could be predicted from the molecular weights of the various alternatively spliced cDNAs (Fagioli *et al.*, 1992). IFNs increase PML expression in a dose-and time-dependent manner with a maximum at 500 units/ml after 14h treatment (Figure 1) (Chelbi-Alix *et al.*, 1995). Kinetics studies indicate that maximum mRNA PML induction by IFNα is at 4h (Figure 2).

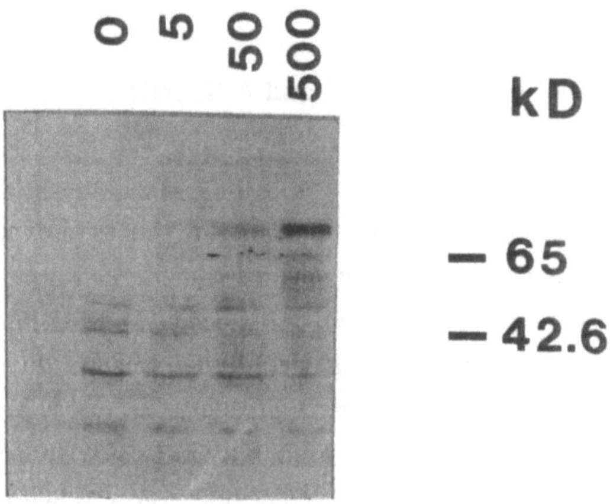

Figure 1: Western blot analysis of PML induction as function of IFN α concentrations. HeLa cells were treated for 18h at the indicated concentrations. Twenty micrograms of whole cell extracts were analysed using rabbit anti-PML antibodies.

We have recently shown that PML mRNA induction by IFN α (Figure 2) and IFN γ was rapid and that the protein synthesis inhibitor cycloheximide failed to antagonise this, which is consistent with a direct activation of the gene (Stadler *et al.*, 1995).

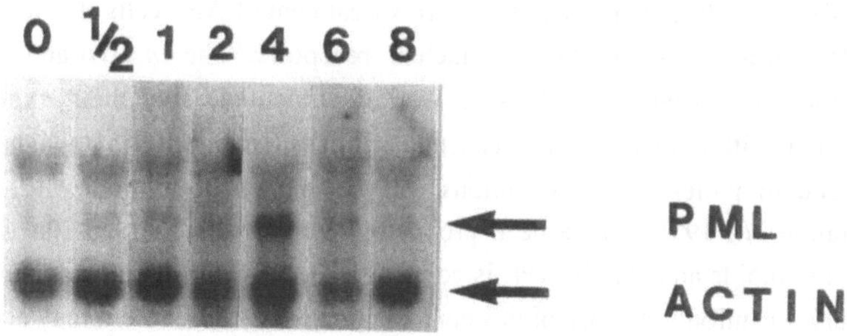

Figure 2: Northern blot analysis of total HeLa RNA treated or not with IFNα. Treatment time is indicated in hours. Northern blot analysis was performed using the radiolabelled *Bgl*II -*Bam*HI fragment of PML cDNA as a probe. Blots were rehybridized with a β actin probe.

Furthermore in the PML promoter we have identified the ISRE and GAS elements, binding of IFN STATs was demonstrated to be weak for the PML GAS, but strong for the PML ISRE which also seemed to contribute subtantially to the IFN γ response (Stadler *et al.*, 1995). Thus, PML is a primary target gene of type I and type II IFNs. The three NB proteins identified to date, Sp100 (Guldner *et al.*, 1992), NDP52 (Korioth *et al.*, 1995) and PML (Chelbi-Alix *et al.*, 1995 ; Stadler *et al.*, 1995) are all induced by IFNs. These results could suggest a role of this nuclear domain in the IFN response.

In NB4 cells, a cell line bearing the t(15;17) translocation (Lanotte *et al.*, 1991), IFNα treatment resulted in an increase of the micropunctuate pattern of PML, Sp100, and PML/RARα without affecting their abnormal microspeckled localization (Chelbi-Alix *et al.*, 1995).The IFN-induced PML/RARα is accompanied by a enhanced sequestration of its heterodimeric partner RXR. This clear increase in the microspeckled RXR protein suggests that not all RXR is PML/RARα associated prior to IFN treatment. IFNα may enhance RXR sequestration and could further impair nuclear receptor function, as previously proposed(Perez *et al.*, 1993; Testa *et al.*, 1994). According to this hypothesis, IFN treatment of APL cells should enhance the differentiation block towards nuclear receptors. The *in vivo* and *in vitro* antitumor effectiveness of IFNs is well documented, and their synergistic combination with retinoic acid in leukaemic cell lines (Bollag, 1991; Gallagher *et al.*, 1987) and in patients with squamous cell carcinomas (Lippman *et al.*, 1992a; Lippman *et al.*, 1992b) have been promising. The effectiveness of the synergy between retinoids and IFN in APL is controversial, however. One report (Warrell, 1993) urges caution in the use of this combinaition in APL because it may accelerate the patient's leukaemia. Our demonstration of PML/RARα induction by IFNα, in APL cells, could provide a molecular explanation to this observation.

The normal PML pattern is disrupted upon infection with DNA viruses but not with RNA viruses.

Previous studies had shown that infection with some DNA viruses such as Herpes- and Adenovirus disrupted the nuclear punctate localisation of Sp100 and PML (Maul and Everett, 1994; Maul *et al.*, 1993; Puvion-Dutilleul *et al.*, 1995). To study the fate of PML during other viral infections, human MRC5 cells which express high endogeneous PML levels, were infected with DNA viruses such as Herpesviridae (Cytomegalovirus, CMV) or RNA viruses such Vesicular stomatitis virus (VSV) (Figure 3). Early after CMV infection the normal speckled PML (and Sp100) patterns were lost (arrows, Figure 3 and not shown), the PML pattern was replaced by a uniform nuclear staining (arrowhead, Figure 3, upper panel) similar to an immediate early CMV protein (Figure, 3 lower panel). Despite the colocalisation of Sp100 and PML found in normal and APL cells, (Koken *et al.*, 1994), Sp100 and PML patterns were not similarly altered by infection with Herpesviridiae: in CMV infection, Sp100 was condensed in the cytoplasm while PML was homogeneously nuclear ; in a later phase, PML appeared in the cytoplasm (data not shown). Infection with RNA viruses such as VSV, Encephalomyocarditis virus and Influenza virus did not alter the PML or Sp100 profiles even until complete cell lysis (Figure 3 and not shown). Thus, during infection with DNA but not RNA, viruses the interactions between PML and nuclear bodies as well as between Sp100 and PML may be disrupted.

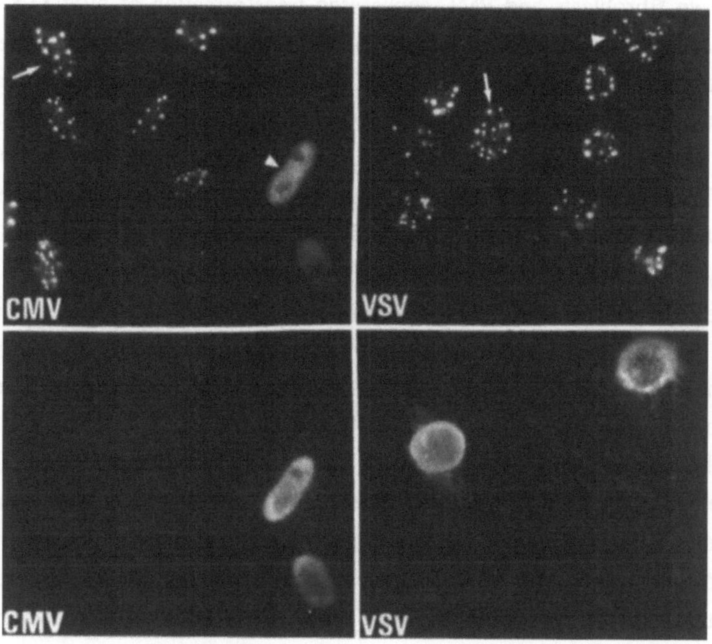

Figure 3: Delocalisation of PML upon infection with DNA viruses but not with RNA viruses. MRC5 primary human fibroblasts were infected with Cytomegalovirus (CMV) or Vesicular Stomatitis Virus (VSV) at a multiplicity of 0.1. Arrows point to uninfected cells with a normal PML pattern, arrowheads point to the PML pattern in infected cells were detected using monoclonal antibodies directed against viral proteins. Double labelling of CMV- and VSV-infected MRC5 cells. Upper panels : PML, lower panels: CMV early protein and VSV antigens. At 6 h p.i. (post infection), immunofluorescence was performed with rabbit anti-PML and mouse monoclonal, E13 (against CMV immediate early antigen) or with monoclonal anti-PML and rabbit anti-VSV antibodies.

PML could mediate some biological effects of IFN

IFNs have well established antiproliferative properties (Gutterman, 1994) and dominant negative mutants in a previously identified IFN-mediator (p68 kinase) are transforming (Koromilas *et al.*, 1992; Meurs *et al.*, 1993). PML expression is cell-cycle regulated, overexpression of PML retards cell-growth (Koken *et al.*, 1995), PML sharply reduces the transforming effects of cooperating oncogenes on primary rat embryo fibroblasts and PML suppressed transformation of NIH/3T3 cells by activated neu oncogene (Liu *et al.*, 1995; Mu *et al.*, 1994). Taken together, these results suggest that the IFN-induced PML protein has anti-oncogenic effects. Note that an altered expression of PML was found in various premalignant or malignant human lesions (Koken *et al.*, 1995). At present, the molecular basis of the antiproliferative effects of PML (and NBs) are not fully understood. However, our demonstration, that PML is IFN-induced, could incorporate PML and NBs into the pathways responsible for IFN-induced tumor suppression.

In some cells, such as MRC5 human fibroblasts, IFNs establish a broad antiviral state without increase in the previously identified antiviral mediators, 2'5'A synthetase, p68 kinase and MxA, (Meurs *et al.*, 1981) and data not shown). The strong induction of PML by IFN in MRC5 cells and the effect of PML on cell growth raised the question of a possible role of this protein in virus resistance. To test a putative antiviral effect of PML, cell lines stably expressing PML were constructed in hamster CHO and mouse GE86 cells. PML expression was demonstrated by immunofluorescence and Western blot analysis (data not shown). Interestingly, clones that express high levels of PML show a reduced rate of proliferation (not shown and (Koken *et al.*, 1995). Note, however, that the amount of PML expresed in these cells is in the range of that induced by high concentrations of IFN in HeLa cells (Figure 1). In the absence of IFN, constitutive overexpression of PML, but not of Sp100, induced resistance to infection by a variety of RNA viruses. While a strong inhibition of viral yield was observed with VSV (rhabdovirus) and influenza A (orthomyxovirus), a weaker inhibition was found for EMCV (picornavirus). The coiled-coil domain of PML, but not the C-terminus is required both for targetting onto NBs and viral resistance (manuscript in preparation). Note that infection with DNA, but not RNA viruses disrupts the speckled PML pattern (Figure 3). By identifying PML as a mediator of some antiproliferative and antiviral IFN-response pathways, our results shed a new light on the mechanism of action of this cytokine. Alterations of PML structure and localisation in APL constitute the first genetic alteration of an IFN effector in a human cancer. Thus, inactivation of PML, an IFN-induced growth suppressor may result in uncontrolled growth in APL.

Acknowledgement: We thank J. Peries for support and encouragment. We acknowledge F. Morinet for providing CMV, and MRC5 cells, and D. Blondel for anti-VSV serum. This work was supported by grants from ARC, Ligue contre le Cancer, Fondation de France.

References

Alcalay, M., Zangrilli, D., Fagioli, M., Pandolfi, P.P., Mencarelli, A., Lo Coco, F., Biondi, A., Grignani, F. and Pelicci, P.G. (1992) Expression pattern of the RAR alpha-PML fusion gene in acute promyelocytic leukemia. Proc Natl Acad Sci U S A 89: 4840-4844.

Barlow, P.N., Luisi, B., Milner, A., Elliott, M. and Everett, R. (1994) Structure of the C3HC4 domain by 1H-nuclear magnetic resonance spectroscopy. A new structural class of zinc- finger. J. Mol. Biol. 237: 201-211.

Bollag, W. (1991) Retinoid and interferon: a new promising combination Brit. Hematol. 79: 87-91.

Chelbi-Alix, M.K., Brouard, A., Boissard, C., Pelaprat, D., Rostene, W. and Thang, M.N. (1994) Induction by Vasoactive Intestinal peptide of interferon synthesis in glial cells but not in neurons J. Cell. Physiol. 158: 47-54.

Chelbi-Alix, M.K. and Chousterman, S. (1992) Ethanol induces 2'5' oligoadenylate synthetase and antiviral activities through interferon β production J. Biol. Chem. 267: 1741-1745.

Chelbi-Alix, M.K., Pelicano, L., Quignon, F., Koken, M.H.M., Venturini, L., Stadler, M.,Pavlovic, J., Degos, L. and de Thé, H. (1995) Induction of PML protein by interferons in normal and APL cells. Leukemia, in press.

Daniel, M., Koken, M., Romagné, O., Barbey, S., Bazarbachi, A., Stadler, M., Guillemin, Degos, L., Chommienne, C., and de Thé, H. (1993) PML protein expression in hematopoietic and acute promyelocytic leukemia cells. Blood 82: 1858-1867.

de Thé, H., Chomienne, C., Lanotte, M., Degos, L. and Dejean, A. (1990) The t(15;17) translocation of acute promyelocytic leukemia fuses the retinoic acid receptor a gene to a novel transcribed locus. Nature 347: 558-561.

de Thé, H., Lavau, C., Marchio, A., Chomienne, C., Degos, L. and Dejean, A. (1991) ThePML-RAR alpha fusion mRNA generated by the t(15;17) translocation in acute promyelocytic leukemia encodes a functionally altered RAR. Cell. 66: 675-84.

Dyck, J.A., Maul, G.G., Miller, W.H., Chen, J.D., Kakizuka, A. and Evans, R.M. (1994) A novel macromolecular structure is a target of the promyelocyte-retinoic acid receptor oncoprotein Cell 76: 333-343.

Fagioli, M., Alcalay, M., Pandolfi, P.P., Venturini, L., Mencarelli, A., Simeone, A., Acampora, D., Grignani, F. and Pelicci, P.G. (1992) Alternative splicing of PML transcripts predicts coexpression of several carboxy-terminally different protein isoforms. Oncogene 7: 1083-1091.

Freemont, P.S. (1993) The RING finger. A novel protein sequence motif related to the zinc finger.Annals N Y Acad. Sci. 684: 174-192.

Gallagher, R.E., Lurie, K.J., Leavitt, R.D. and Wiernik, P.H. (1987) Effects of interferon and retinoic acid on the growth and differentiation of clonogenic leukemic cells from acute myelogenous leukemia patients treated with recombinant leukocyte-αA interferon. Leukemia Res. 11: 609-619.

Goddard, A.D., Borrow, J. and Solomon, E. (1992) A previously uncharacterized gene,PML, is fused to the retinoic acid receptor alpha gene in acute promyelocytic leukaemia. Leukemia 6: 117S-119S.

Guldner, H., Szostecki, C., Grotzinger, T. and Will, H. (1992) IFN enhances expression of Sp100, an autoantigen in primary biliary cirrhosis J. Immunol. 149: 4067-4073.

Gutterman, J. (1994) Cytokine therapeutics: lessons from interferon α.Proc. Natl. Acad. Sci. USA 91: 1198-1205.

Hovanessian, A.G. (1993) Interferon-induced dsRNA-activated protein kinase (PKR) : antiproliferative, antiviral and antitumoral functions. Virology 4: 237-245.

Kastner, P., Perez, A., Lutz, Y., Rochette-Egly, C., Gaub, M.P., Durand, B., Lanotte, M., Berger, R. and Chambon, P. (1992) Structure, localization and transcriptional properties of two classes of retinoic acid receptor alpha fusion proteins in acute promyelocytic leukemia (APL): structural similarities with a new family of oncoproteins EMBO J. 11: 629-642.

Koken, M.H.M., Linares-Cruz, G., Quignon, F., Viron, A., Chelbi-Alix, M.K., Sobczak-Thépot, J., Juhlin, L., Degos, L., Calvo, F. and de Thé, H. (1995) The PML growth-suppressor has an altered expression in human oncogenesis 10: 1315-1324.

Koken, M.H.M., Puvion-Dutilleul, F., Guillemin, M.C., Viron, A., Linares-Cruz, G., Stuurman, N., de Jong, L., Szostecki, C., Calvo, F., Chomienne, C., Degos, L., Puvion, E. and de Thé, H. (1994) The t(15;17) translocation alters a nuclear body in a RA-reversible fashion. EMBO J. 13: 1073-1083.

Korioth, F., Gieffers, C., Maul, G.G. and Frey, J. (1995) Molecular characterization of NDP52, a novel protein of the nuclear domain 10, which is redistributed upon virus infection and interferon treatment. J. Cell Biol. 130 : 1-13.

Koromilas, A.E., Roy, S., Barber, G., Katze, M.G. and Sonenberg, N. (1992) Malignant transformation by a mutant of the IFN-inducible dsRNA-protein kinase. Science 257: 1685-1689.

Lanotte, M., Martin-Thouvenin, V., Najman, S., Balerini, P., Valensi, F. and Berger, R.(1991) NB4, a maturation inducible cell line with t(15;17) marker isolated from a human acute promyelocytic leukemia. Blood. 77: 1080-1086.

Lippman, S.M., Kavanagh, J.J., Paredes-Espinosa, M., Delgadillo-Madrueno, F., Paredes- Casillas, P., Hong, W.K., Holdener, E. and Krakoff, I.H. (1992) 13-cis-Retinoic acid plus Interferon α-2a: highly active systemic therapy for squamous cell carcinoma of the cervix. J. Natl. Cancer Inst. 84: 241-245.

Lippman, S.M., Parkinson, D.R., Itri, L.M., Weber, R.S., Schantz, S.P., Ota, D.M., Schusterman, M.A., Krakoff, I.H., Gutterman, J.U. and Hong, W.K. (1992b) ., 13-cis-Retinoic Acid and Interferon α- 2a: effective combination therapy for advanced squamous cell carcinoma of the skin. J. Natl. Cancer Inst 84: 235-241.

Liu, J.-H., Mu, Z.-M. and Chang, K.-S. (1995) PML suppresses oncogenic transformation of NIH/3T3 cells by activated *neu*.. J. Exp. Med. 181: 1965-1973.

Mahé, D., Roussev, R., Lutz, Y., Puvion-Dutilleul, F. and Fuchs, J.P. (1995) Two hnRNP-associated proteins share common structural features with the adenoviral 72-kDA protein. Exp. Cell Res. 216: 1-12.

Maul, G.G. and Everett, R.D. (1994) The nuclear location of PML, a cellular member of the C_3HC_4 zinc-binding domain protein family, is rearranged during herpes simplex virus infection by the C_3HC_4 viral protein ICP0. J. Gen. Virol. 75: 1223-1233.

Maul, G.G., Guldner, H.H. and Spivack, J.G. (1993) Modification of discrete nuclear domains induced by herpes simplex virus type 1 immediate early gene 1 product (ICPO). J. Gen. Virol. 74: 2679-2690.

Meurs, E., Hovanessian, A.G., Montagnier, L. (1981) Interferon-mediated antiviral state in human MRC5 cells in the absence of detectable levels of 2-5 A synthetase andprotein kinase. J. Interferon Res. 1: 219-232.

Meurs, E., Galabru, J., Barber, G., Katze, M.G. and Hovanessian, A.G. (1993) Tumorsuppressor function of the interferon-induced double-stranded RNA-activated protein kinase. Proc. Natl. Acad. Sci. USA 90: 232-236.

Miki, T., Fleming, T., Crescenzi, M., Molloy, C., Blam, S., Reynolds, S. and Aaronson, S. (1991) Development of a highly efficient expression cDNA cloning system: application to oncogene isolation. Proc. Natl. Acad. Sci. USA 88: 5167-5171.

Mu, Z.M., Chin, K.V., Liu, J.H., Lozano, G. and Chang, K.S. (1994) PML, a growth suppressor disrupted in acute promyelocytic leukemia. Mol. Cell. Biol. 14: 6858-6867.

Pellegrini, S. and Schindler, C. (1993) Early events in signalling by interferons. TIBS 18: 338- 342.

Perez, A., Kastner, P., Sethi, S., Lutz, Y., Reibel, C. and Chambon, P. (1993) PML/RAR homodimers: distinct binding properties and heteromeric interactions with RXR. EMBO J. 12: 3171-3182.

Pestka, S., Langer, J.A., Zoon, K.C. and Samuel, C.E. (1987) Interferons and their actions. Ann.Rev.Biochem 56: 727-777.

Puvion-Dutilleul, F., Chelbi-Alix, M.K., Koken, M., Quignon, F., Puvion, E. and de Thé, H. (1995) Adenovirus infection induces rearrangements in the intranuclear distribution of the nuclear body -associated PML protein. Exp. Cell Res. 218: 9-16.

Reddy, B., Etkin, L. and Freemont, P. (1992) A novel zinc finger coiled-coil domain in a family of nuclear proteins. TIBS. 17: 344-345.

Schindler, C. and Darnell Jr., J.E. (1995) Transcriptional responses to polypeptide ligands: The JAK-STAT pathway. Ann. Rev. Biochem. 64: 621-651.

Sen, G.C. and Ransohoff, R.M. (1993) Interferon-induced antiviral actions and their regulation Advances in Virus Research 42: 57-102.

Stadler, M., Chelbi-Alix, M.K., Koken, M.H. M., Venturini, L., Lee, C., Saïb, A., Quignon, F., Pelicano, L., Guillemin, M.-C., Schindler, C. and de Thé, H. (1995) Transcriptional induction of the PML growth suppressor gene by interferons is mediated through an ISRE and a GAS element. Oncogene in press.

Staeheli, P. (1990) Interferon-induced proteins and the antiviral state. Advances in virus research 38: 147-201.

Staeheli, P., Pitossi, F. and Pavlovic, J. (1993) Mx proteins: GTPases with antiviral activity Trends in Cell. Biol. 3: 268-272..

Szostecki, C., Krippner, H., Penner, E. and Bautz, F. (1987) Autoimmune sera recognize a 100 kD nuclear protein antigen. Clin. Exp. Immunol. 68: 108-115.

Testa, U., Grignani, F., Barberi, T., Fagioli, M., Masciulli, R., Ferrucci, P.F., Seripa, D., Camagna, A., Alcalay, M., Pellici, P.G. and Peschle, C. (1994) PML/RAR alpha+ U937 mutant and NB4 cell lines: retinoic acid restores the monocytic differentiation response to vitamin D3. Cancer Res. 54: 4508-4515.

Tissot, C., and Mechti, N. (1995) Molecular cloning of a new interferon-induced factor that represses human immunodeficiency virus type long terminal repeat expression. J. Biol. Chem 270: 14891-14898.

Visa, N., Puvion-Dutilleul, F., Harper, F., Bachellerie, J.P. and Puvion, E. (1993) Intranuclear distribution of poly A RNA determined by electron microscopy in situ hybridization. Exp.Cell Res. 208: 19-34.

Warrell, R., de Thé, H., Wang, Z. and Degos, L. (1993) Acute promyelocytic leukemia. New Engl. J. Med. 329: 177-189.

Warrell, R.P. (1993) Retinoid resistance. The Lancet 341: 126.

Weis, K., Rambaud, S., Lavau, C., Jansen, J., Carvalho, T., Carmo-Fonseca, M., Lamond, A. and Dejean, A. (1994) Retinoic acid regulates aberrant nuclear localization of PML/RARα in acute promyelocytic leukemia cells. Cell. 76: 345-356.

DIFFERENTIATION THERAPY OF ACUTE PROMYELOCYTIC LEUKEMIA

Laurent DEGOS
Institut d'Hématologie
HOPITAL SAINT LOUIS
1, avenue Claude Vellefaux
75010 PARIS - FRANCE

Abnormal are the malignant cells. The genome is abnormal including chromosomal translocations, gene deletions or mutations. Malignant cell seems to be in an irreversible state. Therefore the only way to treat a patient with cancer is to eliminate the tumor by surgery, radiotherapy, chemotherapy and more recently immunotherapy.

Léo Sachs in Israel demonstrated that cell lines derived from murine leukemia were able by *in vitro* assay to be differentiated into mature granulocytes (Sachs, 1978). Léo Sachs unsettled the dogma of the irreversibility of malignant cells, and he pointed out the impairment of the balance between differentiation and proliferation. Two approaches were thought as rationale for the repair of this disequilibrium : either a pressure on the differentiation or an inhibition of proliferation.

We first studied the inhibition of proliferation by low concentration of antimitotic in vitro and on low dose ARA-C in vivo. We obtained a complete remission in the first three patients treated by low dose ARA-C (Housset et al, 1982).

Therapeutic randomized trial in elderly leukemic patients showed similar survival using low dose ARA-C to conventional chemotherapy while it reduced the frequency of early deaths (Tilly et al, 1990). However it was difficult to distinguish between cytotoxic and differentiation effects.

The cell lines models are artificial, and the differentiation effects of low dose antimitotic drug are mixed with antiproliferative effects. Is it conceivable to find a pure differentiation agent which could be administered to patients ?

All-trans RA is specifically active on acute promyelocytic leukemic cells. Christine Chomienne tested fresh bone marrow cells from 60 leukemic patients and disclosed a specific differentiation effect of retinoic acid in all bone marrow specimens from acute promyelocytic patients using the Nitroblue Tetrazolium

NATO ASI Series, Vol. H 99
Tumor Biology
Regulation of Cell Growth, Differentiation and Genetics in Cancer
Edited by Asterios S. Tsiftsoglou, Alan C. Sartorelli, David E. Housman and T. Michael Dexter

reduction assay (Chomienne et al, 1990). Among the retinoids, all trans retinoic acid (ATRA) gave better results that 13-cis isomer or 4-oxo metabolite. One log difference of concentration for the same magnitude in favor of ATRA was found. The ethyl-ester was not effective and we have recently tested the 9-cis which gave similar results to all-trans.

Acute promyelocytic leukemia is a rare leukemia : 10 % of acute non lymphoblastic leukemia, and is characterized morphologically by the aspects of the abnormal cells classified as M3 subtype, clinically by a bleeding diathesis exacerbated by chemotherapy, and cytogenetically by a specific translocation t(15;17).

ALL TRANS RETINOIC ACID INDUCES A PROGRESSIVE DIFFERENTIATION OF ACUTE PROMYELOCYTIC LEUKEMIC CELLS IN PATIENTS

Usual features of bone marrow acute promyelocytic leukemic patients include homogenous infiltration of large cells, simulating abnormal promyelocytes with several large granules and Auer rods in the cytoplasm. Maturing bone marrow cells appeared during the treatment of APL patients using ATRA (Castaigne et al, 1990). Auer rods were sometimes observed in the mature cells after two weeks, confirming the differentiation process of malignant cells. After 2 to 3 months treatment, bone marrow appeared normal. The differentiation is ascertained by the absence of bone marrow hypoplasia, by the presence of Auer rods in mature cells and by the findings of an intermediate cell population with mature and immature markers.

The progressive differentiation of leukemia cells leads to the disappearance of abnormal cells and the emergence of normal cells. At the time of complete remission the caryotype becomes normal.

ATRA could give complete remission in 95 % of patients without worsening of coagulopathy, and without aplasia in de novo (Huang et al, 1988) or in relapse patients (Degos et al, 1990). However, ATRA has two major side effects : the risk of rapid rise of leukocytes and the appearence of ATRA syndrome (leucocyte activation), and the risk of rapid relapse unless intensive chemotherapy is administered in complete remission.

The French cooperative group started a treatment approach combining ATRA followed by intensive chemotherapy in newly diagnosed APL . We first performed a pilot study between 1990 and 1991 where treatment with ATRA was given until complete remission and followed by a combination of daurorubicin (4 days) and Ara-C (7 days), then three courses of chemotherapy as consolidation, and finally, prolonged continuous maintenance with 6-mercaptopurine and methotrexate. The first daunorubicin-Ara-C course was, however, administered as an emergency treatment if leukocytes were above 6000/μl by day 5 or 10.000/μl by day 10, or 15.000/μl by day 15 of ATRA treatment, in order to prevent ATRA syndrome.

Twenty six patients were included. All had leukocytes under 10 000/μl. Ninety-six per cent achieved complete remission (only one early death) as compared to 76% CR in a historical control treated by chemotherapy alone between 1984 and 1989 (Fenaux et al, 1992). With a minimum follow up of 38 months from CR achievement, event free survival (EFS including as events : resistance, relapse, and death in CR) disease free interval (DFI, taking into account relapse in patients who had reached CR, and censoring deaths in CR) and survival were 62 %, 70 % and 77 % at 4 years, as compared to 28 %, 42 % and 42 % in the historical control group and these differences were all significant.

The prolonged follow up now available in both cohorts of patients shows that the combination of ATRA and chemotherapy mainly reduces the risk of early relapses, within 18 months of CR achievement, whereas the risk of later relapses is similar to that of the chemotherapy group. This suggested that combination therapy did not delay relapses but actually reduced their frequency.

A European trial comparing ATRA followed by chemotherapy and chemotherapy alone in newly diagnosed APL was started in April 1991 after the favorable results obtained in the pilot study. In the chemotherapy group, patients received 3 successive courses of DNR and AraC. In the ATRA group, patients received ATRA until CR, followed by the same 3 chemotherapy courses. The first course was however rapidly administered if WBC were greater than 5000/mm3 at diagnosis, or increased above thresholds defined in the pilot study. The trial was prematuraly closed in December 1992, because the event free survival (EFS) was significantly better in the ATRA group (Fenaux et al, 1993).

The last analysis, performed at the reference date of 1 September 1994, showed actuarial EFS, relapse rate, and survival at two years of 68%, 25 % and 81 % in the ATRA group, as compared to 23 %, 56 % and 51 % respectively in the chemotherapy group ($p=10^{-4}$ $p=10^{-4}$ and $p=0.009$, respectively).

Whatever this additive effect is better achieved by administrating ATRA and chemotherapy together or one after the other is unknown. The current European APL trial (APL 93 trial) is trying to answer this question by randomizing ATRA followed by chemotherapy and ATRA plus chemotherapy (started on day 3 of ATRA) in newly diagnosed APL. Up to now more than 250 patients are included and 95 % of patients obtained a CR in both arms.

In the European APL 93 trial, after consolidation courses, patients in CR are randomized between no maintenance, intermittent treatment with ATRA (15 days every 3 months), continuous low dose chemotherapy (6 mercaptopurine plus methotrexate) or both. (This part of APL 93 trial is organized in cooperation with the GIMEMA Italian group).

The results on more than 1000 patients from France and China confirmed by series of patients from U.S.A., Japan, Australia and European countries (Wang et al, 1993), lead to three common conclusions : 1°) no resistance if the diagnosis of APL is ascertained by the specific rearrangement between the chromosome 15 and 17 ; 2°) no aplasia and thus no severe infection ; 3°) a rapid disappearance of bleeding diathesis in the first week of treatment.
However, all the series accounted 10% of early deaths by hemorraghe mainly during the first week and by a new syndrome of leucocyte activation so called "retinoic acid syndrome".
The bleeding diathesis is correlated with the low concentration of fibrinogen which is due to the degradation by thrombin, plasmin and lysozomial enzymes. Thrombin activation induces a DIC, plasmin activity induces a fibrinolysis and lysozomial enzymes as cathepsine G and elastase are a part of an extended proteolysis (Dombret et al, 1992).
ATRA repairs within a week the fibrinopenia and the fibrinolysis but not the DIC and not the release of lysozomial enzymes which remain during one month. Bleeding diathesis disappears during the first week concomitantly to the normal level of fibrine and then after a thrombosis tendency exists during the first month due to the persistant procoagulant activation.
The second adverse effect is the retinoic acid syndrome (Frankel et al, 1992) generally associated with an hyperleucocytosis (Castaigne et al, 1990) and a leucocyte activation. It affects 35% of newly diagnosed and 15% of first relapse patients, and is chearacterized by fever, dyspnea, respiratory distress, kidney failure and coma. It occurs during the first month of treatment.
Hyperleucocytosis could be due to cell cycles of multiplication during the initial maturation but also to a bone marrow chase concomitant to changes in adhesion properties and in cell deformability. Leukocyte activation could be a consequence of secretion of cytokines (IL1β, IL6, TNFα and IL8).
The syndrome is now well known in western countries but is almost absent in China, for unknown reasons, while more than 700 patients have been treated.
In order to avoid the retinoic acid syndrome we proposed to prevent it by a chemotherapy delivered when white blood cells are above 6000 by day 5, 10.000 by day 10, and 15.000 by day 15. In U.S.A. patients are treated by corticosteroids. Leukopheresis was previously proposed in U.S.A which reduce the hyperleucocytosis but has no effect on the fatal issue of the syndrome.
Sylvie Castaigne, in the hope of reducing the adverse effects, proposed to treat patients with lower doses of ATRA. She treated 30 patients with 25 mg/m2 and even

12 patients with 15 mg/m2 (Castaigne et al, 1993). She obtained the same complete remission rate but no reduction of the frequency of hyperleucocytosis and of retinoic acid syndrome.

No resistance, a differentiation of malignant cells without aplasia, a rapid correction of bleeding diathesis, an easy prevention of adverse effects, a high rate of complete remission, all these spectacular results have been tarnished by a progressive acquired resistance to ATRA which provokes relapses when ATRA is given alone.

The shortness of the duration of complete remission and the weaker activity of subsequent treatment pointed to an acquired resistance. The resistance was explained by a progressive decrease in plasma concentration of ATRA (Warrel et al, 1991). The reasons actually proposed result in an increased catabolytic state for ATRA. On one hand an induction by ATRA itself of the 4-hydroxylase of cytochrome P450, which plays a major role in the catabolism of the drug itself (Muindi et al, 1993). On the other hand a progressive appearance of cytoplasmic binding protein II (CRABP) in the cytoplasm of myeloid cells (Cornic et al, 1992). The appearance of CRABP could sequester free retinoic acid in the cytosol and facilitate its catabolism. CRABP appears after 2 or 3 months of treatment. The reversibility is also progressive. The kinetics of appearance and disappearance of CRABP is correlated with the resistance of patients to ATRA. Patients become again sensitive to ATRA after 6 months to one year out of treatment.

MOLECULAR ABNORMALITIES IN ACUTE PROMYELOCYTIC LEUKEMIA

In 1977 Janet Rowley reported that the t(15;17) translocation was a consistent chromosomal change in APL (Rowley et al, 1977). The specific RA sensitivity of APL on one hand, and the mapping of the retinoic acid receptor alpha close to the breakpoint on chromosome 17 on the other hand, prompted us to investigate the expression of the RA receptor alpha gene in the patients cells.

The specific rearrangement of retinoic acid receptor alpha gene was first recognized by Christine Chomienne (Chomienne et al, 1990). In Northern blot an additional band was found in each case of APL but not in normal individuals or in other types of leukemia (M1, M2, M4, M5). The 15;17 translocation fuses the retinoic acid receptor to a novel transcribed locus, which is called PML for ProMyelocytic Leukemia (de Thé et al, 1990 ; Borrow et al, 1990).

The reciprocal translocation leads to two abnormal chromosome with two reciprocal fusion proteins and to two normal chromosomes (one 17 and one 15). More precisely, the breakpoint on chromosome 17 is constantly within the second intron of retinoic acid receptor excluding the A domain. On the PML gene two major known breakpoints are observed : in the intron 3 and the region intron 6 - exon 6. These two types of breakpoints lead to two major PML-RAR fusion transcripts, called long and

short PML-RAR. The two fusion proteins both contain the functional domains for dimerization and DNA binding. The reciprocal small transcripts are sometimes absent.
The knowledge of these breakpoints allowed to use the amplification of reversed transcribed cDNA sequence of the fusion transcript as a diagnostic laboratory test, and for the assessment of the minimal residual disease (Castaigne et al, 1992). A positive signal at diagnosis ascertained the *in vivo* retinoic acid sensitivity. A positive signal in the follow up of the patients predicts the relapse.

THE RETINOIC ACID RECEPTOR IMPAIRMENT IS RELATED WITH THE ARREST OF MYELOID MATURATION

Retinoic acid receptors belong to the superfamily of ligand inducible transcriptional transactivators. They bind to specific DNA responsive elements in the promotor region of target genes and are able to modulate transcription when the ligand hormone or vitamine is carried. The fusion protein contains the DNA binding and the retinoic acid binding domains.
RAR molecule most easily presents as RAR-RXR heterodimer. RAR has a greater affinity for all-trans and RXR for 9-cis retinoic acid. They bind to the same palindromic DNA sequence spaced by 1 to 5 nucleotides.
Most investigators have focused their efforts upon the impairment of transactivation due to the aberrant product on retinoic acid sensitive genes. In these conditions the endogenous receptors responded to RA, transfected normal RAR increased the signal while the hybrid gene modulated the signal of endogenous and even transfected RAR (de Thé et al., 1991 ; Kakizyka et al, 1991 ; Pandolfi t al, 1991). This could explain the impairment of transactivation and the stop in cell maturation program.
HL60 is a well known myeloid cell line capable of differentiation either into mature granulocytes upon administration of retinoic acid or DMSO, or into monocytes upon phorbol-ester (TPA) treatment. Philippe Rousselot transfected PML-RAR hybrid gene using liposome technique in HL60 (Rousselot et al, 1993). Under these conditions the retinoic acid induced granulocyte differentiation was blocked, while the DMSO or TPA induced terminal differentiation are not inhibited.
The hybrid gene alters the transactivation of normal retinoic acid receptor and blocks the retinoic acid induced maturation of myeloid cells, which could explain at least in part the leukemogenesis. How high doses of all trans retinoic acid could repair these defects is unknown (Chomienne et al, 1991).

THE ABNORMAL PML MOLECULE IS RELATED WITH A DISRUPTION OF A NUCLEAR BODY

The PML molecule contains a ring finger domain which probably does not bind directly to the DNA and also a leucine zipper like region related to the fos family which is responsible for the dimerization. An interference with AP1 transactivation complex (JUN-FOS) was suggested in presence of PML-RAR.

Applying acid phosphatase immunostaining with a rabbit anti PML antibody, Marie Thérèse Daniel described a speckled pattern with large dots inside the nucleus of normal cells (Daniel et al, 1993). The same nuclear body is also recognized by auto antibodies found in primary biliary chirrhosis patients. The cloned gene coding for one of the auto antigens is named sp100.

PML molecules are present on the outer shell of the nuclear body. In APL cases, the PML nuclear body is disrupted leading to a micropunctated pattern in the nucleus and the cytoplasm. The hybrid molecule effect in the disruption is dominant. After ATRA treatment the structure reappears quite rapidly, within a day in *in vitro* treatment or one week in *in vivo* treatment. The conversion should be a key phenomenon to study the genesis of nuclear bodies.

After complete remission, all the cells countain normal nuclear body. In fact PML molecules in acute promyelocyte leukemia reveal an alteration in the nuclear body structure which is possibly kinked to oncogenesis and more strikingly which is reversible under ATRA treatment.

From these investigations one can speculate that PML-RAR hybrid molecules impair the myeloid maturation altering the normal RAR function and disrupt a nuclear body interfering the normal PML function. ATRA restores the myeloid program and rebuilds the nuclear body structure.

In conclusion, all-trans RA in acute promyelocytic leukemia is the first model of differentiation therapy in malignancy, the first specific treatment for a genetic defect due to the translocation t(15;17), the first evidence for a reversible nuclear structure disruption and greatly improves the survival of patients.

REFERENCES

Borrow J., Goddard A.D., Sheer D. and Solomon E. (1990) Molecular analysis of acute promyelocytic leukemia breakpoint cluster region on chromosome 17. Science, 249 : 1577-1580.

Castaigne S., Chomienne C., Daniel MT, Ballerini P., Berger R., Fenaux P.,Degos L. (1990) All trans retinoic acid as a differentiation therapy for acute promyelocytic leukemia. I. Clinical results. Blood, 76 : 1704-1709.

Castaigne S., Lefebvre P., Chomienne C., Suc E., Rigal-Huguet F., Gardin C., Delmer A., Archimbaud E., Tilly H., Janvier M., Isnard F., Travade P., Montfort L., Delannoy A., Rapp MH., Christian G., Montastruc M., Weh H., Fenaux P., Dombret H., Gourmel B. and Degos L. (1993) Effectiveness and pharmacokinetics of low-dose all-trans retinoic acid (25 mg/m2) in acute preomyelocytic leukemia. Blood, 82 : 3560-3563.

Castaigne S., Balitrand N., de The H., Dejean A., Degos L. and Chomienne C. (1992) A PML/RAR alpha fusion transcript is constantly detected by RNA based polymerase chain reaction in acute promyelocytic leukemia. Blood, 79 : 3110-3115.

Chomienne C., Ballerini P., Balitrand N., Daniel MT., Fenaux P., Castaigne S., Degos L. (1990) All trans retinoic acid in acute promyelocytic leukemia. II. In vitro studies : structure function relationship. Blood, 76 : 1710-1717.

Chomienne C., Ballerini P., Balitrand N., Huang ME., Krawice I., Castaigne S., Fenaux P., Tiollais P., Dejean A., Degos L., de The H. (1990) The retinoic acid receptor alpha gene is rearranged in retinoic acid sensitive promyelocytic leukemia. Leukemia, 4 : 802-807.

Chomienne C., Balitrand N., Ballerini P., Castaigne S., de The H., Degos L. (1991) All trans retinoic acid modulates the retinoic receptor alpha in promyelocytic cells. J. Clin. Invest., 88 : 2150-2154.

Cornic M., Delva L., Guidez F., Balitrand N., Degos L., Chomienne C. (1992) Induction of retinoic acid binding protein in normal and malignant human myeloid cells by retinoic acid in AML3 patients. Cancer Res., 52 : 3329-3334.

Daniel MT., Kiken M., Romagne O., Barbey S., Bazarbachi A., Stadler M., Guillemin M.C., Degos L., Chomienne C., de The H. (1993) PML protein expression in haematopoietic and APL cells. Blood, 82 : 1858-1867.

Degos L., Chomienne C., Daniel MT., Berger R., Dombret H., Fenaux P., Castaigne S. (1990) Treatment of frist relapse in acute promyelocytic leukemia with all trans retinoic acid. Lancet, 2 : 1440.

de The H., Chomienne C., Lanotte M., Degos L., Dejean A. (1990) The t(15;17) translocation of acute promyelocytic leukemia fuses the retinoic acid receptor alpha gene to a novel transcribed locus. Nature, 347 : 558-561.

de The H., Lavau C., Marchio A., Chomienne C., Degos L., Dejean A. (1991) The PML-RAR fusion mRNA generated by the t(15;17) trans location in acute promyelocytic leukemia encodes a functionnally altered RAR. Cell, 66 : 675-684.

Dombret H., Sutton L., Duarte M., Daniel MT., Leblond V., Castaigne S., Degos L. (1992) Combinet therapy with all-trans retinoic acid and high-dose chemotherapy in patients with hyperleukocytic acute promyelocytic leukemia and severe visceral hemorrhage. Leukemia, 6 : 1237-1242.

Fenaux P., Castaigne S., Dombret H., Chomienne C., Duarte M., Archimbaud E., Lamy T., Tibeghien P., Tilly H., Dufour P., Cransac M., Guerci A., Sadoun A. and

Degos L. (1991) All-trans retinoic acid in newly diagnosed acute promyelocytic leukemia : a pilot study. Blood (ASH Abstract n° 305 - plenary session) p 79a.

Fenaux P., Castaigne S., Dombret H., Archimbaud E., Duarte M., Morel P., Lamy T., Tilly H., Guerci A., Maloisel F., Bordessoule D., Sadoun A., Tiberghien P., Fegueux N., Daniel MT., Chomienne C., Degos L. (1992) All-trans retinoic acid followed by intensive chemotherapy gives a high complete remission rate and may prolong remissions in newly diagnosed acute promyelocytic leukemia : a pilot study on 26 cases. Blood, 80 : 2176-2181.

Fenaux P., Le Deley MC., Castaigne S. et al (1993) Effect of all trans retinoic acid in newly diagnosed acute promyelocytic leukemia. Results of a multicenter randomized trial. Blood, 82 : 3241-3249.

Frankel SR., Eardley A., Lauwers G., Weiss M. and Warrel RP (1992) The "retinoic acid syndrome" in acute promyelocytic leukemia. Ann. Intern. Med, 117 : 292-296.

Housset M., Daniel MT, Degos L. (1982) Small doses of ARA-C in the treatment of acute myeloid leukemia : differentiation of myeloid leukemia cells ? Brit. J. Haematol. 51 : 125-129.

Huang M., Yu-Chen Y., Shu-Rong C., Lu Mx., Zhoa L., Gu LJ., Wang ZY (1988) Use of all trans retinoic acid in the treatment of acute promyelocytic leukemia. Blood, 73 : 567.

Kakizuka A., Miller WH., Umesono K. et al (1991) Chromosomal translocation t(15;17) in human acute promyelocytic leukemia fuses RAR alpha with a novel putative transcription factor, PML. Cell, 66 : 663-674.

Muindi JF., Young CW (1993) Lipid hydroperoxides greatly increase the rate of oxidative catabolism of all trans retinoic acid by human cell culture microsomes genetically enriched in specified cytochrome p-450 isoforms. Cancer Res, 53 : 1226-1229.

Pandolfi PP, Grignani F., Alcalay M. et al (1991) Structure and origin of the acute promyelocytic leukemia myl/RAR alpha cDNA and characterization of its retinoid-binding and transactivation properties. Oncogene, 6 : 1285-1292.

Rousselot P., Hardas B., Patel A., Guidez F., Gaken J., Castaigne S., Dejean A., de The H., Degos L., Farzaneh F., Chomienne C. (1994) The PML/RAR alpha gene product of the t(15;17) translocation inhibits retinoic acid induced differentiation and mediated transactivation in human myeloid cells. Oncogene, 9 : 545-551.

Rowley JD., Golomb HM., Gougherty C., (1977) 15/17 translocation : a consistent chromosomal change in acute promyelocytic leukaemia. Lancet , 1 : 549-550.

Sachs L. (1978) Control of normal differentiation and the phenotypic reversion of malignancy in myeloid leukemia cells. Nature, 274 : 535.

Tilly H., Castaigne S., Bordessoule D., Casassus P., Le Prise P., Tertian G., Desablens B., Henry-Amar M., Degos L. (1990) Low dose cytarabine versus intensive chemotherapy in the treatment of acute nonlymphocytic leukemia in the elderly. J. Clin. Oncol., 8 : 272-279.

Wang ZY, Degos L. (1993) Retinoic acid in hematological malignancies. International Symposium, Shanghai, Chine 19-21 octobre 1993.

Warrell RP., Muindi J., Frankel SR., Jakubowski A., Dmitrovsky E., Miller WH. (1991) Continuous treatment with all trans retinoic acid progressively decreases plasma drug concentrations : implications for relapse and resistance in acute promyelocytic leukemia. Blood (ASH abstract n° 1063, p. 268a.

BIOLOGY OF HAEMATOPOIETIC CELL GROWTH FACTORS

T.M. Dexter
Christie Hospital, NHS Trust
Paterson Institute for Cancer Research
Wilmslow Road
Manchester M20 9BX
United Kingdom

1. Introduction to the Haematopoietic System

The majority of mature blood cells are destined to live for only a few hours (granulocytes) or weeks (erythrocytes) before being sequestered and destroyed (1,2). In one day, this means that something in the order of 10^{13} myeloid cells must be produced to replace the expended cells. How this is achieved and how homeostasis is maintained are the subjects of this monograph.

Haematopoiesis first becomes apparent in the yolk-sac (3,4). As the embryo develops, the fetal liver becomes the main site of blood cell production and, from there, the haematopoietic cells migrate to the developing bones which then serve as the major site for haematopoiesis throughout adult life. Within the bone marrow, recognisable blood cells in various stages of development are found, as well as more primitive cells which act as the precursors for the various mature cell lineages. It is now known that all these precursor cells are derived from a common cell, the pluripotent stem cell (5-7).

Stem cells are few in number, comprising between 0.01 and 0.05% of the total marrow population, and possess two features that distinguish them from other haematopoietic cells. First is their ability to proliferate to produce more stem cells - a process called self-renewal. Second is their potential to undergo differentiation to produce at least nine highly specialised mature cell types: erythrocytes, neutrophils, eosinophils, basophils, platelets, monocytes/macrophages, osteoclasts and T and B lymphocyctes.

In this process, the pluripotent stem cells produce more developmentally restricted cells such as the multipotent colony-forming cell-mixed (CFC-Mix) that can proliferate, differentiate and develop into mature cells of the various myeloid cell

NATO ASI Series, Vol. H 99
Tumor Biology
Regulation of Cell Growth, Differentiation and
Genetics in Cancer
Edited by Asterios S. Tsiftsoglou, Alan C. Sartorelli,
David E. Housman and T. Michael Dexter

lineages (but not into T or B lymphocytes); bipotent progenitor cells such as granulocyte-monocyte colony forming cell (GM-CFC), that can produce neutrophils, macrophages and osteoclasts; and a variety of unipotent progenitor cells that can proliferate and develop into mature cells of only one cell lineage (8). This progressive restriction in development potential also ensures a tremendous amplification in cell numbers.

For all practical purposes, the stem cells are immortal (6,9,10). Consider, for example, that 5 x 10^4 murine marrow cells are sufficient to reconstitute the haematopoietic system when transplanted into a potentially lethally irradiated sibling. Throughout the recipient animal's lifespan, normal numbers of functionally mature myeloid and lymphoid cells are continuously being produced from the donor cells. Since a normal mouse has a total complement of 3 x 10^8 marrow cells, an individual mouse is capable of contributing sufficient marrow to reconstitute 10,000 siblings. Since each of these recipients is in turn capable of eventually contributing marrow cells to fulfill another round of marrow transplantation (10), it is clear that the most primitive cells in the original graft are capable of an extent of "self-renewal" which far exceeds a normal life-span. It is also clear that the stem cells are the most important cells in the haematopoietic system. These are the cells that are ultimately responsible for regenerating haematopoiesis following bone marrow transplantation, or following severe damage to the haematopoietic system by radiation accident or treatment with chemotherapeutic agents for malignant disease (5,11). If the stem cells are absent, death would rapidly result as a consequence of pancytopenia. If the stem cells are defective, abnormal lineage development leading to a variety of lympho-myelodysplastic syndromes would occur. Thus, a great deal of effort has gone into experiments aimed at understanding how stem cell self-renewal and differentiation are regulated.

In the shorter-term, however, following damage to the haematopoietic system, recovery takes place from the more developmentally-restricted progenitor cells which are `primed' to undergo rapid proliferation and development. Indeed, the short-term recovery (and hence survival) of recipients of marrow grafts probably owes more to these cells than the more primitive stem cells. In other words, when considering morbidity and survival of patients undergoing cytoreductive therapy, all arms of the haematopoietic system have roles to play. How then, is growth and development of stem cells and their progeny regulated and how may this knowledge be applied to the clinic?

2. Characterisation of Haematopoietic Growth Factors

Following the observation, more than a quarter of a century ago, that colonies containing mature neutrophils and macrophages would develop when haematopoietic cells are immobilised in a soft gel matrix and provided with media 'conditioned' by the growth of non-haematopoietic cells (12,13), a great deal of effort has gone into isolating and characterising the factor(s) involved in initiating and stimulating the development of the colony forming cells (now known as the 'GM-CFC'). During the course of this work, the *in vitro* growth of colonies containing mature erythroid cells, megakaryocytes or eosinophils were described (14-16) - and the general term 'colony stimulating factors' (CSFs) was used (particularly by the group in Melbourne led by Dr D Metcalf) to describe the molecules involved in stimulating the proliferation and development of the clonogenic precursor cell populations.

For many years, the nature of these CSFs remained elusive and their target cells were fairly ill-defined. Eventually, however, the widespread use of recombinant DNA technology allowed isolation of the genes coding for the CSFs and the production of large amounts of protein using bacterial, yeast or mammalian cell expression systems (17). The first CSFs to be produced in this way were granulocyte colony stimulating factor (G-CSF), macrophage colony stimulating factor (M-CSF), granulocyte-macrophage colony stimulating factor (GM-CSF) and interleukin-3 (IL-3 or multi-CSF). Concomitantly, the various subpopulations of haematopoietic cells were being analysed using antibodies, lectins and/or fluorescence activated cell sorting, to enrich for the various progenitor clonogenic cell populations (18-22). The combination of large amounts of purified recombinant materials, with the availability of enriched populations of target cells, facilitated further experiments at the mechanistic level. It is now clear that the CSFs do not represent a gene 'family', their structures at the nucleic and amino acid level are quite distinct and receptors are present on the surface of target cells specific for each factor (23). It is worth mentioning however, that the genes coding for IL-3, GM-CSF, M-CSF (as well as genes coding for IL4 and IL5) and the M-CSF receptor are all localised to a fairly small region of human chromosome 5q, while the gene coding for G-CSF is on chromosome 17 (17). The significance of the clustering on chromosome 5 is unknown.

At the same time as the myeloid CSFs were being investigated, various 'interleukins' that were able to modulate proliferation and/or differentiation of T or B lymphocytes and their precursors were also being characterised and molecularly cloned. While to our knowledge the 'classical' CSFs, such as G-CSF, GM-CSF or M-CSF have no effect upon the production or activation of T or B lymphocytes, some of the interleukins, apart from influencing the development of lymphoid cells, can also have an effect upon the production and/or function of myeloid cells. IL-5 for example, is a powerful stimulus for the maturation of eosinophils (24); IL-6 can influence the proliferation of multipotent myeloid progenitor cells (25,26); IL-1, IL-4 and IL-9 can modulate the response of myeloid progenitor cells to the CSFs (25,27,28), and IL-8 can activate neutrophils (29). Clearly, the target cell specificity of these interleukins could be discussed at some length. For brevity, however, this monograph will concentrate mainly on the biology of the CSFs and for further information on the interleukins see references (30) and (31).

3. The Target Cells for Haematopoietic Growth Factors

Using enriched populations of multipotent cells (such as day-12 spleen colony-forming cells, CFU-S) (28) and lineage-restricted progenitor cells such as GM-CFC and CFU-E (32,33), the `target' cells for the haematopoietic growth factors have been determined using both liquid culture and soft gel clonogenic assay systems (Figure 1).

IL-3 directly stimulates proliferation and development of multipotent stem cells, various developmentally-restricted progenitor cells and precursor cells, leading to the formation of various types of mature myeloid cells (34). GM-CSF is more restricted and preferentially promotes growth and development of neutrophil, macrophage and eosinophil progenitor and precursor cells (35). M-CSF acts upon GM-CFC and macrophage precursor cells, leading to the development of monocytes/ macrophages and only occasional neutrophils (36), while G-CSF preferentially stimulates the development of neutrophils from the appropriate precursor cell populations - although a modest effect upon production of macrophages has also been reported (37). Finally, erythropoietin is the primary physiological stimulus for proliferation, haemoglobinisation and terminal maturation of erythroid precursor cells (38).

Recently, there has been much interest in factors that may stimulate very primitive multipotent cells, in particular the newer pleiotropic cytokines like IL-6, Leukaemia Inhibitory Factor (LIF), Stem Cell Factor (SCF) and IL-1. Their effects on haematopoiesis may be better observed when they act together with other factors (39-41). For examples, IL-6 may induce cells to move from G_0 state to the G_1 phase of the cell cycle, in which they become responsive to other factors such as IL-3, thus showing a synergistic effect on colony formation (40). Also, SCF synergises with other factors to enhance colony formation from a range of progenitor cells (42), and allows cells to respond to low concentrations of the other growth factors (such as IL-3 or G-CSF). This effect is relevant to our understanding of haemopoiesis in the bone marrow, where it is difficult to detect appreciable levels of the classical CSFs. LIF has been shown to influence the maintenance of haemopoietic stem cells (43), although it is not necessary for their terminal differentiation.

It should be stressed that these stimulatory effects of the CSFs are best seen using highly-enriched target cell populations - essentially free of accessory cells. This is particularly important in view of the reported synergistic interactions or additive effects between growth factors.

4. Synergistic Interactions of Haematopoietic Growth Factors

When enriched populations of multipotent haematopoietic cells are exposed to M-CSF or to G-CSF, they die (44). Lineage-restricted GM-CFC and their progeny, on the other hand, proliferate and develop in response to these growth factors (45). What has recently been found is that, if M-CSF and G-CSF are used in combination, they can also recruit more primitive multipotent cells and allow these cells to proliferate and develop into mature neutrophils and macrophages (Figure 2) (46,47). In this case, therefore, synergy is seen: the combination of growth factors can recruit cell populations that are unresponsive to either factor used alone.

Since synergy is observed using a wide range of growth factor combinations (25,48), the effects may be particularly relevant in the context of how regulation of haematopoiesis takes place in the bone marrow. For example, IL-1 (which is not a growth stimulus for multipotent cells) can synergise with IL-3, with GM-CSF and with M-CSF (48). IL-6 is also a powerful stimulus when used in combination with IL-3 or G-CSF (40). Also SCF synergises with G-CSF and with other factors (49). Since IL-1 and IL-6 are constitutively produced by marrow stromal cells (50), they may well be important components in regulating blood cell production.

Similarly, SCF is also produced by marrow stromal cells, but what is particularly interesting here is that this cytokine is produced both as soluble material and as an integral component of the cell membrane (51). The latter may have a dual role to play as a growth factor and as an adhesion molecule.

5. Stromal Cell-mediated Haematopoiesis

In the bone marrow, haematopoiesis occurs in association with stromal cells which have been shown to produce a variety of cytokines either constitutively or following stress. As discussed above, IL-1, IL-6 and SCF can be detected in marrow stromal cells maintained *in vitro*. In addition, M-CSF, GM-CSF and G-CSF can also be produced by stromal cells either constitutively or following induction by a variety of stimuli (52,53).

It may be significant that, so far, production of IL-3 has not been detected either at the messenger RNA or the protein level. While the role of IL-3 in bone marrow haematopoiesis is unclear, it is worth emphasising that many (perhaps all) of the known target cells for IL-3 can also be recruited using other growth factors, either alone or in combination. In other words, there may well be no absolute requirement of haematopoietic cells for IL-3. Since the other haematopoietic growth factors show overlapping biological activities and target cells, it is likely that this idea also holds true for some of the other growth factors. Studies in mice, where the genes coding for the growth factors have been compromised by genetic engineering, will eventually answer these questions. It has already been shown that mice in which the gene coding for LIF is not functional are apparently normal except for a defect in embryo implantation (54). These observations emphasise the large degree of redundancy that exists in the regulation of haematopoiesis.

Meanwhile, the question remains as to how these cytokines may work together to regulate steady-state haematopoiesis and ensure that the sizes of the various cell compartments are maintained in homeostasis. What prevents too many stem cells from undergoing either self-renewal or differentiation? How are the relative sizes of the mature cell populations regulated?

A possible solution lies in the intimate cell to cell interactions that occur between stromal cells and haematopoietic cells (55). In the absence of added growth factors, haematopoiesis can be maintained *in vitro* for several months if marrow stromal `adherent' cells are present in long-term marrow cultures (56,57). In this case, the maintenance of haematopoietic stem cell self-renewal, differentiation and development into mature cells requires physical contact between the stromal cells and the haematopoietic cells: if this contact is prevented, haematopoiesis rapidly declines (58,59). This suggests that `free' cytokines, produced by and secreted from the stromal cells, do not have a major role to play in haematopoiesis and that the required stimuli for haematopoietic cell growth and development are primarily localised to the surface (membrane) of the stromal cells.

What is the nature of these membrane associated molecules? Although SCF can exist as a membrane-bound growth factor (51), this is not apparently true for many other growth factors. However, a clue that other growth factors may be localised, was provided in the observation by Gordon and colleagues that extracellular matrix (ECM) molecules can sequester growth factors (60). Subsequent studies showed that an important ECM component in this process was heparan sulphate: heparan sulphate isolated from the membranes of stromal cells (that could support haematopoiesis) could bind IL-3 and GM-CSF and present these growth factors to target haematopoietic cells, in a biologically active form (61). Thus, it seems likely that various cytokines (either produced by the marrow stromal cells themselves or `sequestered' from a circulating pool) are localised at the membrane of stromal cells.

Since the marrow stromal cells themselves are heterogeneous - comprising fibroblasts, reticular cells, adipocytes, endothelial cells and 'macrophages' (55), the possibility that these different populations can produce/sequester specific types of growth factors needs to be considered (45).

Recent work (De Wynter, Flavell and Dexter, unpublished) supports this premise: immunostaining of stromal cells of long-term bone marrow cultures has shown that neither GM-CSF nor G-CSF are uniformly distributed. Rather, they are localised to specific stromal cell populations and then only to restricted regions of the cell. The idea, then, is that within the bone marrow, there exist anatomically discrete stromal cell environments where particular growth factors are localised and where they exert their effects upon haematopoietic stem and progenitor cells. In other words, there are probably cell environments that preferentially promote the development of erythroid cells, other environments that stimulate the development of neutrophils and so on. This concept is supported by the observation that multipotential cells with different capacity for self-renewal and progenitor cells committed to different lineages are found at increased concentrations in well defined areas of the murine bone marrow (62). Similar findings on selective localisation of both progenitor cells and some factor producing cells, have been reported for human bone marrow (63). Some recent work also suggests that the range and concentration of growth factors can also influence the probability of self-renewal or differentiation of multipotent 'stem' cells *in vitro* (45), supporting the idea that self-renewal of stem cells only takes place in privileged microenvironments ('niches') in the bone marrow (64).

Thus the process of stem cell self-renewal, differentiation and lineage-restriction, as well as the growth and development of the mature haematopoietic cells, can all be modulated (at least in part) by growth factors. Clearly, however, growth factors are not the only regulatory component in haematopoiesis: adhesion molecules and 'homing' molecules also have a role to play in determining interactions between stromal cells and haematopoietic cells and are presumably a prerequisite for enabling the haematopoietic cells to lodge in specific regions and then to respond to the localised developmental stimuli (65,66).

6. How do Haematopoietic Cells Respond to Growth Factors?

When haemopoietic cells are removed from the bone marrow and cultured *in vitro* in the absence of growth factors, they rapidly die (44). Recent data suggests that death is mediated via activation of a mechanism (67), involving breakdown of DNA into the typical nucleosomal fragments associated with `programmed cell death' or apoptosis. The finding that the continued presence of haematopoietic growth factors is absolutely essential for the maintenance of viability of the haematopoietic stem and progenitor cell has led to some interesting studies at the mechanistic level.

The haematopoietic growth factors elicit their effects on cells by binding to high affinity cell surface receptors. Each growth factor has a specific receptor subunit or complete receptor molecule which binds the growth factor permitting the vectorial flow of "information" across the cell surface membrane. Whilst the receptors are all different, they do fall into families, so for example IL-3 and GM-CSF bind to an αIL-3 and αGM-CSF receptor subunit, and these interact with other subunits to form heterodimeric complexes (68). M-CSF, on the other hand, binds to a cell surface receptor which oligonises upon occupation, thereby activating an intrinsic tyrosine kinase which causes the phosphorylation of other proteins.

Using a variety of cell lines that depend upon various haematopoietic growth factors for their continued survival and proliferation, it has been demonstrated that removal of the growth factor leads to a decrease in the rate of glucose transport (69) and a fall in the levels of intracellular ATP (70), indicating that the growth factors have a major role to play in influencing the primary metabolism of target cells. There is also a change in membrane fluidity (71), as yet of unknown significance. Following re-addition of the growth factor to deprived cells (and, indeed, also to freshly isolated GM-CFC progenitor cell populations), there is a stimulation of protein phosphorylation on tyrosine residues, also serine or threonine specific protein kinases such as protein kinase C (PKC) or MAP kinase are activated by the addition of growth factor (72-74). All these processes appear to be important components of a cascade mechanism leading to the subsequent proliferative events set in motion upon addition of the haematopoietic growth factor.

Perhaps this is not too surprising, since activation of some of these systems appears to be common to most (perhaps all) of the classical growth factors.

What is intriguing, however, is that the response of haematopoietic cells to IL-3, GM-CSF and M-CSF is not associated with rapid changes in intracellular calcium levels (75-77) - suggesting that the activation of PKC is not mediated via the `classical' breakdown patterns of the inositol lipid system. Indeed, this was confirmed in several systems: treatment of cells with IL-3 or M-CSF does not result in rapid hydrolysis of phosphatidyl inositol 4,5 bis phosphate (PIP2) or a change in the proportion of the other inositol lipids (69,74,76,77). Clearly, the haematopoietic growth factors are exerting their effects via alternative pathways - yet to be fully understood, but certainly not involving the other second messenger system, cyclic AMP (78).

One particularly interesting aspect to emerge from this work is that IL-3, GM-CSF and M-CSF (and probably also G-CSF) have a similar effect upon target cells in terms of Na+/H+ activation and intracellular alkalinisation (33). But, the outcome of the response of the same target cells to these agents is quite different - GM-CSF, for example, will promote the development of both neutrophils and macrophages from GM-CFC, whereas M-CSF preferentially stimulates only macrophage production. Thus, it is possible that the biochemical events seen so far, following addition of these growth factors to target cells represent 'house-keeping' functions - necessary to allow the cells to survive and to prime them for proliferation - and that the more specific changes leading to the production of developmentally-restricted progeny have yet to be determined.

Figure 1: Target cells for haematopoietic growth factors. Effects best or only seen using combinations of factors.

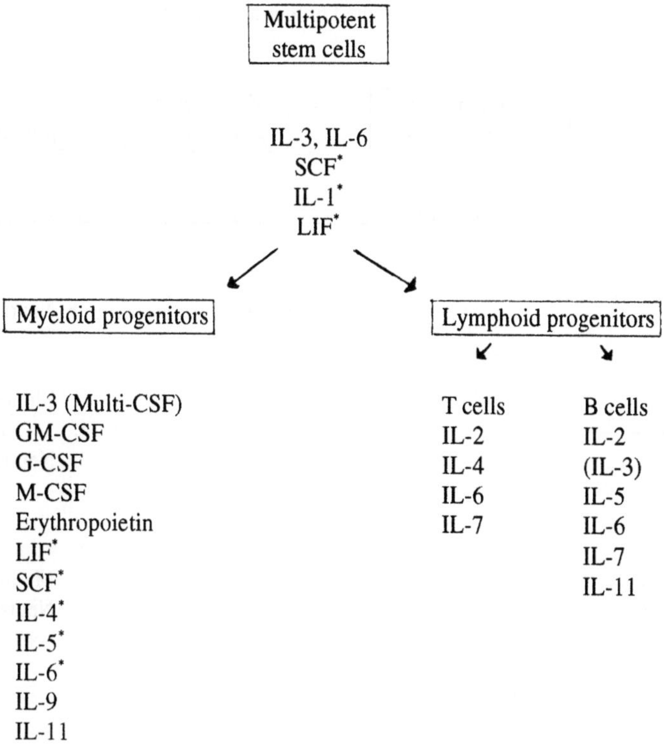

* Effects best or only seen with combinations of factors

Figure 2: Synergistic interactions between haematopoietic growth factors.

Target cells: highly enriched multipotent d12 CFU-S
highly enriched progenitor cells, GM-CFC

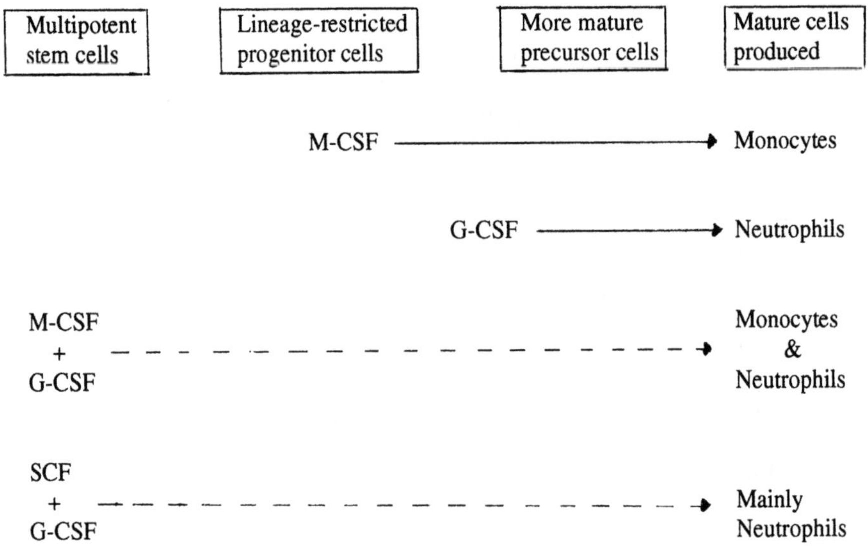

References

1. Cronkite, E.P. and Feinendegen, L.E. (1976). Notions about human stem cells. *Nouv. Rev. Fr. Hematol. Blood Cells* **17(1-2)**: 269-84.

2. Goldman, J.M. In: *Blood and its Disorders*. (Hardisty, R.M. and Weatherall, D.J., eds.). Blackwell, London (1982), p.601-27.

3. Moore, M.A. and Metcalf, D. (1970). Ontogeny of the haemopoietic system: yolk sac origin of *in vivo* and *in vitro* colony forming cells in the developing mouse embryo. *Br. J. Haematol.* **18(3)**: 279-96.

4. Wood, D.G. In: *Blood and its Disorders*. (Hardisty, R.M. and Weatherall, D.J., eds.). Blackwell, London (1982), p.75-98.

5. Metcalf, D. and Moore, M.A.S. (1971). Haemopoietic Cells. North Holland, Amsterdam.

6. Abramson, S., Miller, R.G. and Phillips, R.A. (1977). The identification in adult bone marrow of pluripotent and restricted stem cells of the myeloid and lymphoid systems. *J. Exp. Med.* **145(6)**: 1567-79.

7. Snodgrass, R. and Keller, G. (1987). Clonal fluctuation within the haematopoietic system of mice reconstituted with retrovirus-infected stem cells. *EMBO J.* **6(13)**: 3955-60.

8. Metcalf, D. (1988). The Molecular Control of Blood Cells. Harvard University Press, London.

9. Metcalf, D., Johnson, G.R. and Burgess, A.W. (1980). Direct stimulation by purified GM-CSF of the proliferation of multipotential and erythroid precursor cells. *Blood* **55(1)**: 138-47.

10. Harrison, D.E. (1973). Normal production of erythrocytes by mouse marrow continuous for 73 months. *Proc. Natl. Acad. Sci. USA.* **70(11)**: 3184-8.

11. Ford, C.E., Hamerton, J.L., Barnes, D.W.H. *et al.* (1956). Cytological identification of radiation-chimaeras. *Nature* **177**: 452-4.

12. Bradley, T.R. and Metcalf, D. (1966). The growth of mouse bone marrow cells *in vitro*. *Aust. J. Exp. Biol. Med. Sci.* **44(3)**: 287-99.

13. Pluznik, D.H. and Sachs, L. (1966). The induction of clones of normal mast cells by a substance from conditioned medium. *Exp. Cell Res.* **43(3)**: 553-63.

14. Stephenson, J.R., Axelrad, A.A., McLeod, D.L. and Shreeve, M.M. (1971). Induction of colonies of hemoglobin-synthesizing cells by erythropoietin *in vitro*. *Proc. Natl. Acad. Sci. USA.* **68(7)**: 1542-6.

15. Metcalf, D., MacDonald, H.R., Odartchenko, N. and Sordat, B. (1975). Growth of mouse megakaryocyte colonies *in vitro*. *Proc. Natl. Acad. Sci. USA.* **72(5)**: 1744-8.

16. Metcalf, D., Nossal, G.J., Warner, N.L., Miller, J.F., Mandel, T.E., Layton, J.E. and Gutman, G. (1975). Growth of B-lymphocyte colonies *in vitro*. *J. Exp. Med.* **142(6)**: 1534-49.

17. Nicola, N.A. (1989). Hemopoietic cell growth factors and their receptors. *Ann. Rev. Biochem.* **58**: 45-77.

18. Ploemacher, R.E. and Brons, N.H. (1988). Cells with marrow and spleen repopulating ability and forming spleen colonies on day 16, 12 and 8 are sequentially ordered on the basis of increasing rhodamine 123 retention. *J. Cell Physiol.* **136(3)**: 531-6.

19. Berenson, R.J., Bensinger, W.I., Kalamasz, D. and Martin, P. (1986). Elimination of Daudi lymphoblasts from human bone marrow using avidin-biotin immunoadsorption. *Blood* **67(2)**: 509-15.

20. Spangrude, G.J., Heimfeld, S. and Weissman, I.L. (1988). Purification and characterization of mouse hematopoietic stem cells [published erratum appears in Science 1989, June 2:244(4908): 1030]. *Science* **241(4861)**: 58-62.

21. Visser, J.W., Bauman, J.G., Mulder, A.H., Eliason, J.F. and de Leeuw, A.M. (1984). Isolation of murine pluripotent hemopoietic stem cells. *J. Exp. Med.* **159(6)**: 1576-90.
22. Lord, B.I. and Spooncer, E. (1986). Isolation of haemopoietic spleen colony forming cells. *Lymphokine Res.* **5(1)**: 59-72.
23. Nicola, N.A. and Metcalf, D. (19??). Subunit promiscuity among hemopoietic growth factor receptors. Minireview. *Cell* 1-4.
24. Sanderson, C.J. (1990). Eosinophil differentiation factor (Interleukin-5). In: *Colony-stimulating factors.* (Dexter, T.M., Garland, J.M. and Testa, N.G., eds.). Marcel Dekker Inc., New York, p.231-256.
25. Ikebuchi, K., Ihle, J.N., Hirai, Y., Wong, G.G., Clark, S.C. and Ogawa, M. (1988). Synergistic factors for stem cell proliferation: Further studies of the target stem cells and the mechanism of stimulation by interleukin-1, interleukin-6 and granulocyte colony-stimulating factor. *Blood* **72(6)**: 2007-14.
26. Caracciolo, D, Clar, S.C. and Rovera, G. (1989). Human interleukin-6 supports granulocytic differentiation of hematopoietic progenitor cells and acts synergistically with GM-CSF. *Blood* **73(3)**: 666-70.
27. Rennick, D., Yang, G., Muller-Sieburg, C., Smith, C., Arai, N., Takabe, Y. and Gemmell, L. (1987). Interleukin-4 (B-cell stimulatory factor 1) can enhance or antagonize the factor-dependent growth of hemopoietic progenitor cells. *Proc. Natl. Acad. Sci. USA.* **84(19)**: 6889-93.
28. Baggiolini, M., Imboden, P. and Detmers, P. (1992). Neutrophil activation and the effects of interleukin-8/nuetrophil-activating peptide 1 (IL-8/NAP-1). *Cytokines* **4**: 1-17.
29. Schaafsma, M.R., Falkenburg, J.H., Duinkerken, N., Van-Snick, J., Landegent, J.E., Willemze, R. and Fibbe, W.E. (1993). Interleukin-9 stimulates the proliferation of enriched human erythroid progenitor cells: additive effect with GM-CSF. *Ann. Hematol.* **66(1)**: 45-9.
30. Hamblin, A.S. Lymphokines, IRL Press, Oxford. (1988).
31. Quesniaux, V.F. (1992). Interleukins 9, 10, 11 and 12 and kit ligand: A brief overview. *Res. Immunol.* **143(4)**, 385-400.
32. Heyworth, C.M., Ponting, I.L. and Dexter, T.M. (1988). The response of haemopoietic cells to growth factors: Developmental implications of synergistic interactions. *J. Cell Sci.* **(Pt.2)**: 239-47.
33. Williams, D.E., Straneva, J.E., Cooper, S., Shadduck, R.K., Waheed, A., Gillis, S., Urdal, D. and Broxmeyer, H.E. (1987). Interactions between purified murine colony-stimulating factors (natural CSF-1, recombinant GM-CSF, and recominant IL-3) on the *in vivo* proliferation of purified murine granulocyte-macrophage progenitor cells. *Exp. Hematol.* **15(10)**: 1007-12.
34. Cook, N., Dexter, T.M., Lord, B.I., Cragoe, E.J. Jr. and Whetton, A.D. (1989). Identification of a common signal associated with cellular proliferation stimulated by four haemopoietic growth factors in a highly enriched population of granulocyte/macrophage colony-forming cells. *EMBO J.* **8(10)**: 2967-74.
35. Morris, C.F., Young, I.G. and Hapel, A.J. (1990). Molecular and cellular biology of interleukin-3. In: *Colony Stimulating Factors.* (Dexter, T.M., Garland, J.M. and Testa, N.G., eds.). Marcel Dekker Inc., New York, p.177-214.
36. Gough, N.M. and Nicola, N.A. (1990). Granulocyte-macrophage colony-stimulating factor. In: *Colony Stimulating Factors.* (Dexter, T.M., Garland, J.M. and Testa, N.G., eds.). Marcel Dekker Inc., New York, p.111-53.
37. Kawasaki, E.S. and Ladner, M.B. (1990). Molecular biology of macrophage colony-stimulating factor. In: *Colony Stimulating Factors.* (Dexter, T.M., Garland, J.M. and Testa, N.G., eds.). Marcel Dekker Inc., New York, p.155-76.

38. Nicola, N.A. (1990). Granulocyte colony-stimulating factor. In: *Colony Stimulating Factors.* (Dexter, T.M., Garland, J.M. and Testa, N.G., eds.). Marcel Dekker Inc., New York, p.77-109.
39. Goldwasser, E., Beru, N. and Smith, D. (1990). Erythropoietin. In: *Colony Stimulating Factors.* (Dexter, T.M., Garland, J.M. and Testa, N.G., eds.). Marcel Dekker Inc., New York, p.257-76.
40. Gough, N.H. and Williams, R.L. (1989). The pleiotrophic actions of leukaemia inhibitory factor. *Cancer Cells* **1**: 77-80.
41. Ikebuchi, K., Wong, G.G., Clark, S.C., Ihle, J.N., Hinai, Y. and Ogawa, M. (1987). Interleukin 6 enhancement of interleukin-3-dependent proliferation of multipotent haemopoietic progenitors. *Proc. Natl. Acad. Sci. USA.* **84**: 9035-39.
42. Witte, O.N. (1990). Steel locus defines new multipotent growth factor. *Cell* **63**: 5-6.
43. Heyworth, C.M., Whetton, A.D., Nicholls, S., Zsebo, K. and Dexter, T.M. (1992). Stem cell factor directly stimulates the development of enriched granulocyte and macrophage colony forming cells and promotes the effects of other colony stimulating factors. *Blood* **80**: 2230-2236.
44. Escary, J-L., Perreau, J., Dumenil, D., Ezine, S. and Brulet, P. (1993). Leukaemia inhibitory factor is necessary for maintenance of haematopoietic stem cells and thymocyte stimulation. *Nature* **363**: 361-364.
45. Metcalf, D. (1977). Hemopoietic Colonies. Springer-Verlag, Berlin/Heidelberg.
46. Dexter, T.M., Heyworth, C.M., Spooncer, E. and Ponting, I.L.O. (1990). The role of growth factors in self-renewal and differentiation of haemopoietic stem cells. *Phil. Trans. R. Soc. Lond.* **B237**: 85-98.
47. Whetton, A.D. and Dexter, T.M. (1989). Myeloid haemopoietic growth factors. *Biochim. Biophys. Acta.* **989(2)**: 111-32.
48. Dexter, T.M. (1989). Haemopoietic growth factors. *Br. Med. Bull.* **45(2)**, 337-49.
49. Moore, M.A.S., Warren, D.J. and Souza, L. (1987). In: *UCLA Symposium on Molecular and Cellular Biology. Recent Advances in Leukaemia and Lymphoma.* (Gale, R.P. and Golde, D.W., eds.). Alan R. Liss, New York, p.445-6.
50. Molineux, G., Migdalska, A., Szmitkowski, M., Zsebo, K. and Dexter, T.M. (1991). The effects upon haemopoiesis of recombinant stem cell factor (ligand for c-kit) administered *in vivo* to mice either alone or in combination with G-CSF. *Blood* **78**, 961-6.
51. Testa, N.G. and Dexter, T.M. (1993). Cytokines in haemopoiesis, their applications in oncology. In: *Cancer Chemotherapy.* (Hickman, J.A. and Tritton, T.R., eds.). Blackwell, Oxford, p.344-62.
52. Miossec, P., Cavender, D. and Ziff, M. (1986). Production of interleukin 1 by human endothelial cells. *J. Immunol.* **136(7)**: 2486-91.
53. Flanagan, J.G. and Leder, P. (1990). The kit ligand: A cell surface molecule altered in steel mutant fibroblasts. *Cell* **63**: 185-94.
54. Fibbe, W.E., van Damme, J., Billiau, A., Goselink, H.M., Voogt, P.J., van Eeden, G., Ralph, P., Altrock, B.W. and Falkenburg, J.H. (1988). Interleukin 1 induces human marrow stromal cells in long-term culture to produce granulocyte colony-stimulating factor and macrophage colony-stimulating factor. *Blood* **71(2)**:430-5.
55. Fibbe, W.E., van Damme, J., Billiau, A., Voogt, P.J., Duinkerken, N., Kluck, P.M. and Falkenburg, J.H. (1986). Interleukin-1 (22-K factor) induces release of granulocyte-macropjage colony-stimulating activity from human mononuclear phagocytes. *Blood* **68(6)**:1316-21.

56. Stewart, C.L., Kaspar, P., Brunet, L.J., Bhatt, H., Gadi, In., Kontgen, F. and Abbondanzo, S.J. (1992). Blastocyst implantation depends on maternal expression of leukaemia inhibitory factor. *Nature* **359**: 76-79.

57. Dexter, T.M., Spooncer, E., Simmons, P. and Allen, T.D. (1984). Long-term marrow culture: An overview of techniques and experience. In: *Long-term Marrow Culture*, Alan R. Liss, p.58-96.

58. Dexter, T.M., Allen, T.D. and Lajtha, L.G. (1977). Conditions controlling the proliferation of haemopoietic stem cells *in vitro*. *J. Cell Physiol.* **91(3)**: 335-44.

59. Dexter, T.M. (1982). Stromal cell-associated haemopoiesis. *J. Cell Physiol.* **(Suppl.1):** 87-94.

60. Bentley, S.A., Tralka, T.S. and Alabaster, O. (1981). Phagocytic properties of bone marrow fibroblasts. *Exp. Hematol.* **9(4)**: 313-8.

61. Kierney, P.C. and Dorshkind, K. (1987). B lymphocyte precursors and myeloid progenitors survive in diffusion chamber cultures but B cell differentiation requires close association with stromal cells. *Blood* **70(5)**: 1418-24.

62. Gordon, M.Y., Riley, G.P., Watt, S.M. and Greaves, M.F. (1987). Compartmentalization of a haematopoietic growth factor (GM-CSF) by glycosaminoglycans in the bone marrow microenvironment. *Nature* **326(6111)**: 403-5.

63. Roberts, R., Gallagher, J, Spooncer, E., Allen, T.D., Bloomfield, F. and Dexter, T.M. (1988). Heparan sulphate bound growth factors: A mechanism for stromal cell mediated haemopoiesis. *Nature* **332(6162)**: 376-8.

64. Lord, B.I. and Testa, N.G. (1988). The haemopoietic system: Structure and regulation. In: *Hematopoiesis: Long-term Effects of Chemotherapy and Radiation.* (Testa, N.G. and Gale, R.P., eds.). Marcel Dekker, New York, p.1-26.

65. Testa, N.G., Hendry, J.H. and Molineux, G. (1985). Long-term bone marrow damage in experimental systems and in patients after radiation or chemotherapy. *Anticancer Res.* **5**: 101-10.

66. Schofield, R. (1978). The relationship between the spleen colony-forming clel and the haemopoietic stem cell. *Blood Cells* **4(1-2)**: 7-25.

67. Farrar, W.L., Thomas, T.P. and Anderson, W.B. (1985). Altered cytosol/membrane enzyme redistribution on interleukin-3 activation of protein kinase C. *Nature* **315(6016)**: 235-7.

68. Whetton, A.D., Vallance, S.J., Monk, P.N., Cragoe, E.J., Dexter, T.M. and Heyworth, C.M. (1988). Interleukin-3-stimulated haemopoietic stem cell proliferation. Evidence for activation of protein kinase C and Na+/H+ exchange without inositol lipid hydrolysis. *Biochem. J.* **256(2)**: 585-92.

69. Whetton, A.D., Monk, P.N., Consalvey, S.D., Huang, S.J., Dexter, T.M. and Downes, C.P. (1988). Interleukin-3 stimulates proliferation via protein kinase C activation without increasing inositol lipid turnover. *Proc. Natl. Acad. Sci. USA.* **85(10)**: 3284-8.

70. Coffey, R.G., Davis, J.S. and Djeu, J.Y. (1988). Stimulation of guanylate cyclase activity and reduction of adenylate cyclase activity by granulocyte-macrophage colony-stimulating factor in human blood neutrophils. *J. Immunol.* **140(8)**: 2695-701.

71. Sullivan, R., Griffin, J.D., Simons, E.R>, Schafer, A.I., Meshulam, T., Fredette, J.P., Maas, A.K., Gadenne, A.S., Leavitt, J.L. and Melnick, D.A. (1987). Effects of recombinant human granulocyte and macrophage colony-stimulating factors on signal transduction patheays in human granulocytes. *J. Immunol.* **139(10)**: 3422-30.

72. Whetton, A.D., Monk, P.N., Consalvey, S.D. and Downes, C.P. (1986). The haemopoietic growth factors interleukin-3 and colony-stimulating factor-1 stimulate proliferation but do not induce inositol lipid breakdown in murine bone-marrow-derived macrophages. *EMBO J.* **5(12)**: 3281-6.

73. Lord, B.I., Molineux, G., Testa, N.G., Kelly, M., Spooncer, E. and Dexter, T.M. (1986). The kinetic response of haemopoietic precursor cells, *in vivo*, to highly purified, recombinant interleukin-3. *Lymphokine Res.* **5(2)**: 97-104.

74. Metcalf, D., Begley, C.G., Johnson, G.R., Nicola, N.A., Lopez, A.F. and Williamson, D.J. (1986). Effects of purified bacterially synthesized murine multi-CSF (IL-3) on hematopoiesis in normal adult mice. *Blood* **68(1)**: 46-57.

75. Shadduck, R.K., Waheed, A., Boegel, F., Pigoli, G., Porcellini, A. and Rizzoli, V. (1987). The effect of colony stimulating factor-1 *in vivo*. *Blood Cells* **13(1-2)**: 49-63.

76. Garnich, M.B., Stoudeam, R.E.J. and Donahue, R.S. (1989). *Proc. Am. Soc. Clin. Oncol.* **8**, Abstract 717.

77. Tamura, M., Hattori, K., Nomura, H., Oheda, M., Kubota, N., Imazeki, I., Ono, M., Ueyama, Y., Nagata, S., Shirafuji, N. *et al.* Induction of neutrophilic granulocytosis in mice by administration of purified human native granulocyte colony-stimulating factor (G-CSF). *Biochem. Biophys. Res. Commun.* **142(2)**: 454-60.

78. Molineux, G., Pojda, Z. and Dexter, T.M. (1990). A comparison of hematopoiesis in normal and splenectomized mice treated with granulocyte colony-stimulating factor. *Blood* **75(3)**: 563-9.

Mos, Meiosis and Cellular Transformation

Kenji Fukasawa, Taesaeng Choi, and George F. Vande Woude
ABL-Basic research Program
NCI-Frederick Cancer research & Development Center
Frederick, MD 21702, USA

Mos and meiotic interphase.

Mos was initially identified as the cellular sequence acquired by Moloney murine sarcoma virus (MuSV) produced in sarcomas developed in new born mice (Moloney, 1966; Frankel and Fischinger, 1976). Mos is a serine/threonine kinase and is highly oncogenic *in vitro* and *in vivo* (reviewed by Yew et al., 1993). High levels of c-*mos* mRNA are present in ovaries (Propst and Vande Woude, 1985) and c-Mos has been shown to function in oocytes during meiotic maturation (Sagata et al., 1988,1989a; Freeman et al., 1989; O'Keefe et al., 1989; Paules et al., 1989; Barrett et al., 1990; Zhao et al., 1991). In *Xenopus*, oocytes initiate meiotic maturation after hormonal stimulation and arrest at metaphase II as unfertilized eggs (Masui and Markert, 1971). In *Xenopus* oocytes, Mos is required for both initiation of meiotic maturation (Sagata et al., 1989a) and metaphase II-arrest (Sagata et al., 1989b; Colledge et al., 1994; Hashimoto et al., 1994), functions that were attributed to maturation promoting factor (MPF) or $p34^{cdc2}$ kinase and cyclin B (Gautier et al., 1988,1990; Draetta et al., 1989). MPF activation is an absolute requirement for the onset of meiosis, and is implicated in series of M-phase events (Lohka and Maller, 1985; Murray and kirschner, 1989), including chromosome condensation, germinal vesicle breakdown (GVBD), and reorganization and formation of meiotic spindles.

Mitogen activated protein kinase (MAPK), which functions in somatic cells as a key molecule in signal transduction pathway in response to a mitogenic stimuli (reviewed by Crews et al, 1992; Thomas, 1992; Pelech and Sanghera, 1992), has been shown to be activated thoughout meiosis (Ferrell et al., 1991; Fukasawa et al., 1994; Verlhac et al., 1994; Fig. 1). Moreover, Mos was identified as an activator of MAPK pathway through directly phosphorylating and activating MAPK kinase (MEK) (Posada et al., 1993; Nebreda and Hunt, 1993; Shibuya and Ruderman, 1993; Resing et al., 1995). Fig. 1A illustrates the kinetics of MPF and MAPK activities during meiotic maturation of mouse oocytes. Soon after oocytes were released from GV-stage, both MPF and MAPK are

NATO ASI Series, Vol. H 99
Tumor Biology
Regulation of Cell Growth, Differentiation and Genetics in Cancer
Edited by Asterios S. Tsiftsoglou, Alan C. Sartorelli, David E. Housman and T. Michael Dexter

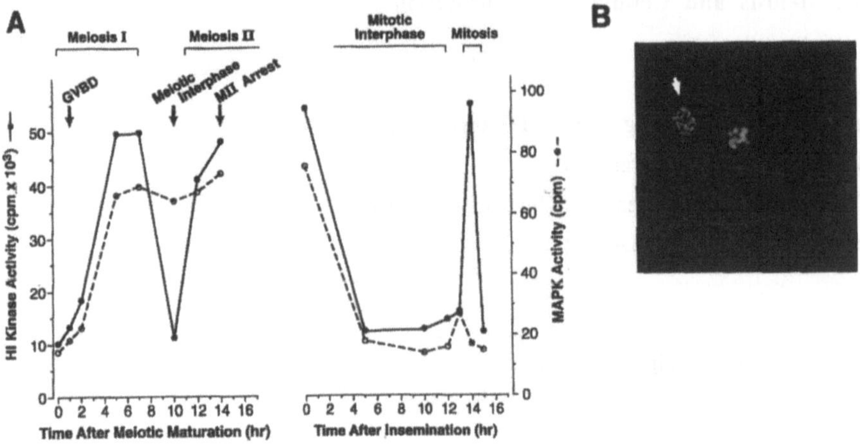

Fig. 1. MAPK and MPF activities during meiosis and the first mitotic cycle.

A. Maturing oocytes or one-cell embryos were assayed for MAPK and MPF activities at the indicated time points. MPF activity was measured in total cell extracts from 10 oocytes or 10 embryos using histone H1 as a substrate. MAPK activity was measured by the in-gel kinase assay method using myelin basic protein (MBP) as a substrate.

B. The oocyte during meiotic interphase was stained with 4'-6-diamino-2-phenylindol (DAPI) DNA dye. The arrow indicates the first polar body.

MPF activity oscillates during meiotic maturation. In particular, MPF activity rapidly decreases at or near the end of meiosis I. By contrast, MAPK, once activated, remains active thoughout meiosis. In mitotic cell cycle (of one-cell embryo), MPF activity increases at the onset of mitosis. However, no notable activation of MAPK was observed during mitosis.

concomitantly activated. However, during metaphase-anaphase transition in meiosis I, MPF activity abruptly decreases, while MAPK activity remains high. After extrusion of first polar body (Fig. 1B), oocytes temporally enter a phase often referred to as "meiotic interphase". Unlike mitotic interphase, no DNA replication is observed, chromosomes remain condensed and nuclear membrene does not reform despite of the absence of MPF activity (Fig. 1B). The persistence of these phenotypes during meiotic interphase in the absence of MPF activity has been enigmatic, but our initial clue as to how it may occur came from examination of somatic cells overexpressing Mos (see the legend to Fig. 2, Fig. 2A). In those cells, inappropriate chromosomal condensation (Fig. 2D) and partial nuclear

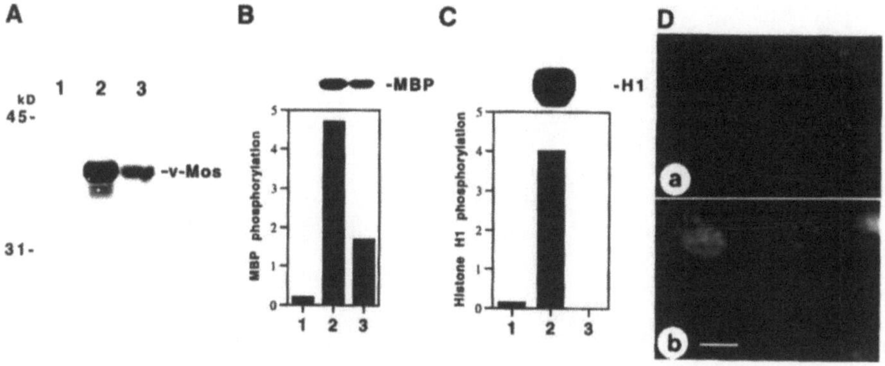

Fig. 2. Characterization of mouse fibroblasts (Swiss 3T3) overexpressing v-Mos.

A. Swiss 3T3 cells were infected with MuSV (clone 124) (Ball et al., 1973). The multiplicities of infection were >5 focus-foming units per cell to ensure all the cells were infected. As a control, cells were infected with murine leukemia virus (MuLV). At 60 - 70 hours post-infection (P.I.), >70% of the MuSV-infected cells rounded-up and floated into the media (floating cells). No phenotypic changes were observed in control MuLV-infected cells. The majority of the floating cells were cell cycle-arrested at 2N- or 4N-DNA content (Fig. 5). These viable floating cells were examined for Mos expression by immunoblot using anti-mouse Mos. No Mos-expression was observed in the control MuLV-infected cells (lane 1). The MuSV-infected floating cells expressed high levels of Mos (lane 2), which was 3-fold more than the MuSV-infected cells that remained attached to the dish surface at 70 hours P.I. (lane 3).

B. Anti-MAPK immunoprecipitates prepared from the control MuLV-infected cells (lane 1), the MuSV-infected floating cells (lane 2), and the MuSV-infected cells that remained attached to the dish surface at 70 hours P.I. (lane 3) were examined for the MBP-phosphorylation activity. Consistent with the high level expression of Mos, the MuSV-infected floating cells display high MAPK activation.

C. $p34^{cdc2}$ kinases precipitated using suc1-beads from the control MuLV-infected cells (lane 1), the mitotic cells harvested after nocodazole treatment (lane 2), and the MuSV-infected floating cells (lane 3) were assayed for histone H1 kinase activity. The control MuLV-infected cells display low H1 kinase activity (~3% of that observed in mitotic cells), consistent with the fact that they are proliferating and 3-5% of the cell population are at mitosis. By contrast, no H1 kinase activity was observed in the MuSV-infected floating cells. Thus, the MuSV-infected floating cells contain highly activated MAPK in the absence of $p34^{cdc2}$ kinase activity.

D. DAPI-staining of the MuSV-infected floating cells (panel b) shows that the majority of the cells contain partially condensed chromosomes. Panel a shows DAPI-staining of the control proliferating MuLV-infected cells. Bar, 10μm.

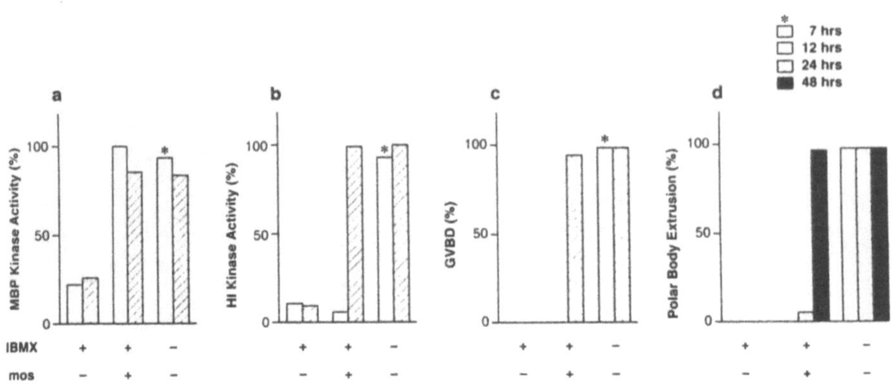

Fig. 3. MAPK, MPF, GVBD and first polar body extrusion are induced by Mos in the presence of IBMX.

Immature mouse oocytes were either maintained in IBMX, or injected with *mos* RNA and maintained in IBMX or the IBMX was removed to allow meiotic maturation to occur. Oocytes were selected after 7, 12, 24, and 48 hours: a) MAPK activity was measured as in Fig. 1A) histone H1 kinase (MPF) activity was measured in total cell extracts; c) GVBD and; d) extrusion of the first polar body were determined. Kinase activities were normalized to the highest activity. GVBD, histone H1 kinase, and MAPK activation were prevented when oocytes were cultured with IBMX without *mos* RNA-injection. However, *mos* RNA-injected oocytes, at 12 hours after culture in the presence of IBMX, had high MAPK activity in the absence of histone H1 kinase activity. At 24 hours, GVBD had occurred in the *mos* RNA-injected oocytes and histone H1 kinase and MAPK activities were elevated to the control maturing oocyte levels. These oocytes also extruded the first polar body, and arrested at metaphase II by 48 hours.

membrene were observed (Fukasawa et al., 1995a) in the absence of $p34^{cdc2}$ kinase-associated histone H1 kinase activity (Fig. 2C). However, these cells displayed high constitutive activation of MAPK (Fig. 2B), suggesting that MAPK may be responsible for the chromosome condensation during meiotic interphase. This prediction was further supported by microinjecting *mos* or gain-of-function mutant*mek* RNA into mouse oocytes in the presence of IBMX. IBMX is a phosphodiesterase inhibitor, and thus stabilizes protein kinase A (PKA). We have previously shown that PKA activation efficiently block MPF activation, but does not interfere with the activation of MAPK pathway (Matten et al., 1994). The *mos* or *mek* RNA injection into mouse oocytes under such conditions triggers activation of MAPK in the absence of MPF activity (Fig. 3), and oocytes show both condensation of chromosomes as well as partial nuclear membrene breakdown (Choi et al., 1996), and could explain why, during meiotic interphase, chromosomes remain ciondensed and the nuclear envelope does not reform.

Meiosis vs. mitosis: functions of Mos/MAPK in early stages of meiosis.

In *Xenopus* oocytes, activation of the Mos/MAPK pathway appears to be required for early stages of meiosis I including initiation of meiotic maturation upon hormonal stimulation (Sagata et al., 1989a). Mouse oocytes microinjected with *mos* RNA in the presence of IBMX mature slower than normal oocytes, and MAPK is activated in the absence of MPF under these conditions (Choi et al., 1996; Fig. 3). This allowed us to examine the potentional functions of the Mos/MAPK pathway in the early stages of meiosis. At 12 hours after injection of *mos* RNA, we observed multiple nuclear asters (Fig. 4A, a-a"), and many of the asters appeared to associate with condensed chromosomes (Fig. 4A, a"). These phenotypes differ from mitosis, in which two centrosomes orthogonally position outside the nucleus at G2/M boarder, nucleate microtubules to form asters, and capture the condensed chromosomes after nuclear envelope breakdown (Fig. 4B). These data implicate the Mos/MAPK pathway in the early events of meiotic maturation. After 14 hours, MPF activation is triggered probably through the newly-synthesized cyclin B combining with dephosphorylated $p34^{cdc2}$ (Choi et al., 1996). Concomitant with MPF activation, GVBD occurs and the chromosome-associated asters congregate to form a pre-srpindle (Fig. 4A, b and c). Centrosomal proteins such as 5051 antigen [pericentrin (Doxsey et al., 1994)] and γ-tubulin (Oakley , 1990; Stearns et al., 1991; Joshi et al., 1992) (not shown) are present in this complex (Fig. 4A, c), and migrate to the poles as the spindles polewardly elongates (Fig.4A, d-f). The differences in the formation of spindle apparatus between mitosis and meiosis are schematically depicted in Fig. 4B. These observations suggest that at least some of the differences are mediated by the Mos/MAPK pathway.

Possible functions of Mos/MAPK in the control of polar body formation.

Requirement of the Mos/MAPK pathway in meiosis-specific microtubule-reorganization has been suspected (Gotoh et al., 1991; Verlhac et al., 1993; Fukasawa et al., 1994,1995b). Upon introduction of high levels of Mos into 3T3 cells by MuSV-infection, the majority of cells either become growth-arrested or undergo apoptosis (see the legend to Fig. 2). Among the arrested cells, ~30% were arrested with 4N-DNA content and binucleated, indicating that the cells failed to undergo cytokinesis (Fig. 5). Immunocytological examination of cells prior to the cytokinesis block showed that their

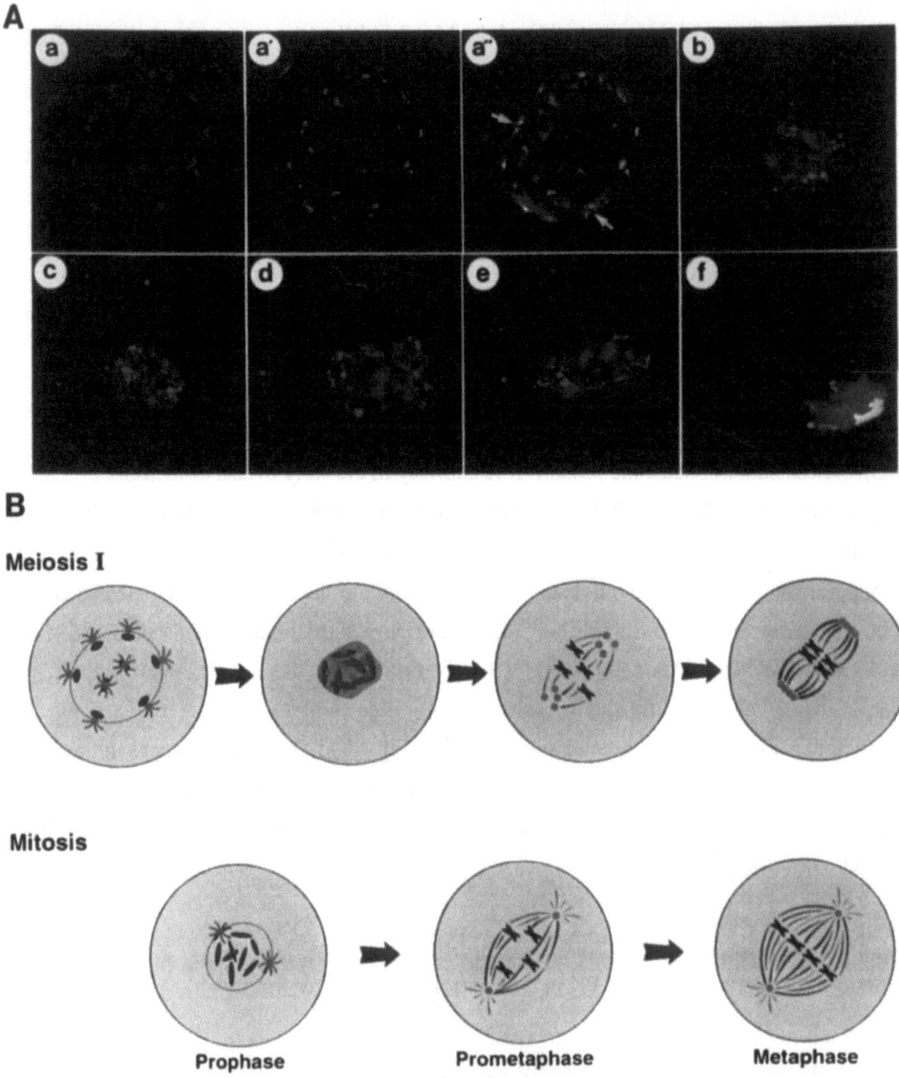
A
a
a'
a''
b
c
d
e
f
B
Meiosis I
Mitosis
Prophase
Prometaphase
Metaphase

Fig. 4. Confocal laser scanning microscopy analysis of meiotic spindle formation.

(A) Oocytes maintained in IBMX were selected at 12 hours through 48 hours after *mos* RNA-injection. Each panel shows a representative oocyte from each time point stained for tubulin (green), centrosome [5051 antigen (red)], and DNA (blue). Panels a-a" show the oocyte stained at 12 hours for tubulin (panel a), 5051 antigen (panel a'), and overlay analyses with DNA staining (panel a"). The remaining panels (b-f) are the overlay analyses of representative oocytes examined at 16 hours (panels b and c), 18 hours (panels d and e), and 20 hours (panel f). At 12 hours, the centrosomal staining is directly localized with the microtubule arrays and often colocalize with condensed chromatin. The 5051 staining is more punctated, and frequently bivalent staining patterns were observed (indicated by arrows in panel a"). After MPF activation and GVBD, the 5051 antigens redistribute to perimeter of the "pre-spindle" and is less colocalized with microtubules (panel b). At 18 hours (panels d and e), during early stages of spindle pole formation, 5051 antigens preferentially localize at the emerging spindle poles.

(B) Schematic description of differences in spindle formation between meiosis I and mitosis. At the onset of meiosis I, numerous asters form in and around the nucleus and physically associate with centrosomal proteins (5051 antigens, shown in red) and condensed chromosomes (shown in blue). Upon GVBD, the "pre-spindle" forms. During this stage, association of microtubules, 5051 antigens and chromosomes are less distinctive. The "pre-spindle" then elongate to the poles to form metaphase spindles, and 5051 antigens move polewardly along with the elongating spindles and eventually localize at the spindle poles. In mitotic cells, at G2/M boarder, two well-defined centrosomes (shown in red) translocate around the nucleus to position orthogonally. After the breakdown of nuclear membrene, the microtubules radiating from centrosomes capture condensed chromosomes to form metaphase spindles.

spindle apparatus displayed some features specific to meiotic spindles, in particular diminished astral microtubules (Fig. 6c). In mitosis, astral microtubules radiating from spindle poles (Fig. 6a) are involved in positioning of the apparatus in the center of the cell, ensuring proper formation of the cleavage furrow (Dan et al., 1990). In the cells overexpressing Mos prior to cytokinesis block, mal-positioning of the spindles, in which one (Fig. 6d) or both spindle poles (Fig. 6f) juxtaposed to the cell surface, was observed. This is perhaps responsible for the failure to undergo cytokinesis in these cells.

During meiosis I, a metaphase spindle forms in the center of the oocyte before translocating and juxtapositioning to the cell surface. This process is prerequisite for the subsequent asymmetric cell divisions to produce two daughter cells which differ greatly in size. This is referred to as a polar body formation and extrusion. The polar body (Fig. 1B, indicated by an arrow) is small in size, and especially the first polar body which is formed at the end of meiosis I rapidly disintegrates. The mal-positioning of spindle apparatus observed in 3T3 cells overexpressing Mos and containing highly activated MAPK may represent the inappropriate polar body formation processes in a somatic cell. These studies suggest that the Mos/MAPK pathway may be involved in the processes of oocyte meiosis-specific polar body formation and extrusion.

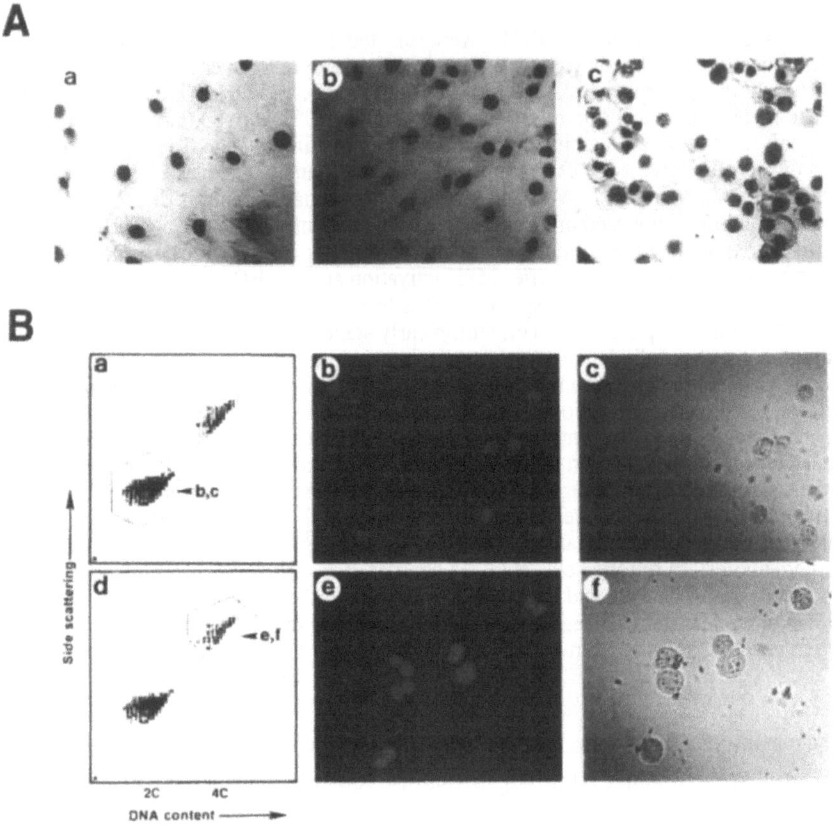

Fig. 5. Microscopic and FACS analyses of 2N- and 4N-arrested MuSV-infected floating cells.

A. The floating cells collected from the media at 70 hours after MuSV-infection (see the legend to Fig. 2) were Giemsa-stained (panel c). As controls, mock-infected cells (panel a) and MuLV-infected cells (panel b) were stained. ~30% of the MuSV-infected floating cells are binucleated (panel a).

B. The MuSV-infected floating cells were sorted into 2N- (panel a) and 4N-arrested (panel d) populations. Panels b and e show propidium iodide-stained 2N- and 4N-arrested floating cells and panels c and f show the phase-contrast images of panels b and e, respectively. >90% of 4N- arrested cells are binucleated, suggesting the failure of those cells to undergo cytokinesis.

The Mos/MAPK pathway and cellular transformation.

We have proposed that the activation of the Mos/MAPK pathway in somatic cells constitutively imposes molecular changes normally reserved only for oocyte maturation.

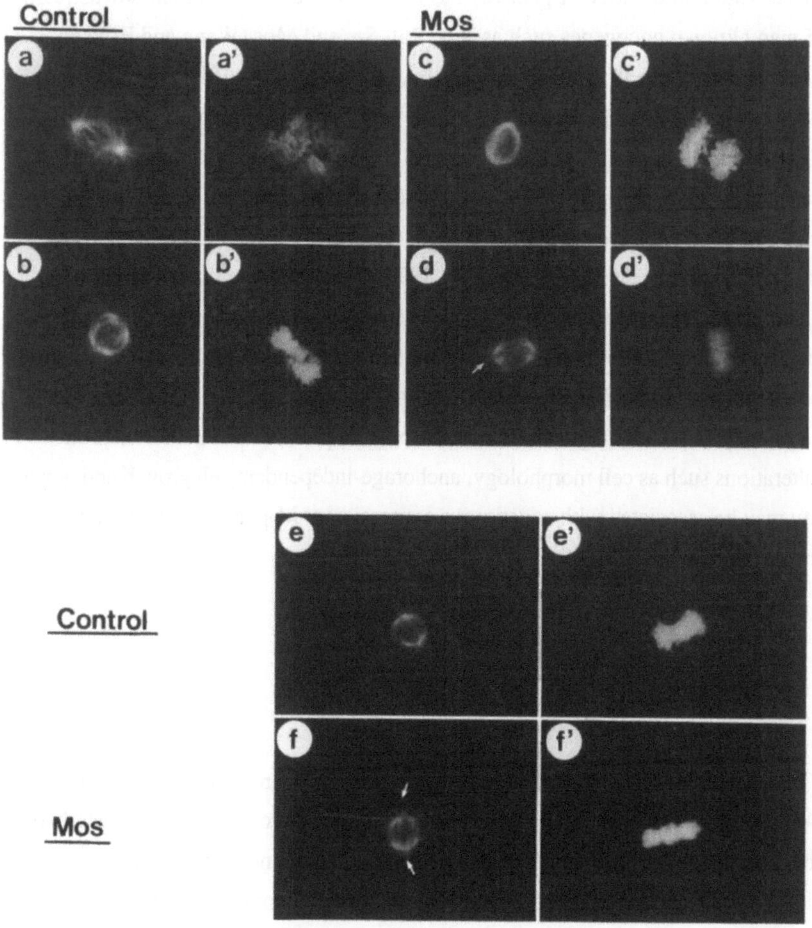

Fig. 6. The floating cell precursors contain mitotic spindle poles lacking astral microtubules and asymmetrically (or aberrantly) positioned spindle apparatus.

The control mock-infected Swiss 3T3 cells and MuSV-infected cells that remained attached to the culture slides at 40 hours P.I. were immunostained with anti-ß-tubulin (panels a-f) and were also stained with DAPI (panels a'-f'). The asters in the MuSV-infected cells were undetectable (panel c) or do not radiate from centrosomes as observed in prophase through metaphase control cells (panel a). In the MuSV-infected cells, the spindle apparatus was asymmetrically positioned with one spindle pole juxtaposed to the cell surface (panel d). The attachment of the spindle pole to the membrene was further revealed by a cell in which both spindle poles were associated with the antipodal cell membrenes, resulting in spindle apparatus and chromosomes oriented perpendicular to the longitudinal axis of the cell (panel f). For comparison, mock-infected cells with a similar cell shape show the spindle apparatus oriented properly, parallel to the longitudinal axis of the cell (panel e).

Since the activation of the MAPK pathway has been shown to be a common downstream target of many known oncogenes such as Ras, Raf, Src and Mos (Wang and Erikson, 1992; Leevers and Marshall, 1992, Howe et al., 1992, Dent et al., 1992, Kyriakis et al., 1992; Gupta et al., 1992) and is essential for their cellular transforming activities (Cowley et al., 1994; Mansour et al., 1994; Fukasawa et al., 1994; Okazaki and Sagata, 1995; Fukasawa et al., 1995c), the inappropriate alterations imposed by MAPK activities in oocyte maturation may be more general. Indeed, oncogenic Ras and Src have also been shown to exert M-phase activities (Daar et al., 1991; Chackalaparampil and Shalloway, 1988). At present, it is not clear how M-phase-associated activities of those oncogenes contribute to induction and/or maintenance of transformed phenotypes. Considering the profound effects of the Mos/MAPK pathway on the dynamic behavior of macromolecules observed in oocytes undergoing meiotic maturation, it is possible that transformation-related alterations such as cell morphology, anchorage-independent cell growth and genetic instability may be associated with constitutive expression of M-phase events through the activation of the MAPK pathway.

Acknowledgements

This research was sponsored in part by the National Cancer Institute, DHHS, with ABL. Additional research funding was provided by the G. Harold and Leila Y. Mathers Charitable Foundation. The contents of this publication do not necessarily reflect the views or policies of the Department of Health and Human Services, nor does mention of trade names, commercial products, or organizations imply endorsement by the U.S.

References

Barrett CB, Schroetke RM, Van der Hoorn FA, Nordeen SK, and Maller JL (1990) Ha-ras^{val12thr59} activates S6 kinase and p34^{cdc2} kinase in *Xenopus* oocytes: evidence for c-*mos*^{xe}-dependent and independent pathways. Mol. Cell. Biol. 102:310-315

Chackalaparampil I and Shalloway D (1988) Altered phosphorylation and activation of pp60^{src} during fibroblast mitosis. Cell 52:801-810

Choi T, Rulong S, Resau J, Fukasawa K, Matten W, Kuriyama R, Mansour SJ, Ahn NG, and Vande Woude GF (1996) Mos/MAPK can induce early meiotic phenotypes in the absence of MPF: A novel system for analyzing spindle formation during meiosis I. Proc. Natl. Acad. Sci. USA, in press.

Colledge WH, Carlton MBL, Udy GB, and Evans MJ (1994) Disruption of c-*mos* causes parthenogenetic development of unfertilized mouse eggs. Nature 370:65-68

Cowley S, Paterson H, Kemp P, and Marshall CJ (1994) Activation of MAP kinase kinase is necessary and sufficient for PC12 differentiation and for transformation of NIH3T3 cells. Cell 77:841-852

Crews CM, Alessandrini A, and Erikson RL (1992) Erks: their fifteen minutes has arrived. Cell Growth & Diff. 3:135-142

Daar I, Nebreda AR, Yew N, Sass P, Paules R, Santos E, Wigler M, and Vande Woude GF (1991) The ras oncoprotein and M-phase activity. Science 253:74-76

Dan K and Tanaka Y (1990) Attachment of one spindle pole to the cortex in unequal cleavage. In Cytokinesis: Mechanisms of furrow formation during cell division. pp. 108-119

Dent P, Haser W, Haystead TAJ, Vincent LA, Roberts TM, and Sturgill TW (1992) Activation of mitogen-activated protein kinase by v-Raf in NIH/3T3 cells and *in vitro.* Science 257:1404-1407

Doxsey SJ, Stein P, Evans L, Calarco PD, and Kirschner M (1994) Pericentrin, a highly conserved centrosome protein involved in microtubule organization. Cell 76:639-650

Draetta G, Luca F, Westendorf J, Brizuela L, Ruderman J, and Beach D (1989) cdc2 protein kinase is complexed with both cyclin A and B: evidence for proteolytic inactivation of MPF Cell 56:829-838

Ferrell J, Wu M, Gerhart J, and Martin G (1991) Cell cycle tyrosine phosphorylation of $p34^{cdc2}$ and a microtubule-associated protein kinase homolog in *Xenopus* oocytes and eggs. Mol. Cell. Biol. 11:1965-1971

Frankel AE and Fischinger PJ (1976) Nucleotide sequences in mouse DNA and RNA specific for Moloney sarcoma virus. Proc. Natl. Acad. Sci. USA 73:3705-3709

Freeman RS, Pickham KM, Kanki JP, Lee BA, Pena SV, and Donoghue DS (1989) Xenopus homolog of the *mos* proto-oncogene transforms mammalian fibroblasts and induces maturation of *Xenopus* oocytes. Proc. Natl. Acad. Sci. USA 86:5805-5809

Fukasawa K, Murakami MS, Blair DG, Kuriyama R., Hunt T, Fichinger P. and Vande Woude GF (1994) Similarities between somatic cells overexpressing the *mos* oncogene and oocytes during meiotic interphase. Cell Growth & Diff. 5:1093-1103.

Fukasawa K, Rulong S, Resau J, Pinto da Silva P and Vande Woude GF (1995a) Overexpression of *mos* oncogene product in Swiss 3T3cells induces apoptosis preferentially during S-phase. Oncogene 10:1-8.

Fukasawa K and Vande Woude GF (1995b) Mos overexpression in Swiss 3T3 cells induces meiotic-like alterations of the mitotic spindle. Proc. Natl. Acad. Sci. USA 92:3430-3434

Fukasawa K, Zhou R, Matten WT, Armstrong AJ, Darr I, Oskarsson M, Sathyanarayana BK, Maclvor L, Wood TG, and Vande Woude GF (1995c) Mutagenic analysis of functional domains of the *mos* protooncogene and identification of the sites important for MAPK activation and DNA binding. Oncogene 11:1447-1457

Gautier J, Norbury C, Lohka M, Nurse P, and Maller J (1988) Purified maturation-promoting factor contains the product of a *Xenopus* homolog of the fission yeast cell cycle gene *cdc2*. Cell 54:433-439

Gautier J, Minshull j, Locka M, Glotzer M, Hunt T, and Maller J (1990) Cyclin is a component of maturation-promoting factor from Xenopus. Cell 60:487-494

Gotoh Y, Nishida E, Matsuda S, Shiina N, Kosako H, Shiokawa K, Akiyama T, Ohta K, and Sakai H (1991) *In vitro* effects on microtubule dynamics of purified *Xenopus* M-phase-activated MAP kinase. Nature 349:251-254

Gupta SK, Gallego C, Johnson GL, and Heasley LE (1992) MAP kinase is constitutively activated in *gip2* and *src* transformed Rat 1a fibroblasts. J. Biol. Chem. 267:7987-7990

Hashimoto N, Watanabe N, Furuta Y, Tamemoto H, Sagata N, Yokoyama M, Ozaki K, Nagayoshi M, Takeda N, Ikawa Y, and Aizawa S (1994) Parthenogenetic activation of oocytes in c-*mos*-deficient mice. Nature 370:68-71

Howe LR, Leevers SJ, Gomez N, Nakielny S, Cohen P, and Marshall C (1992) Activation of MAP kinase pathway by the protein kinase raf. Cell 71:335-342

Joshi HC, Palacios MJ, McNamara L, and Cleveland DW (1992) γ-tubulin is a centrosome protein required for cell cycle-dependent microtubule nucleation. Nature 356:80-83

Kyriakis JM, App H, Zhang X-F, Banerjee P, Brautigan DL, Rapp UR, and Avruch J (1992) Raf-1 activates MAP kinase-kinase. Nature 358:417-421

Leevers SJ and Marshall C (1992) Activation of extracellular signal-regulated kinase, ERK2, by $p21^{ras}$ oncoprotein. EMBO J. 11:569-574

Lohka MJ and Maller JL (1985) Induction of nuclear envelope breakdown, chromosome condensation and spindle formation in cell free extracts. J. Cell Biol. 101:518-523

Mansour SJ, Matten WT, Hermann AS, Candia JM, Rong S, Fukasawa K, Vande Woude GF, and Ahn NG (1994) Transformation of mammalian cells by constitutively active MAP kinase kinase. Science 265:966-970

Masui Y and Markert CL (1971) Cytoplasmic control of nuclear behavior during meiotic maturation in frog oocytes. J. Exp. Zool. 117:129-146

Matten W, Daar I, and Vande Woude GF (1994) Protein kinase A acts at multiple points to inhibit *Xenopus* oocyte maturation. Mol. Cell. Biol. 14:4419-4426

Moloney JB (1966) A virus-induced rhabdomyosarcoma of mice. Natl. Cancer Inst. Monogr. 22:139-142

Murray AW and Kirschner MW (1989) Dominoes and clocks: the union of two views of the cell cycle. Science 246:614-621

Nebreda AR and Hunt T (1993) The c-*mos* proto-oncogene protein kinase turns on and maintains the activity of MAP kinase, but not MPF, in cell-free extracts of *Xenopus* oocytes and eggs. EMBO J. 12:1979-1986

Oakley BR, Oakley CE, Yoon Y, and Jung MK (1990) γ-tubulin is a component of the spindle pole body that is essential for microtubule function in *Aspergillus nidulans*. Nature 338:662-664

Okazaki K and Sagata N (1995) MAP kinase activation is essential for oncogenic transformation of NIH3T3 cells by Mos. Oncogene 10:1149-1157

O'Keefe SJ, Wolfes H, Kiessling AA, and Cooper GM (1989) Microinjection of antisense c-mos oligonucleotides prevents meiosis II in the maturing mouse egg. Proc. Natl.Acad. Sci. USA 86:7038-7042

Paules RS, Buccione R, Moschel RC, Vande Woude GF, and Eppig JJ (1989) The mouse c-*mos* proto-oncogene product is present and functions during oogenesis. Proc. Natl. Acad. Sci. USA 86:5395-5399

Pelech SL and Sanghere JS (1992) MAP kinase: charting the regulatory pathways. Science 257:1355-1356

Posada J, Yew N, Ahn NG, Vande Woude GF and Cooper JA (1993) Mos stimulates MAP kinase in *Xenopus* oocytes and activates a MAP kinase kinase *in vitro*. Mol. Cell. Biol. 13:2546-2553

Propst F and Vande Woude GF (1985) c-*mos* proto-oncogene transcripts are expressed in mouse tissues. Nature 335:516-518

Resing KA, Mansour SJ, Hermann AS, Johnson RS, Candia JM, Fukasawa K, Vande Woude GF, and Ahn NG (1995) Determination of v-Mos-catalyzed phosphorylation sites and autophosphorylation sites on MAP kinase kinase by ESI/MS. Biochemistry 34:2610-2620

Sagata N, Oskarsson M, Copeland T, Brumbaugh J, and Vande Woude GF (1988) Function of c-*mos* proto-oncogene product in meiotic maturation in *Xenopus* oocytes. Nature 335:519-625

Sagata N, Darr I, Oskarsson M, Showalter S, and Vande Woude GF (1989a) The product of the *mos* proto-oncogene as a candidate "initiator" for oocyte maturation. Science 245:643-646

Sagata N, Watanabe N, Vande Woude GF, and Ikawa Y (1989b) The c-*mos* proto-oncogene product is a cytostatic factor responsible for meiotic arrest in vertebrate eggs. Nature 342:512-518

Shibuya E and Ruderman J (1993) Mos induces the *in vitro* activation of mitogen-activated protein kinases in lysates of frog oocytes and mammalian somatic cells. Mol. Biol. Cell 4:781-790

Stearns T, Evans L, and Kirschner M (1991) γ-tubulin is a highly conserved component of the centrosome. Cell 65:825-836

Thomas G (1992) MAP kinase, by any other name smells just as sweet. Cell 68:3-6

Verlhac MH, Pennart HD, Maro B, Cobb MH, and Clark HJ (1993) MAP kinase becomes stably activated at metaphase and is associated with microtubule-organizing centers during meiotic maturation of mouse oocytes. Dev. Biol. 158:330-340

Verlhac MH, Kubiak JZ, Clarke HJ, and Maro B (1994) Microtubule and chromatin behavior follow MAP kinase activity but not MPF activity during meiosis in mouse oocytes. Development 120:1017-1025

Wang H-CR and Erikson RL (1992) Activation of protein serine-threonine kinases p42, p63, and p37 in Rous sarcoma virus-transformed cells: signal transduction/transformation dependent MBP kinases. Mol. Biol. Cell 3:1329-1337

Yew N, Strobel M, and Vande Woude GF (1993) Mos and the cell cycle: the molecular basis of the transformed phenotype. Curr. Opi. Gen. Dev. 3:19-25

Zhao X, Singh B, and Batten BE (1991) The role of c-*mos* proto-oncogene in mammalian meiotic maturation. Oncogene 6:43-49

Cell Cycle Regulation in Intestinal Epithelial Cells

Andrei L. Gartel, Medeia Gartel, Eugene Goufman, and Angela L. Tyner
Dept. of Genetics
University of Illinois
Chicago, IL 60612
USA

Abstract

To investigate the role that cell cycle regulation may play during differentiation of intestinal epithelial cells, we examined the expression of p53, p21, cyclin D1, cyclin E and E2F in differentiating Caco-2 cells. Caco-2 cells were established from a human colon adenocarcinoma, and differentiate spontaneously after reaching confluence. p53 was undetectable in undifferentiated Caco-2 cells, while significant levels of p21 were observed. p53 transcripts could be detected by day 14 and levels increased slightly toward day 21. Steady state levels of p21 mRNA increased approximately four fold during process of Caco-2 differentiation. The levels of cyclins D1, E and transcription factor E2F-1 mRNA increased early during the differentiation of Caco-2 cells, but decreased as the cells were maintained in culture in a differentiated state. Our data suggest that the Caco-2 cell line will provide a good model system for studying the function of p21 and other cell cycle regulators in the process of terminal differentiation.

Introduction

Major transitions in the cell cycle are regulated by the cyclin-dependent protein kinases (Cdks). Cdk activity is modulated by the cyclins which bind and activate the Cdks (Morgan, 1995). Each Cdk interacts with a particular subset of cyclins, and each mammalian cyclin can interact with multiple Cdks. Cyclin function is primarily controlled by changes in cyclin levels. Cdks are also regulated by phosphorylation of threonine and tyrosine residues (Kato et al., 1994). Regulators of G1 progression in mammalian cells include D-type cyclins (D1, D2, D3), and cyclin E (Sherr and Roberts, 1995). D-type cyclins are associated with either cdk4 or cdk6, and cyclin E is

NATO ASI Series, Vol. H 99
Tumor Biology
Regulation of Cell Growth, Differentiation and Genetics in Cancer
Edited by Asterios S. Tsiftsoglou, Alan C. Sartorelli, David E. Housman and T. Michael Dexter

associated with cdk2 (Bates et al., 1994). These cyclins play important, but probably distinct roles in the regulation of the G1-S phase transition (Dulic et al., 1992).

Substrates of the G1 CDKs include the retinoblastoma protein (pRb) and other pRb-related proteins. pRb binds and negatively regulates transcription factors such as E2F, whose activities are required for the G1-S-phase transition (Hinds and Weinberg, 1994). pRb phosphorylation results in the release of E2F, enabling it to activate genes important for S-phase entry. However, it was shown that cyclin E, though not D, is essential for entry into S phase even in cells lacking functional pRb (Ohtsubo et al., 1995).

Cyclin kinase inhibitors (CKIs) were recently identified, and these proteins bind to and inhibit the activity of Cdks. p21, also known as WAF1 (wildtype p53 activated factor-1) or CIP1 (CDK interacting protein-1) (El-Deiry et al., 1993; Harper et al., 1993; Xiong et al., 1993) was the first CKI that was identified, and it can bind and inhibit G_1 cyclin/ Cdk complexes (Harper et al., 1993; Xiong et al., 1993). It was shown that multiple p21 molecules are required for inhibition of cyclin/Cdk kinase activity (Harper et al., 1993; Zhang et al., 1994). p21 overexpression results in the arrest of the cell cycle in G_1 (Xiong et al., 1993; Luo et al., 1995) and in the suppression of tumor cell growth (El-Deiry et al., 1993). In contrast, expression of antisense p21 RNA in G_0 -arrested cells results in cell cycle progression (Nakanishi et al., 1995). p21 also binds to proliferating cell nuclear antigen (PCNA) (Waga et al., 1994) and affects DNA replication through this interaction (Li et al., 1994). Expression of p21 has been found to be regulated primarily at the transcriptional level (Morgan, 1995). p21 is induced during differentiation of a number of cell types in vitro (Steinman et al., 1994; Jiang et al., 1994; Halevy et al., 1995; Parker et al., 1995). p21 induction has also been correlated with terminal differentiation of muscle and terminal differentiation in a variety of fetal and adult mouse tissues (Halevy et al., 1995; Parker et al., 1995). Although the p21 gene is transcriptionally activated by p53 in response to DNA damaging agents, p21 induction occurs independently of p53 expression in most of the differentiating systems (Steinman et al., 1994; Jiang, Lin et al., 1994; Halevy et al., 1995; Parker et al., 1995; Macleod et al., 1995; Missero et al., 1995; Datto et al., 1995). Mice with targeted disruption of p21 gene developed normally, but were defective in G1 checkpoint control (Brugarolas et al., 1995; Deng et al., 1995).

We examined the expression of p21 and a variety of other cell cycle regulators in the colon adenocarcinoma Caco-2 cell line. Preconfluent Caco-2 cells are not polarized and appear undifferentiated. After reaching confluence and being maintained in culture, these cells polarize and undergo enterocyte-like differentiation (Pinto et al., 1983). They provide a useful in vitro system for studying intestinal epithelial cell differentiation, and for elucidating the role of p21 in this process.

MATERIALS AND METHODS

Cell line and RNA preparation

The human colon cancer cell line Caco-2 was obtained from the American Type Culture Collection and grown in DMEM/F12 with 20% fetal calf serum. The cells were harvested at 2 days (70% confluent), 10 days, 14 days and 21 days after plating and total RNA was prepared. Total RNA was prepared from Caco-2 cells by a single-step method using RNAzol ™ B (TEL-TEST, INC).

Quantitative RT-PCR

For synthesis of cDNA we used "SuperScript ™ Preamplification System" (Gibco-BRL). 2 μg of total cellular RNA was annealed with 50 ng of random hexadeoxynucleotide primer at 70°C for 10 min. cDNA was synthesized in 20 μl of solution containing 20 mM Tris-HCl (pH 8.4), 50 mM KCl, 10 mM DTT, 500 μM dNTP, and 200 units of SuperScript II reverse transcriptase at 42°C for 50 minutes, after which we added 2 units of RNAse H. 2.5 μl of cDNA product was added to 30 μl of PCR reaction mixture that contained 200 μM dNTP, 2.5 mM $MgCl_2$, 75 pmol primers (Table 1), 1.5 units of AmpliTaq Polymerase (Gibco-BRL). Before the reaction AmpliTaq Polymerase was preincubated 10 min with TaqStart ™ Antibody (Clontech).

Table 1. PCR primers for the detection of cell cycle related gene transcripts

PCR product	Sense primer(5' to 3')	Antisense primer (5' to3')	Product size (bp)
p21	ACTGTGATGCGCTAATGGC	ATGGTCTTCCTCTGCTGTCC	233
p53	TGGTACAGTCAGAGCCAACC	AGCAGTCACAGCACATGACG	201
E2F1	TATGGTGGCAGAGTCAGTGG	TGCAGAGCAGATGGTTATGG	220
cyclin D1	ATGTGTGCAGAAGGAGGTCC	GTAGATGCACAGCTTCTCGG	244
cyclin E	CTGTGTCAAGTGGATGGTTCC	TGCTTCTTACCGCTCTGTGG	231
sucrase-isomaltase	CAGCCTTATCCAAGTCTGG	CAACAGCAGGTGTCTTGTTGC	165
S14	GGCAGACCGAGATGAATCCTCA	CAGGTCCAGGGGTCTTGGTCC	143

Each cycle of PCR included 45 sec of denaturation at 94°C, 1 min of primer annealing at different temperatures for each primer pair, and 1 min of extension/synthesis at 72°C. For each combination of primers the kinetics of PCR amplification was studied, the number of cycles corresponds to the plateau were determined, and PCR was performed at the exponential range. We used amplification of transcripts of human S14 ribosomal protein gene as internal control (Foley et al., 1993). PCR products were separated on a 2% agarose gel and stained with ethidium bromide. Gels were photographed in UV light using a Hitachi KPM1-U camera, and bands corresponding to each specific PCR product were quantitated using NIH Image (available by ftp from zippy.nimh.nih.gov).

Results and Discussion

Differential expression of cell-cycle dependent genes during Caco-2 cell differentiation

To analyze expression levels of cell cycle-related genes during differentiation we isolated RNA from Caco-2 cells undergoing differentiation and performed quantitative RT-PCR with an endogenous control (human S14 ribosomal protein) and primers specific for these genes (Table 1). Ubiquitously expressed RNA that was coamplified with target RNA was used as an internal control. After normalization with the respect to the internal control the RNA levels in different samples were compared directly. This method permits one to examine the extent of degradation and recovery of RNA. Expression of any gene with a published sequence could be studied very conveniently after synthesizing primers specific for the gene of interest. Using this method we determined the relative levels of expression of cell-cycle genes in Caco-2 cells at day 2,10,14 and 21 after plating (Fig.1). Sucrase-isomaltase mRNA which is expressed in differentiated enterocytes (from day 10 to 21), was undetectable in preconfluent undifferentiated cells (day 2). Expression of this differentiation marker increases throughout the process of Caco-2 cell differentiation (Fig. 1). In parallel with an increase in sucrase-isomaltase expression, levels of p21 RNA increase approximately four fold (Fig. 1). Following DNA damage, the p21 gene is transcriptionally activated by p53 in many tissues (El-Deiry et al., 1993; Macleod et al., 1995). In differentiating Caco-2 cells, p53 is virtually undetectable at day 2, while significant levels of p21 are observed. Steady state levels of p53 transcripts are observed on day 14 with slight increase to day 21 (Fig. 1). p53-independent expression of p21 has been observed during the process of cell differentiation in vitro (Steinman et al., 1994; Jiang, et al., 1994; Halevy et al., 1995; Parker et al., 1995). Expression of p21 during mouse development and in various tissues including the intestine, has also been demonstrated in the absence of p53 (Parker et al., 1995; Macleod et al., 1995; El-Deiry et al., 1995). In contrast to what was seen in other tissues, p21 expression did not noticeably increase in the intestine as a whole, with irradiation of wild-type mice

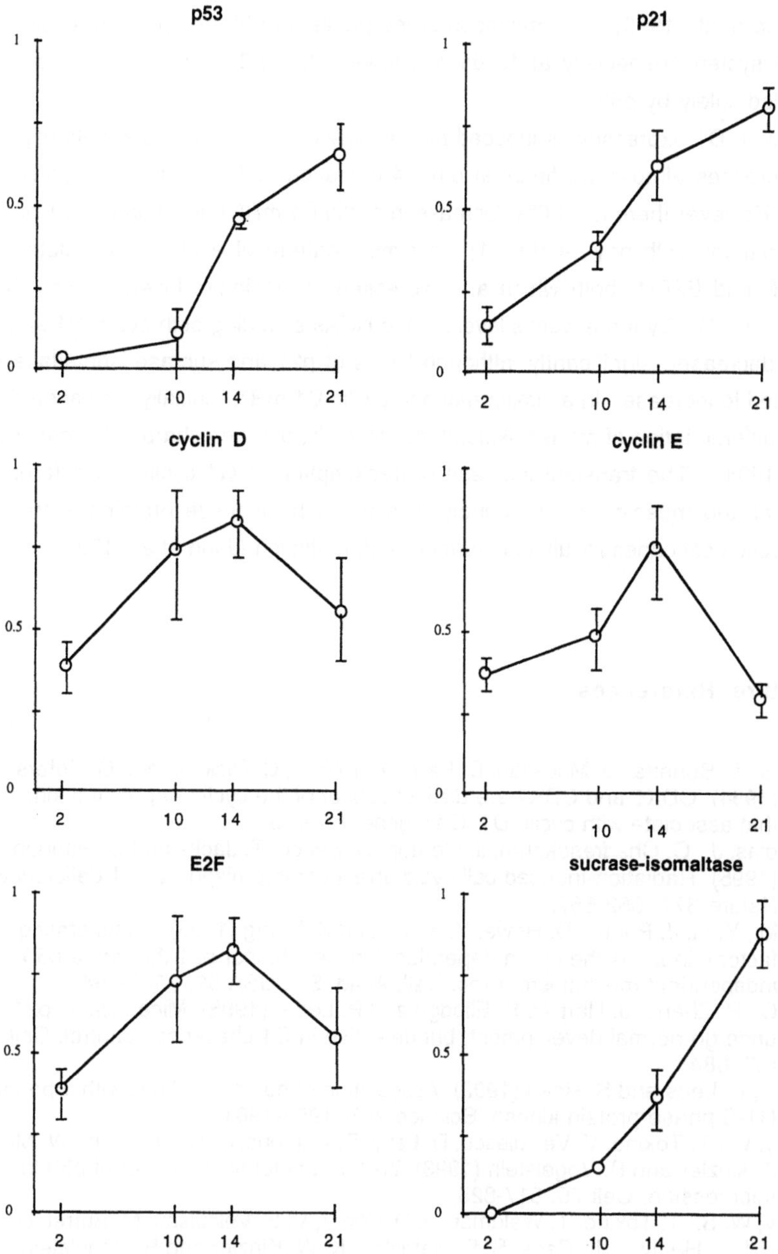

Figure 1. Expression patterns of several cell-cycle regulatory genes in differentiating Caco-2 cells. The results of 4-6 experiments are shown with bars indicating standard deviations.

(Macleod et al., 1995). A comparison of the profile for p53 and p21 expression in the Caco-2 system, especially at 10 days, suggests that p21 expression is not linearly controlled solely by p53.

Cyclin D1 expression is induced by a variety of growth factors and its expression also increases almost two times at day 14 of Caco-2 cell growth in comparison with day 2. However there is a 30% decrease in cyclin D1 mRNA level on day 21 of culture in comparison with day 14 (Fig. 1). A similar pattern of expression is detected for cyclin E and E2F-1, both which are expressed at maximum levels on day 14 post plating (Fig. 1). By three weeks levels of the RNAs encoding both cyclin D1 and cyclin E had decreased significantly, although levels of p21 and sucrase-isomaltase RNAs continued to increase. In a similar manner, cyclin D1 mRNA initially increased twofold during differentiation of mouse skeletal muscle cells, but then sharply decreased (Jahn et al., 1994). The transient increase of transcription of G1 cyclins in differentiating intestinal and muscle cells may indicate a function for these proteins in the signal transduction cascades resulting in terminal differentiation (Jahn et al., 1994).

Literature References

Bates, S., L. Bonetta, D. Macallan, D. Parry, A. Holder, C. Dickson and G. Peters (1994). CDK6 and CDK4 are distinct subset of the cyclin-dependent kinases that associate with cyclin D1. Oncogene 9: 71-79.

Brugarolas, J., C. Chadrasekaran, J. Gordon, D. Beach, T. Jacks and G. Hannon (1995). Radiation-Induced cell cycle arrest compromised by p21 deficiency. Nature 377: 552-557.

Datto, M., Y. Li, J. Panus, D. Howe, Y. Xiong and X. Wang (1995). Transforming growth factor β induces the cyclin-dependent kinase inhibitor p21 through a p53-independent mechanism. Proc. Natl. Acad. Sci. USA 92: 5545-5549.

Deng, C., P. Zhang, J. Harper, S. Elledge and P. Leder (1995). Mice lacking p21 undergo normal development, but defective in G1 checkpoint control. Cell 82: 675-684.

Dulic, V., E. Lees and S. Reed (1992). Association of human cyclin E with a periodic G1-S phase protein kinase. Science 257: 1958-1961.

El-Deiry, W., T. Tokino, V. Velculescu, D. Levy, R. Parsons, J. Trent, D. Lin, W. Mercer, K. kinzler and B. Vogelstein (1993). WAF1, a potential mediator of p53 tumor suppression. Cell 75: 817-825.

El-Deiry, W. S., T. Tokino, T. Waldman, J. D. Oliner, V. E. Velculscu, M. Burrel, D. E. Hill, E. Healy, J. L. Rees, S. R. Hamilton, K. W. Kinzler and B. Vogelstein (1995). Topological control of p21 (WAF1/CIP1) expresison in normal and neoplastic tissues. Cancer Res. 55: 2910-2919.

Foley, K., M. Leonard and J. Engel (1993). Quantitation of RNA using the polymerase chain reaction. Trends Genet. 9: 380-385.

Halevy, O., B. Novitch, D. Spicer, S. Skapek, J. Rhee, G. Hannon, D. Beach and A. Lassar (1995). Correlation of terminal cell cycle arrest of skeletal muscle with induction of p21 by MyoD. Science 267: 1018-1021.
Harper, J. W., G. Adami, N. Wei, K. Keyomarsi and S. Elledge (1993). The p21 cdk-interacting protein Cip1 is a potent inhibitor of G1 cyclin-dependent kinases. Cell 75: 805-816.
Hinds, P. W. and R. A. Weinberg (1994). Tumor suppressor genes. Curr. Opin. Genet. Dev. 4: 135-141.
Jahn, L., J.-i. Sadoshima and S. Izumo (1994). Cyclins and cyclin-dependent kinases are differentially regulated during terminal differentiation of C2C12 muscle cells. Exp. Cell Res. 212: 297-307.
Jiang, H., J. Lin, Z. Su, F. Collart, E. Huberman and P. Fisher (1994). Induction of differentiation in human promyelocytic HL-60 leukemia cells activates p21, WAF1/CIP1, expression in the absense of p53. Oncogene 9: 3397-3406.
Kato, J., M. Matsuoka, D. Strom and C. Sherr (1994). regulation of cyclin D-dependent kinase 4 (cdk4) by cdk4-activating kinase. Mol. Cell. Biol. 14: 2713-2721.
Li, R., S. Waga, G. Hannon, D. Beach and B. Stillman (1994). Differential effects by the p21 CDK inhibitor on PCNA dependent DNA replication and repair. Nature 371: 534-537.
Luo, Y., J. Hurwitz and J. Massague (1995). Cell-cycle inhibition by independent CDK and PCNA binding domains in p21. Nature 375: 159-161.
Macleod, K., N. Sherry, G. Hannon, D. Beach, T. Tokino, K. Kinzler, B. Vogelstein and T. Jacks (1995). p53-dependent and independent expression of p21 during cell growth, differentiation, and DNA damage. Genes & Dev. 9: 935-944.
Missero, C., E. Calautti, R. Eckner, J. Chin, L. Tsai, D. Livingston and G. P. Dotto (1995). Involvement of the cell-cycle inhibitor Cip1/WAF1 and the E1A-associated p300 protein in terminal differentiation. Proc. Natl. Acad. Sci. USA 92: 5451-5455.
Morgan, D. O. (1995). Principles of CDK regulation. Nature 374: 131-134.
Nakanishi, M., G. Adami, R. Robetorye, A. Noda, S. Venable, D. Dimitrov, O. Pereira-Smith and J. Smith (1995). Exit from Go and entry into the cell cycle of cells expressing p21 antisense RNA. Proc.Natl. Acad.Sci.USA 92: 4352-4356.
Ohtsubo, M., J. Theodoras, J. Schumacher, J. Roberts and M. Pagano (1995). Human cyclin E, a nuclear protein essential for the G1 to S transition. Mol. Cell. Biol. 15: 2612-2624.
Parker, S., G. Eichele, P. Zhang, A. Rawls, A. Sands, A. Bradley, E. Olson, J. Harper and S. Elledge (1995). p53-independent expression of p21 in muscle and other terminally differentiating cells. Science 267: 1024-1027.
Pinto, M., S. Robine-Leon, M. D. Appay, M. Kedinger, N. Triadou, E. Dussaulx, L. B., P. Simmon-Assmann, K. Haffen, J. Fogh and A. Zweibaum (1983). Enterocyte-like differentiation and polarization of the human colon carcinoma cell line Caco-2 in culture. Biol. Cell 47: 323-330.
Reznitzky, D. and S. Reed (1995). Different roles for cyclins D1 and E in regulation of the G1-to-S transition. Mol. Cell. Biol. 15: 3463-3469.
Sherr, C. and J. Roberts (1995). Inhibitors of mammalian G1 cyclin-dependent kinases. Genes & Dev. 9: 1149-1163.
Steinman, R., B. Hoffman, A. Iro, C. Guillouf, D. Liebermann and M. El-Houseini (1994). Induction of p21(WAF-1/CIP1) during differentiation. Oncogene 9: 3389-3396.

Waga, S., G. Hannon, D. Beach and B. Stillman (1994). The p21 inhibitor of cyclin-dependent kinases controls DNA replication by interaction with PCNA. Nature 369: 574-578.

Xiong, Y., G. Hannon, H. Zhang, D. Casso, R. Kobayashi and D. Beach (1993). p21 is a universal inhibitor of cyclin kinases. Nature 366: 701-704.

Zhang, H., G. Hannon and D. Beach (1994). p21-containing cyclin kinases exist in both active and inactive states. Genes&Dev. 8: 1750-1758.

MG-160, a Sialoglycoprotein of the Medial Cisternae of the Golgi Apparatus, Is Closely Related to a Receptor of Fibroblast Growth Factors and to a Ligand for E-Selectin. Functional Implications

Jacqueline O.Gonatas, Y-J Chen, A. Stieber, Zissimos Mourelatos and Nicholas K.Gonatas.

Department of Pathology and Laboratory Medicine

University of Pennsylvania School of Medicine

418 Johnson Pavillion, Philadelphia, PA, 19104

USA.

Summary.

MG-160, a membrane sialoglycoprotein of the medial cisternae of the rat Golgi apparatus, was identified and isolated with the use of an organelle specific antibody (Gonatas et al.1989). A polyclonal antiserum against MG-160, purified by immunoaffinity chromatography from rat brain, reacted with the Golgi apparatus of several species including human (Croul et al. 1990).

Complementary DNAs for MG-160 were cloned and sequenced and the deduced amino acid sequence revealed a type I membrane polypeptide consisting of 1,171 amino acids with an M_r of 133,403. The protein displayed a cleavable signal peptide, followed by a lumenal domain of a few Gln-Pro repeats and 16 contiguous, approximately 60-residue-long, regularly spaced cysteine-rich repeats showing sequence identities

NATO ASI Series, Vol. H 99
Tumor Biology
Regulation of Cell Growth, Differentiation and Genetics in Cancer
Edited by Asterios S. Tsiftsoglou, Alan C. Sartorelli, David E. Housman and T. Michael Dexter

ranging from 15% to 35%, a single transmembrane domain and a 13 amino acid carboxy-terminal cytoplasmic tail(Gonatas et al. 1995). MG-160 binds basic fibroblast growth factor and shows high levels of amino acid sequence identities with a chicken fibroblast growth factor receptor (CFR), and ESL-1, a ligand to E-selectin (Burrus et al. 1992; Steegmeier et al. 1995). The protein is ubiquitously expressed early during the development of chicken embryos and in adult animals (Stieber et al. 1995)

The human gene for MG-160 (GLG1), has been assigned to chromosome 16q22-q23, within a region containing markers showing loss of heterozygocity in several neoplasms (Mourelatos et al. 1995). These findings are consistent with the hypothesis that MG-160 is a multiligand binding protein involved in the regulation of the secretion of certain fibroblast growth factors and ligands mediating interactions between cells.

Introduction.

The Golgi apparatus is key to the orthograde or centrifugal traffic involving the transport, processing and targeting of proteins destined for regulated or constitutive secretion, the plasma membrane and lysosomes (Kornfeld,1992; Mellman and Simons,1992; Rothman and Orci,1992). In addition, a system of cisternae and tubules at the trans aspect of the Golgi apparatus, previously named GERL by Novikoff and Novikoff (for Golgi-Endoplasmic Reticulum-Lysosome, 1977), and now referred to as TGN (for Trans Golgi Network, Griffiths and Simons, 1986), is involved in the endocytosis, or retrograde traffics, of immunoglobulins, certain toxins ,lectins and antigens, and of mannose-6-phosphate receptors (Antoine et al. 1974; Gonatas et al. 1975; Gonatas et al. 1977; Joseph et al. 1978, 1979; Harper et al. 1980; Gonatas and Stieber 1980; Gonatas et al. 1980; Trojanowski et al.

1981; Gonatas, 1982; Sandvig et al. 1992; Hickey et al. 1983; Gonatas et al. 1984; Kornfeld, 1992; Gonatas, 1994).

Virtually all intrinsic membrane proteins of the cisternae of the Golgi apparatus which have been sequenced are enzymes with a type II configuration, i.e. they contain a single transmembrane domain, a lumenal carboxy-terminal catalytic domain, and a cytoplasmically oriented amino-terminus (Paulson and Colley, 1989; Kleene and Berger, 1993).

TGN38, is the only type I membrane protein which has been localized in the *trans*-Golgi network (Luzio et al. 1990). However, TGN, the network of cisternae and tubules at the trans aspect of the Golgi apparatus, is structurally and functionally different from the compact cisternae of the organelle (Griffiths and Simons, 1986). One of the most striking morphologic and functional differences between the Golgi cisternae and TGN is their differential response to the action of the secretion blocker Brefeldin A (BFA). Thus, while in BFA treated cells the Golgi cisternae disappear and their membrane proteins are distributed back into the rough endoplasmic reticulum, the TGN aggregates around the centriole (Reaves and Banting, 1992; Johnston et al. 1994). Another important difference between membrane proteins of the cisternae of the Golgi apparatus and the TGN, is that a protein of TGN, the TGN38/41, recycles between the *trans* Golgi network and the cell surface, a property not shared by the proteins of the Golgi cisternae (Stanley and Howell, 1993).

The sialoglycoprotein MG-160 is the only type I intrinsic membrane protein of the Golgi cisternae described so-far, i.e. the protein displays a single transmembrane domain, an intralumenal amino-terminus, and a carboxy-terminus cytoplasmic tail (Gonatas et al 1989). Most likely, MG-160 is not an enzyme, since an antibody against the protein did not inhibit the activities of certain Golgi enzymes or the uptake of nucleotide sugars by intact Golgi vesicles (Gonatas et al. 1989).

The biochemical and molecular properties of MG-160 will be reviewed and speculations on its function will be discussed.

Materials and Methods.

The methods of preparation of enriched fractions of the Golgi apparatus from rat brain neurons, the production of organelle-specific monoclonal and polyclonal antibodies, and for immunocytochemistry and molecular biology used in these studies have been reported in detail (Gonatas et al. 1985,1989,1995; Croul et al. 1990; Mourelatos et al. 1995). Crucial to these studies has been the original description by Louvard et al. of a method for the preparation of organelle specific antibodies which have been exploited in the identification of novel organelle specific proteins (1982).

Results and Discussion.

Morphologic and Biochemical properties of MG-160.

By immuno-electron microscopy, MG-160 was localized predominantly in the medial cisternae of the Golgi apparatus of rat brain neurons, astrocytes, anterior pituitary, and pheomochromocytoma cells, PC12. By light microscopic immunocytochemistry, a stain consistent with the Golgi apparatus was observed in rat kidney, parathyroid, thyroid, pancreatic islet cells and adrenal. Unreduced MG-160 had a greater electrophoretic mobility (130 kDa), than the reduced protein (160 kDa), consistent with the presence of intrachain disulfide bonds. The protein contained N-linked carbohydrates including sialic acid. The yield of MG-160 purified by immunoaffinity chromatography was 0.9 μg/g of rat brain, and represented approximately 3% of proteins of the Golgi apparatus (Gonatas et al. 1989).

MG-160 is a primordial and conserved protein of the Golgi apparatus.

The monoclonal antibody 10A8, used in the isolation of MG-160, reacted only with rat tissues. However, a polyclonal antibody raised against purified MG-160 reacted by western blotting and immunocytochemistry with the Golgi apparatus of frog, chicken,mouse, rabbit, bovine and human brain cells (Croul et al.1990). Furthermore, developmental studies of the chicken homologue for MG-160 revealed that the protein is ubiquitously and exclusively localized in the Golgi apparatus and appeared early in development within the endoblast and ectoblast prior to the formation of the primitive streak (Stieber et al. 1995). The localization of MG-160 within neuroectodermal cells of the neural tube, and the notochord was particularly prominent. However, although isolated MG-160 binds bFGF, we were not able to localize bFGF with the Golgi apparatus of chicken embryo cells (Stieber et al. 1995).

MG-160 and/or sialyltransferases probably recycle between the TGN and medial Golgi.

The sialic acid containing MG-160 was found in the medial Golgi cisternae, while sialyltransferases have been localized in the distal trans Golgi cisternae or TGN. In order to resolve the question whether MG-160 acquires sialic acid residues in the trans Golgi cisternae or in the TGN and then undergoes retrograde transport in the medial Golgi, experiments were conducted in PC-12 cells treated with BFA. The results of these studies enhanced the hypothesis that under physiological conditions a retrograde traffic of membrane proteins is operational in the Golgi apparatus (Johnston and Gonatas, 1993; Johnston et al. 1994).

Cloning and sequencing of cDNAs for rat MG-160.

Rat cerebral cortex and hypothalamus cDNA libraries were screened with partially degenerate oligonucleotide probes, designed and synthesized on the basis of the amino acid sequences of four peptides derived from MG-160 (Gonatas et al. 1995). The deduced amino acid sequence revealed a type I protein with a single transmembrane domain , an amino terminus lumenal domain, and a short cytoplasmic carboxy terminus. The protein contains 5 potential NXT glycosylation sites and 16 contiguous approximately 60 residue long cysteine-rich segments (Gonatas et al. 1995). It has been proposed by Russell et al. that ".. a multiplicity of cysteine-rich repeats may allow a single protein to bind different protein ligands by employing different combinations of repeats" (1989). The recognition of an approximately 90% identity between MG-160 and several fibroblast growth factor receptors, and a ligand to E-selectin is consistent with the above hypothesis and strongly suggests that MG-160 may bind to additional ligands (Burrus et al. 1992; Zhou and Olwin, 1994; Steegmaier et al. 1995). The functional implications of a Golgi protein binding to a number of apparently unrelated ligands remains to be determined. One plausible hypothesis is the MG-160 regulates the secretion of certain FGFs, such as FGF 3 which is retained in the Golgi apparatus-complex (Kiefer et al. 1993).

Chromosomal localization of *GLG1*, the gene for MG-160.

Insights into the function of a protein may be gained from the chromosomal localization of its gene. The gene of the human homologue of MG-160, named *GLG1*, has been assigned to chromosome 16q22-q23. It may be of interest that loss of heterozygocity for loci in the distal arm of chromosome 16 has been found in a variety of human neoplasms, and deletion of 16q:16q22 is critical for the development of the 16q- syndrome (quoted in Mourelatos et al. 1995). Studies on the organization of the *GLG1* gene and analysis of this gene in patients with neoplasms and loss of heterozygocity on chromosome 16q and in patients with the 16q- syndrome might reveal a role of MG-160 in these disorders.

MG-160 as a marker of the Golgi apparatus in neurodegenerative disorders.

Using a polyclonal antiserum against the human homologue of MG-160, an unusual lesion of the neuronal Golgi apparatus was identified in certain neurons in Amyotrophic Lateral Sclerosis (ALS), in transgenic mice with a mutation of the Superoxide Dismutase I gene (SODI), found in certain familial cases of ALS, and in Alzheimer's Disease (Gonatas, 1994; Mourelatos et al. 1995; Stieber et al. 1995). This lesion resembles the fragmentation of the Golgi apparatus induced by microtubule depolymerizing agents and suggests that in the above disorders either the neuronal microtubules or proteins involved in linkages between microtubules and Golgi membranes are affected.

Conclusions.

The studies, summarized in this paper, strongly suggest that MG-160 is a multi-ligand binding protein and that the extensive cysteine-rich repeats may be involved in the binding between MG-160 and apparently unrelated proteins such as FGFs and E-selectin. In view of the importance of FGFs in development, the appearance of abundant MG-160 during the early stages of the development of chicken embryos may be significant and relevant to its proposed regulatory function on the secretion of certain FGFs (Burgess and Maciag, 1989; Sensenbrenner, 1993). This hypothesis remains to be tested. In addition, the study of the organization of the *GLG1* gene, the human homologue for MG-160, will contribute to the assessment of its function and its probable involvement in human diseases.

Acknowledgments: Supported by a Senator Javits Neuroscience Investigator award (NS 05572) to Dr.Gonatas.

References

Antoine, J-C., Avrameas, S., Gonatas, N.K., Stieber, A., and Gonatas, J.O. (1974). Plasma membrane and internalized immunoglobulins of lymph node cells studies with conjugates of an antibody or its Fab fragments with horseradish peroxidase. J.Cell Biol. 63: 12-23.

Burgess, W.H., and Maciag,T.(1989). The heparin-binding (fibroblast) growth factor family of proteins. Annu.Rev.Biochem. 58, 575-606.

Burrus, L.W., Zuber, M.E., Lueddecke, B.A., and Olwin,B.B.(1992). Identification of a cysteine-rich receptor for fibroblast growth factors. Mol. and Cell Biol. 12, 5600-5609.

Croul, S., Mezitis, S.G.E., Stieber, A., Chen, Y., Gonatas, J.O., Goud, B., and Gonatas, N.K (1990) Immunocytochemical visualization of the Golgi apparatus in several species, including human, and tissues with an antiserum against MG-160, a sialoglycoprotein of rat Golgi apparatus. J. Histochem. Cytochem. 38:957-963.

Gonatas, N.K., Stieber, A., Kim, S.U., Graham, D.I., and Avrameas, S. (1975). Internalization of plasma membrane ricin receptors into the Golgi apparatus. Exp. Cell Res. 94: 426-431.

Gonatas, N.K., Kim, S.U., Stieber, A., and Avrameas, S. (1977). Internalization of lectins in neuronal GERL. J.Cell Biol. 73:1-13.

Gonatas, N.K., and Stieber, A. (1980). Surface distribution and endocytosis of the NS3 antigen of neuroblastoma. Exp. Cell Res. 128: 237-248.

Gonatas, J., Stieber, A., Olsnes, S., and Gonatas, N.K. (1980). Pathways involved in fluid phase and adsorptive endocytosis in neuroblastoma. J. Cell Biol. 87:579-588.

Gonatas, N.K. (1982). The role of the neuronal Golgi apparatus in a centripetal membrane vesicular traffic. J.Neurop.Exp.Neurol. 41:6-17.

Gonatas, J.O., Gonatas, N.K., Stieber, A., and Fleischer, B. (1985). Isolation and characterization of an enriched Golgi fraction from neurons of developing rat brains.J.Neurochem. 45:497-507.

Gonatas, J.O., Mezitis, S.G.E., Stieber, A., Fleischer, B. and Gonatas, N.K (1989) MG-160: A novel sialoglycoprotein of the medial cisternae of the Golgi apparatus. J. Biol. Chem. 264(1):646-653.

Gonatas,N.K.(1994).Contributions to the Physiology and Pathology of the Golgi apparatus. Rous-Whipple award lecture. Am.J.Path. 145, 751-761.

Gonatas, N.K., Stieber, A., Hickey, W.F., Herbert, S.H., and Gonatas, J.O. (1984). Endosomes and Golgi vesicles in adsorptive and fluid phase endocytosis. J. Cell Biol. 99:1379-1390.

Gonatas, J.O., Mourelatos, Z., Stieber, A., Lane, W.S., Brosius, J., and Gonatas, N.K (1995) MG-160, a membrane sialoglycoprotein of the medial cisternae of the rat Golgi apparatus, binds basic fibroblast growth factor and exhibits a high level of sequence identity to a chicken fibroblast growth factor receptor. J.CellSci. 108, 457-467, 1995.

Griffiths, G., and Simons K. (1986). The trans Golgi network:sorting at the exit site of the Golgi complex. Science 234:438-443.

Harper, C.G., Gonatas, J.O., Stieber, A., and Gonatas, N.K. (1980). In vivo uptake of wheat germ agglutinin-horseradish peroxidase conjugates into neuronal GERL and lysosomes. Brain Res. 188:465-472.

Hickey, W.F., Stieber, A., Hogue-Angeletti, R., Gonatas, J., and Gonatas, N.K. (1983). Nerve growth factor induced changes in the Golgi apparatus of PC-12 rat pheochromocytoma cells as studied by ligand endocytosis, cytochemical and morphometric methods. J. Neurocyt. 12:751-766.

Johnston, P.A., and Gonatas, N.K (1993) Effect of Brefeldin A treatment on the resident Golgi protein, MG160. NATO ASI Series, Vol H74. Molecular Mechanisms of Membrane Traffic. Eds. D. J. Morré, K.E. Howell, and J.J.M. Bergeron, Springer-Verlag Berlin Heidelberg.

Johnston, P.A., Stieber, A., and Gonatas N.K. (1994). A hypothesis on the traffic of MG-160, a medial Golgi sialoglycoprotein from the *trans*-Golgi network to the Golgi cisternae. J.Cell Science 107:529-537.

Joseph, K.C., Kim, S.U., Stieber, A., and Gonatas, N.K. (1978). Endocytosis of cholera toxin into neuronal GERL. Proc.Natl.Acad.Sci. U.S.A. 75:2815-2819.

Joseph, K.C., Stieber, A., and Gonatas, N.K. (1979). Endocytosis of cholera toxin in GERL-like structures of murine neuroblastoma cells pretreated with GM_1 ganglioside. J. Cell Biol.81: 543-554.

Kiefer, P., Peters, G., and Dickson, C. (1993). Retention of fibroblast growth factor 3 in the Golgi complex may regulate its export from cells. Mol. Cell Biol. 13:5781-5793.

Kleene, R., and Berger, E.G. (1993). The molecular and cell biology of glycosyltransferases. Biochim.Biophys. Acta 1154: 283-325.

Kornfeld, S. (1992). Structure and function of the mannose 6-phosphate/insulin like growth factor II receptors. Annu. Rev. Biochem. 61: 307-330.

Louvard, D., Reggio, H., Warren G. (1982). Antibodies to the Golgi complex and the rough endoplasmic reticulum. J. Cell Biol. 92:92-107.

Luzio,J.P., Brake,B., Banting,G., Howell,K.E., Braghetta,P.,and Stanley,K.K. (1990). Identification,sequencing and expression of an integral membrane protein of the *trans*-Golgi network (TGN38). Biochem.J.270:97-102.

Mellman, I., and Simons, K. (1992). The Golgi complex: in vitro veritas ? Cell 68:829-840.

Mourelatos, Z., Gonatas, J.O., Nycum, L.M., Gonatas, N.K., and Biegel, J.A (1995). Assignment of the gene for MG-160, a Fibroblast Growth Factor binding membrane sialoglycoprotein of the Golgi apparatus, to Chromosome 16q22-q23 by fluorescence in situ Hybridization. Genomics. 28,354-355.

Mourelatos, Z., Gonatas, N.K., Gurney, M.E., and Dal Canto M.C. The Golgi apparatus of spinal cord motor neurons in transgenic mice expressing mutant Cu,Zn superoxide dismutase (SOD) becomes fragmented in early, preclinical stages of the disease.(1996). In Press, Proc. Natl.Acad.Sci. USA.

Novikoff, A.V., and Novikoff, P.M.(1977). Cytochemical contributions to differentiating GERL from the Golgi apparatus. Histochem.J. 9:525-551.

Paulson, J.C. and Colley,K.J.(1989). Glycosyltransferases. Structure, localization, and control of cell type-specific glycosylation. J.Biol.Chem. 264:17615-17618.

Reaves, B., and Banting,G.(1992). Perturbation of the morphology of the *trans*-Golgi network following brefeldin A traetment: redistribution of a TGN-specific integral membrane protein ,TGN38.J.Cell Biol.116:85-94.

Rothman, J.E., and Orci, L. (1992). Molecular dissection of the secretory pathway. Nature 355:409-415.

Russell, D.W., Brown, M.S., and Golstein,J.L. (1989). Different combinations of cysteine-rich repeats mediate binding of low density lipoprotein receptor to two different proteins. J.Biol.Chem. 264:21682-21688.

Sandvig, K., Prydz, K., Hansen, S.H., and Van Deurs, B. (1992). Retrograd transport of endocytosed Shiga toxin to the endoplasmic reticulum. Nature 358: 510-512.

Stieber, A., Mourelatos, Z., Chen, Y-J, Le Douarin, N., and Gonatas, N.K. (1995). MG160, a membrane protein of the Golgi apparatus which is homologous to a fibroblast growth factor receptor and to a ligand for

E-selectin, is found only in the Golgi apparatus and appears early in chicken embryo development. Exp.Cell Res. 219,562-570,1995.

Steegmaier, M., Levinovitz, A., Isenmann, S., Borges, E., Lenter, M., Kocher, H.P., Kleuser, B., and Vestweber, D. (1995). The E-selectin ligand ESL-1 is a variant of a receptor for fibroblast growth factor. Nature. 373,615-620.

Stieber, A., Mourelatos, Z., and Gonatas, N.K. In Alzheimer's Disease the Golgi apparatus of a population of neurons without neurofibrillary tangles is fragmented and atrophic (1996). In Press,Am.J.Path.

Sensenbrenner, M. (1993). The neurotrophic activity of fibroblast growth factors. Progr. in Neurobiol. 41,683-704.

Stanley, K.K., and Howell, K.E. (1993). TGN38/41: a molecule on the move. Trends in Cell Biol. 3:252-255.

Trojanowski, J.Q., Gonatas, J.O., and Gonatas, N.K.(1981). A light and electron microscopic study of the intraneuronal transport of horseradish peroxidase and wheat germ agglutinin-peroxidase conjugates in the rat visual system. J.Neurocyt. 10:441-1981:

Zhou, Z., and Olwin,B.B.(1994). Characterization of the FGF binding domain of the cysteine-rich FGF receptor. Mol.Biol. of Cell. 5, 136a.

The Regulation of Human β Globin Gene Expression: The Dynamics of Transcriptional Competition in the Human β-Globin Locus

Frank Grosveld[1,2], Niall Dillon[2], Peter Fraser[1], John Strouboulis[2] and Mark Wijgerde[1].

[1]MGC-Department of Cell Biology and Genetics, Erasmus University, Rotterdam, The Netherlands.

[2]Laboratory of Gene Structure and Expression, National Institute for Medical Research, The Ridgeway, Mill Hill, London, UK.

Introduction

The genes of the human ß-globin locus are arranged in the order in which they are expressed during development (fig. 1). This arrangement, and the involvement of the LCR in activating all of the genes gives the locus an intrinsic structural polarity. It has been suggested that this polarity could play a direct role in the regulation of the locus by affecting competition between the genes for the activating function of the LCR (Hanscombe et al., 1991).

Why would the expression of the upstream genes prevent expression of those located downstream but not vice versa? Existing competition models are based on the idea that the high-level expression of the genes in the ß-globin locus is potentiated by direct interaction of each gene with the LCR. There is strong evidence that enhancers work through such interactions (Muller et al., 1989; Bickel and Pirotta, 1990 and references therein). The polar competition that appears to operate in the locus would be explained if the LCR had an intrinsic preference for interaction with a more proximally located gene. A mechanism by which the LCR searches along the chromosome and interacts with the first gene it encounters, would have this property although this would probably require energy and a complex machinery. A second possibility would be the formation of chromatin structures that would bring gene and LCR together, but these would require specific and developmentally regulated functions for the spacer regions in the locus. A comparison of the phenotypes of different deletions in the locus (for review, see Poncz et al., 1988), together with the transgenic mouse data, argues against such a role. An alternative and much simpler possibility is that the determining parameter is the relative

NATO ASI Series, Vol. H 99
Tumor Biology
Regulation of Cell Growth, Differentiation and Genetics in Cancer
Edited by Asterios S. Tsiftsoglou, Alan C. Sartorelli, David E. Housman and T. Michael Dexter

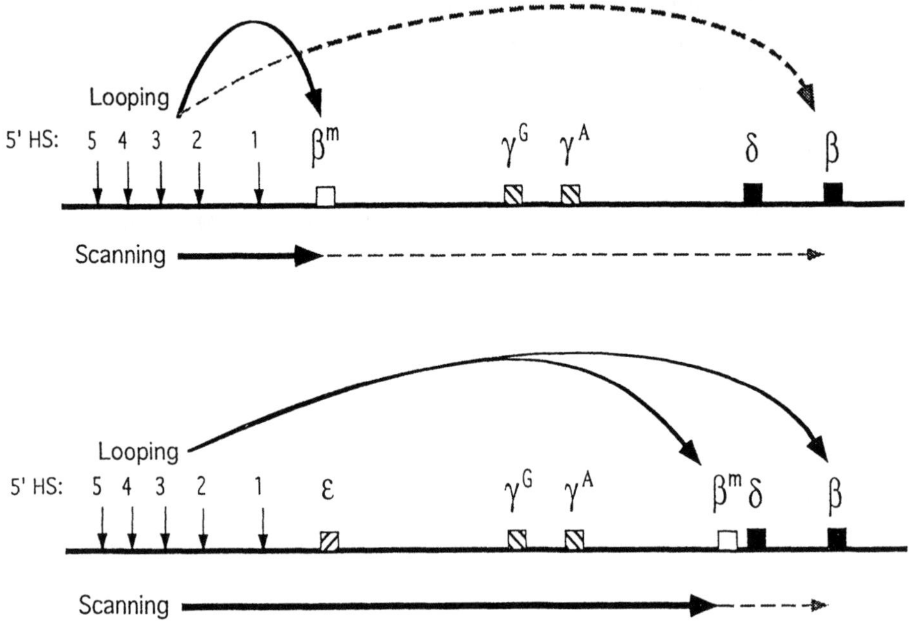

Fig.1 - Schematic representation of ML1 (top) and ML2 (bottom). ML1 contains a marked ß globin gene (ßm) replacing the ε globin gene. ML2 contains the same ßm but now positioned just 5' to the δglobin gene. The solid arrows indicate either a direct contact between the LCR and genes (looping) or a chromatin spreading effect (scanning). The dashed arrows indicate the same processes but taking place at a much lower level.

frequency of contact between genes and LCR (Hanscombe et al., 1991). Where two genes each retain a significant capacity to form stable interactions with the LCR, a difference in their frequency of contact with the LCR would dramatically affect competition between them. During the fetal stage, the interaction of the γ genes with the LCR would be much more frequent than that of the ß gene and, owing to the action of stage-specific factors, would also be stronger. The combination of these two parameters acting together would allow the γ genes to largely compete out ß expression. In the adult stage, although the ß gene would now have the stronger interaction with the LCR, its much lower frequency of contact would make it difficult for it to compete out a γ gene that retained a significant capability to form such interactions. It is possible that this effect could be achieved by a very strong ß interaction, but it seems unlikely that this would have evolved specifically to silence γ expression. The more likely alternative mechanism would be one of promoter-mediated silencing of the early genes by stage-specific factors; the transgenic mouse data indicate that this is the one which has in fact evolved (Dillon et al., 1991; Raich et al., 1990). Such a competition mechanism may have evolved to ensure that the switchover from γ- to ß-globin expression is smooth and that the total output of the ß gene locus is kept constant and in balance with the output from the α locus.

Recently developed technology allowing the complete human ß-globin locus to be introduced into mice (Strouboulis et al., 1992) has given us the opportunity to directly test this model by placing a second ß-globin gene at two different positions in the locus. Analysis of the steady state levels of transcripts from the mutated loci shows that a proximally located ß-globin gene has a strong competitive advantage over a more distally located gene but that the magnitude of this effect depends on the relative distance of the two genes from the LCR. A critical component of the competition model is the notion that the genes would actually be transcribed separately (and compete) rather than be transcribed simultaneously by interaction with the LCR. We therefore developed an in situ hybridization protocol to measure whether the genes are transcribed alternately or simultaneously by detecting the primary transcripts of the ϵ, γ and ß globin genes in a transgenic mouse carrying a single copy of the human ß globin locus (Strouboulis et al., 1992). The results show that the genes are transcribed alternately in a type of dynamic flip flop mechanism. Taken together our results our results provide strong support for a mechanism of transcriptional activation by the LCR involving direct contact with the promoters through looping out of intervening DNA.

Generation of transgenic mice carrying mutant ß-globin loci

To test the effect of gene position on expression within the locus, we made use of a technique which involves ligating two cosmid inserts to generate a 70 kb fragment containing the complete locus (Strouboulis et al., 1992). Conventional cloning methodology carried out on the individual cosmids was used to generate two modified loci each containing a second ß-globin gene (ßm) which was marked so that its transcript could be distinguished from that of the wild-type gene. In mutant locus 1 (fig. 1), the marked gene replaced the ϵ gene while in mutant locus 2 it was inserted close to the cap site of the δ gene (fig. 1). Marking of the gene was achieved by replacing exon 1 and part of intron 1 with the equivalent sequences from the $^{A}\gamma$ gene. These sequences are known not to have any function in the regulation of the γ globin gene.

Transgenic mice were generated by microinjection of the mutant loci into oocytes. Since a multi-copy tandem array would place an LCR close to the 3' end of the wild-type ß-globin gene it was necessary to analyse animals that carried the modified loci at single copy. In order to do this we generated a large number of transgenic founder animals and then used Southern blotting, particularly of the end to identify single copy transgenics.

A total of 57 founder transgenics were generated (23 for ML1 and 34 for ML2). Tail DNA from these founders was digested with BamH1 and probed with fragments from either end of the locus. Founders that showed a joining fragment indicative of a multi-copy tandem repeat were discarded while putative single copy animals were bred to generate transgenic lines. To date we have analyzed two single copy lines for ML1 and 2 for ML2. The integrity of the locus in each of these lines was tested by probing blots of EcoR1 digested DNA with the complete cosmids used to generate the locus (Strouboulis et al., 1992). All of the lines had the locus fully intact (data not shown).

Effect of position in the locus on transcriptional function

RNA was prepared from whole adult blood from each line and subjected to S1 analysis to determine the relative proportions of marked ß and wild-type ß mRNA (Fig. 2). Adult blood RNA from transgenic line 72 carrying the wild-type locus was also included in the analysis (Strouboulis et al., 1992). A number of conclusions can be made from the comparison of the levels of human γ and ß RNA and mouse globin RNA.

Although the ratio between ßm and ß varies considerably, the total output (ß + ßm) of the mutated loci is similar to that of the wild type normal locus. This suggests a balanced competition between the genes for activation by the LCR and implies that the LCR would interact with one gene at a time, rather than multiple genes at the same time (Breshnik and Felsenfeld, 1994; Furukawa et al., 1994). It would argue strongly against the type of model that envisions different HS interacting with different genes at the same time (Engel, 1993).

The expression of the normal ß gene is decreased in both the ML1 and ML2 mice, due to the presence of the ßm gene. The two ML1 lines which carry the marked ßm gene at the position of the ϵ-globin gene show a much greater reduction in the wild type ß expression than the ML2 lines which have the ßm gene inserted close to the δ gene (Fig. 2).

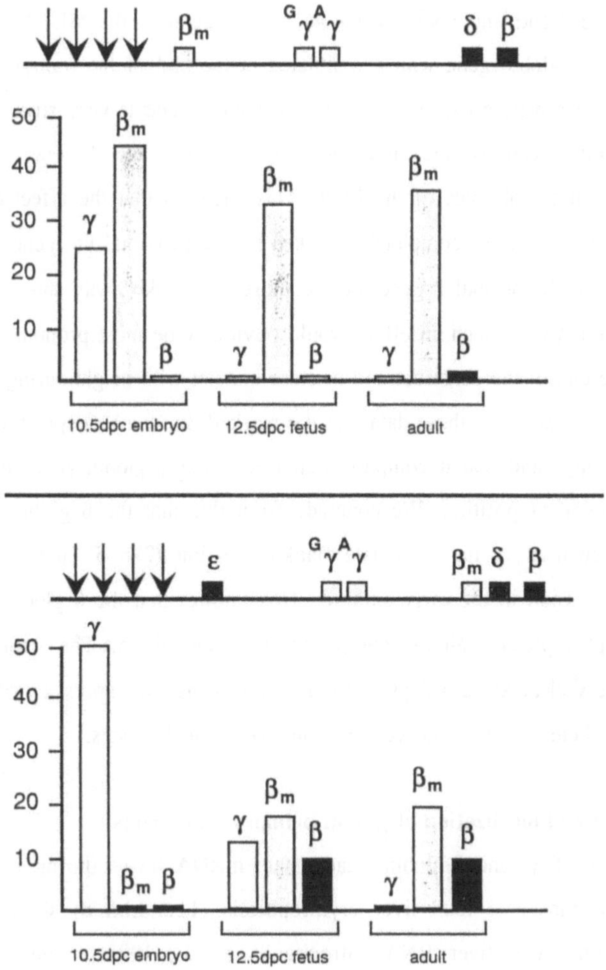

Fig.2 - Bar diagram representing the expression levels of the human γ, ß and ßm genes from the ML1 and LM2 loci in transgenic mice at different times of development. ML1 is shown on top, ML2 on bottom. The developmental time points are indicated below each set of three bars.

This result directly demonstrates the existence of polar competition in the complete ß globin locus. The polar effect is strongest when the proximal gene is close to the LCR and there is a large difference in the relative distance of the two genes from the LCR. The single copy integrated loci show a difference in the level of expression (i.e. line 1 versus line 2 and line 3 versus line 4), in particular in the ML1 lines with respect to the wild type ß globin gene which is situated at the end of the transgenic locus. This low level position sensitive expression of the normal ß gene is very striking in the ML1 locus and is reminiscent of the situation observed when the ß gene is expressed in transgenic mice in the absence of the LCR. This suggests that the effect of the LCR in this normal ß gene is almost completely blocked by the proximal ßm gene. It is therefore not surprising that the normal ß gene located more than 50kb away shows some position of integration sensitivity. Such an effect would obviously be most pronounced when the ß gene is near the end of the construct and in close contact with neighbouring mouse DNA.

It is also clear from these data that the marked ßm can be expressed effectively at the embryonic stage and that it competes better with the γ globin genes than does the ϵ gene in the equivalent position. We conclude from this that the ß globin gene does not contain any sequences in its immediate flanking region (2kb 5' and 3') that actively suppress its expression in the early embryo. This implies that the ß globin in the in vivo situation is kept repressed almost completely by competition. The strong competitive effect is also very clear at 12.5 days in ML1, suppressing the expression of γ completely as would be predicted by the polar competition model for the locus.

Insitu detection and localization of globin primary transcripts

Analysis of γ- and ß-globin steady state mRNA levels during switching in the early stages of transgenic fetal liver erythropoiesis (11.5-13.5 d) show that both are detectable in total fetal liver RNA (Strouboulis et al., 1992). γ-gene expression is decreasing and ß-globin gene expression is increasing. Both γ- and ß-globin polypeptides occur in most cells of the transgenic fetal liver (Strouboulis, PhD Thesis 1994) indicating that commitment to γ or ß gene expression does not occur before the onset of globin transcription. The detection of globin mRNA or polypeptides are not reliable indicators of concurrent transcriptional activity in developing cells due to the long half-lives of such molecules. Alternatively, primary transcripts are both temporally and spatially associated with the transcriptional event and have short half-lives due to rapid splicing into mature mRNA. The half-life of the mouse ß-major and human globin primary transcripts have been calculated to be approx. 5 minutes in transgenic mouse foetal livers (Wijgerde et al.1995). Detection of primary transcripts in situ is therefore an accurate indicator of ongoing or very recent transcription. If transcription complexes last longer than the half-life of the primary transcript, two types of signals may occur; single gene transcription signals or multigene signals. Whereas the explanation for the single gene signal is obvious, the multi gene signal could be the result of two possibilities. It could be the result of concurrent transcription from two genes, or the result of a recent switch, causing a signal overlap between transcription from one gene and the decay of the pre mRNA (processing) from the other gene.

We have used in situ hybridization with gene-specific intron probes to localize human globin primary transcripts in nuclei of individual embryonic and fetal transgenic erythroid cells (Wijgerde et al.,1995). The signals appear in the nucleus as fluorescent foci at the location of an actively transcribing gene (Lawrence et al., 1989, Zhang et al., 1994). Probe penetration is almost complete as > 97 % of the active mouse α-globin genes can be detected. Heterozygous cells show single foci, whereas homozygous cells show two foci, demonstrating detection of each locus (not shown). To determine if the LCR can activate the γ- and ß-globin genes simultaneously we performed in situ hybridizations with gene-specific intron probes for γ- and ß-globin primary transcripts in homozygous day 12 transgenic mouse fetal liver cells (Fig.3). Two types of signals are obtained. Primary transcript signals for both γ- (red) and ß-globin (green) from each chromosome suggesting that the LCR can activate both genes simultaneously from a single chromosome as would be expected for co-transcription. Alternatively, the double signal could be due to an overlap between transcription of one gene and the decay of the pre mRNA signal from the other gene, as would be expected for single gene transcription. The latter explanation would also fit the cells with the other type of signal namely single gene signals from each chromosome, ie only γ-globin or only ß-globin per chromosome (Fig.4). This suggests that the LCR activation mechanism is mono-gene specific. Considering that these nuclei contain the transacting factors required for both γ- and ß-globin transcription, co-transcription would be unlikely if the γ only, ß only cell occurs frequently given the high probe penetration conditions (> 97%). We therefore quantitated the occurrence of the different cell types (Fig.3).

Day 12 homozygous transgenic fetal liver cells show all possible primary transcript signal combinations of γ- and ß-globin. 58 % of globin transcribing nuclei have only ß-globin transcription on both chromosomes while 9% have γ-globin only. The remaining one third of the erythroid cells contain combinations of γ- and ß-globin transcription and are therefore involved in the switching process. However, fewer than half (34%) of the chromosome in these "switching cells" have both γ- and ß-globin signals on the same chromosome. The majority have single gene signals only, suggesting that the LCR interaction is largely mono-gene specific. The same distribution is found at different stages of the cell cycle (not shown). Of particular interest, is the fact that in greater than 90% of the switching cells the two chromosomes are responding differently to the same trans-acting factor environment. This suggests a dynamic, continuously changing system in which the individual loci respond stochastically to changes in the factor environment.

Two forms of such a dynamic process can be envisioned. Either the switch in interaction is progressive from γ to ß, or it switches back and forth between the genes in a kind of flip-flop mechanism. γ- and ß-globin signals on the same chromosome would then be indicative of a recent switch producing a (brief) period of overlap in which the decaying "old" primary transcripts are detected in the presence of the newly synthesized transcripts. The time length of this overlap period would be dependent on the half-life of the primary transcript. The progressive switching mechanism would predict that heterozygous γ gene transcribing cells should not have ß mRNA in their cytoplasm, whereas the flip-flop mechanism would result in a proportion of γ (but not ß) transcribing nuclei, which contain ß mRNA in their cytoplasm.

We therefore added a third probe which specifically recognizes the human ß-globin mRNA (exon probe) and could show that there are many heterozygous transgenic cells that contain a strong γ transcription signal, while the ß-gene signal is absent, indicating that the γ-gene is currently being transcribed, but the ß gene is not. The cytoplasm however, shows an accumulation of ß-mRNA indicating previous transcriptional activity of the ß-gene in that cell (data not shown). This observation strongly supports a dynamic flip-flop between the genes. Many of the chromosomes with both γ- and ß-globin primary transcript signals (Fig. 3) would therefore be the result of flip-flop rather than co-transcription.

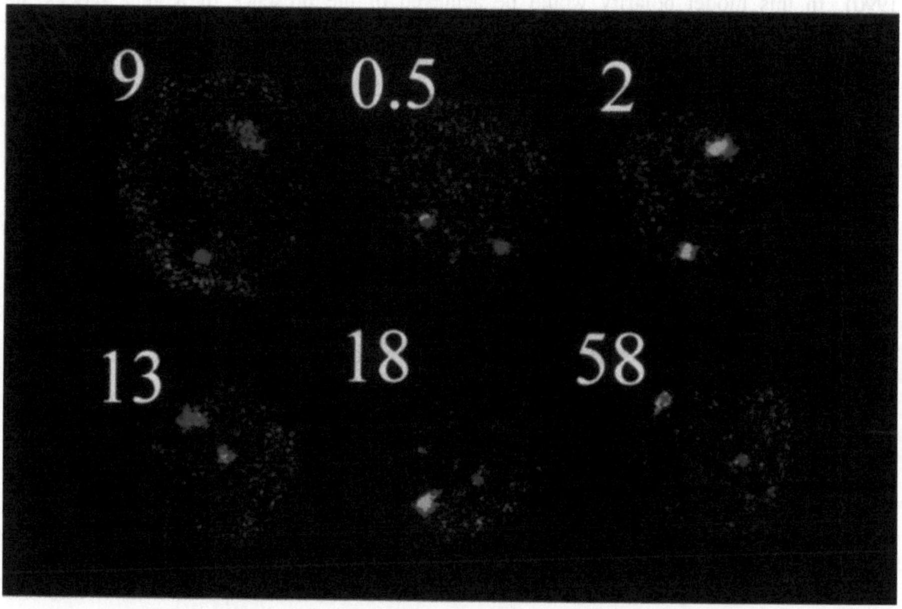

Fig.3 In situ hybridization of transgenic and non transgenic fetal liver cells. Quantitation of the different types of transgenic mouse 12.5 day fetal liver cells. Red (Texas red) repesents γ signal, green (FITC) represents ß signal and yellow the combination of both. Scoring was done by epifluores-cence on a minimum of 400 cells per liver sample and at least two livers. The frequency of occurrence of each cell type is given as a percentage.

Discussion

Long range activation of transcription can be explained by two completely different types of models. In one model the LCR initiates a chromatin change that spreads in a linear fashion along the chromatin until it reaches a gene (scanning fig. 1). The fact that the LCR and in particular HS3 works only efficiently when integrated into chromatin (Tuan et al., 1989; Moon & Ley, 1991) supports this model. In addition the status of the chromatin structure is dependent on the presence of the LCR in cis (Forrester et al.,

1990). In this model polarity would be achieved of the presence of a gene blocks the further progression of the active chromatin structure. However, the data we present here cannot be explained by such a mechanism. The effect of a proximal gene on a distal gene would not be dependent on their relative distance to the LCR. The alternative model postulates direct contact between the LCR and the sequence around each of the genes and the looping out of the intervening DNA (Fig. 1 looping). Competition would be the result of exclusive engagement of the LCR by one of the genes and the formation of loops would be dependent on the frequency of contact between the genes and the LCR. Polarity would be the result of a higher frequency of contact between the LCR and the more proximal gene. Since this frequency would depend on the relative distance between the genes and the LCR, moving the proximal gene away from the LCR and closer to the distal gene would be expected to reduce its ability to compete and give rise to an increase in the absolute level of expression of the distal gene. Our results are consistent with this type of model and in agreement with the in situ hybridization experiments (Wijgerde et al., 1995).

The in situ hybridization experiments show the dynamic nature of the interactions between the human ß-globin genes and the ß-globin LCR throughout development and in addition that the LCR-gene interactions are highly stable. All of our data can be explained by a mechanism in which the LCR activates only a single gene in the locus at any given time. The LCR, although made up of a series of hypersensitive sites would act as a functional unit or holocomplex activating multiple genes from the same chromosome via a stochastic mechanism of dynamic rather than static interactions (Fig.4). Although a low level of cotranscription cannot be excluded, such a mechanism cannot account for our observations.

A very interesting extra piece of information could be obtained from the in situ flip flop experiments. Measurement of all the populations of cells and the RNA precursor half lives also allowed us for the first time to measure individual periods of transcription time in vivo which varies from 4 to 60 minutes (Wijgerde et al.,1995). The flip flop mechanism operates both in the G1 and G2 periods of the cell cycle implying that replication is not required for a change in gene expression. On the basis of these results we postulate that chromatin interactions are dynamic and that transcriptional frequency and duration is regulated by the frequency and stability of direct contact between the LCR and the genes. Our results also imply that the interaction between the LCR and gene is required for the entire duration of a transcriptional period. We intend to test this hypothesis by measuring the transcriptional events in ß globin loci, which have altered expression patterns or which are prone to the position effects resulting from mutations in the LCR or their introduction in mice that lack specific transcription factors.

In conclusion the two types experiments, which use entirely different approaches provide strong evidence of direct interaction between the LCR and single genes by a dynamic looping mechanism of transcriptional activation. It does not show whether some process spreading in a linear fashion along the chromatin is involved in the first activation of the locus. The balance of eg. γ- vs ß-gene transcription would gradually change during a switching period as the factor composition changes in the population.

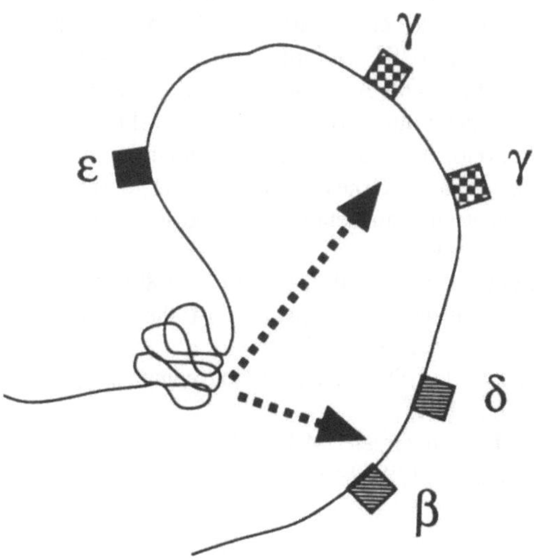

Fig.4 - A model of the LCR interactions with the genes via a stochastic looping mechanism. The LCR is indicated as a squiggly line to indicate that the different H S regions of the LCR could act together or even form a holocomplex to establish an interaction of the LCR with one of the genes.

References

- Bickel, S., and Pirotta, V. (1990) Self association of the Drosophila zeste protein is responsible for transvection effects. EMBO Journal, 9:2959-2967.
- Bresnick, E.H. and Felsenfeld, G. (1994) Dual promoter activation by the human ß-globin locus control region. Proc. Natl. Acad. Sci. Vol. 91:1314-1317.
- Dillon, N. and Grosveld F. (1991) Human γ-globin genes silenced independently of other genes in the ß-globin locus. Nature, 350:252-254.
- Engel, D. (1993) Developmental regulation of human ß-globin gene transcription: a switch of loyalties? Trends in Genetics, vol. 9, no. 9.

- Forrester, W.C., Epner, E., Driscoll, M.C. et al. (1990) A deletion of the human ß-globin locus activation region causes a major alteration in chromatin structure and replication across the entire ß-globin locus. Genes & Development, 4:1637-1649.
- Furukawa, T., Zitnik, G., Leppig, K., Papayannopoulou, Th. and Stamatoyannopoulos, G. (1994) Coexpression of γ and ß globin mRNA in cells containing a single human ß globin locus: results from studies using single-cell reverse transcription polymerase chain reaction. Blood, Vol. 83, no. 5:1412-1419.
- Hanscombe, O., Whyatt, D., Fraser, P. et al. (1991) Importance of globin gene order for correct developmental expression. Genes & Development, 5:1387-1394.
- Lawrence, J.B., Singer, R.H. and Marselle, L.M. (1989) Highly localized tracks of specific transcripts within interphase nuclei visualized by in situ hybridization. Cell, 57:493-502.
- Moon, A.M. and Ley, T.J. (1991) Functional properties of the ß-globin locus control region in K562 erythroleukemia cells. Blood, vol. 77, no. 10:2272-2284.
- Muller, H., Sogo, J. and Schaffner, W. (1989) An enhancer stimulates transcription in trans when attached to the promoter via a protein bridge. Cell, 58:767-777.
- Poncz, M., Henthorn, P., Stoeckert, C. and Surrey S. (1988) Globin gene expression in hereditary presistence of fetal hemoglobin and δß thalassemia. In McLean N (eds), Oxford Surveys on Eukaryotic Genes, vol 5:163-203. Oxford: Oxford University Press.
- Raich, N., Enver, T., Nakamoto, B. et al. (1990) Autonomous developmental control of human embryonic globin switching in transgenic mice. Science, 250:1147-1149.
- Strouboulis, J., Dillon, N. and Grosveld, F. (1992) Developmental regulation of a complete 70-kb human ß-locus in transgenic mice. Genes & Development, 6:1857-1864.
- Strouboulis, J. Regulation of the complete human beta-globin gene locus in transgenic mice. Ph.D. thesis, Open University, UK, 1994.
- Tuan, D.Y.H., Solomon, W.B., London, I.M. and Lee, D.P. (1989) An erythroid specific, developmental-stage-independent enhancer far upstream of the human 'ß-like-globin' genes. Proc. Natl. Acad. Sci., USA, 86:2554-2558.
- Wijgerde, M., Grosveld, F., and Fraser, P. (1995) Transcription complex stability and chromatin dynamics in vivo. Nature, 377:209-213.
- Zhang, G., Taneja, K.L., Singer, R.H. and Green, M.R. (1994) Localization of pre mRNA splicing in mammalian nuclei. Nature, 372:809-812.

Post-Transcriptional Regulation of Iron Metabolism

Matthias W. Hentze
Gene Expression Programme
European Molecular Biology Laboratory
Meyerhofstrasse 1
D-69117 Heidelberg

Germany
Tel.: 0049-6221-387 501
FAX: 0049-6221-387 518
e-mail: Hentze@EMBL-Heidelberg.DE

ABSTRACT

Regulation of iron metabolism is a fundamental process with profound medical implications. Cell growth and erythropoiesis are most critically affected by iron deficiency, while iron overload predominantly causes hepatic and degenerative disease. The underlying molecular control mechanisms are becoming increasingly well understood. Cellular iron uptake via the transferrin receptor and intracellular storage within a molecule consisting of 24 ferritin H- and L subunits is posttranscriptionally regulated through iron-responsive elements (IREs) contained in the untranslated regions of the respective mRNAs. IREs are also found in the messages encoding a rate-limiting enzyme for erythroid heme synthesis, 5-aminolevulinate synthase (eALAS), as well as mitochondrial aconitase, a Krebs' cycle enzyme of critical importance for cellular metabolic activity. These IRE-containing mRNAs are regulated by binding of a cytoplasmic protein, iron regulatory protein (IRP-1). IRP-1 binding reduces the translation rates of ferritin, eALAS and aconitase mRNAs and indirectly increases the synthesis of transferrin receptors by stabilising the mRNA against degradation. IRP-1 binding to IREs is regulated by multiple cellular signals: first by cellular iron levels, second by nitric oxide (NO), and third by oxidative stress (in the form of hydrogen peroxide, H_2O_2). Iron deficiency, NO and H_2O_2 independently

NATO ASI Series, Vol. H 99
Tumor Biology
Regulation of Cell Growth, Differentiation and Genetics in Cancer
Edited by Asterios S. Tsiftsoglou, Alan C. Sartorelli, David E. Housman and T. Michael Dexter

activate IRE binding by inducing the posttranslational conversion of 4Fe-4S IRP-1 to, apparently, apoIRP-1. Thus, a regulatory network that affects iron homeostasis, heme synthesis and the Krebs' cycle is co-ordinately operated by at least three independent signals via IRP-1, providing a framework for understanding physiological and pathological responses that affect cell growth and metabolism as well as erythropoiesis.

All eukaryotic cells need iron for growth. As a component of ribonucleotide reductase iron is required for DNA synthesis, as part of iron sulphur- or hemo-proteins of the Krebs cycle and the respiratory chain for energy metabolism, and in hemoglobin for oxygen transport. Inspite of these beneficial roles that iron plays, its reactivity with intermediates of aerobic metabolism forms the basis of its serious cytotoxicity that becomes clinically apparent in iron overload syndromes. Moreover, iron is essentially insoluble at physiologic pH in an oxygen atmosphere, which necessitates specialised mechanisms for solubilizing and taking up iron. Mammalian cells have solved this problem in the form of the serum protein transferrin that binds two atoms of iron per molecule and is taken up into cells via a specific cell membrane receptor, the transferrin receptor (TfR). Iron is released from transferrin within the endosome and, after crossing the endosomal membrane by as yet undefined mechanisms, can be used for cellular metabolism or be stored in ferritin, a hollow sphere consisting of 24 subunits of ferritin L- (light) and H- (heavy) chains.

In mammals (including man), iron uptake (via the TfR) [Casey et al., 1988; Müllner et al., 1989], iron storage (in ferritin) [Hentze et al., 1987; Aziz and Munro, 1987], erythroid heme synthesis (via 5-aminolevulinate synthase [eALAS]) [Cox et al., 1991; Dandekar et al., 1991; Melefors, et al., 1993] as well as the Krebs cycle (via mitochondrial aconitase) [Dandekar et al., 1991] all appear to be under the regulatory influence of the IRE/IRP system, which is the subject of this summary. IREs are well defined, ~30 nucleotide long RNA structures that were identified in the 5' untranslated regions (UTRs) of ferritin H- and L-chain mRNAs, eALAS mRNA and mitochondrial aconitase mRNA as well as in the 3' UTR of TfR mRNA (Fig.1).

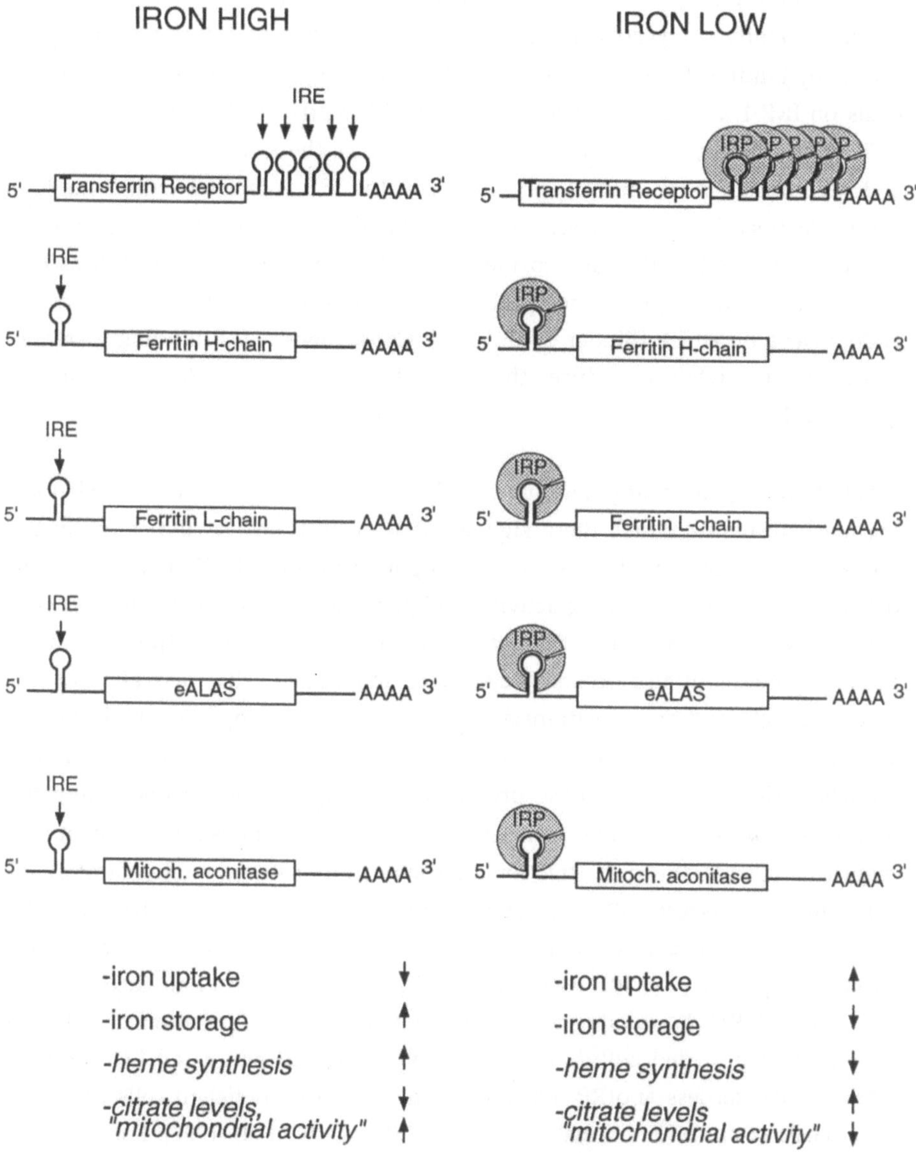

Fig. 1: Messenger RNAs that are regulated by the IRE/IRP system. Note that regulation of mitochondrial aconitase expression has not yet been demonstrated in vivo.

IREs are binding sites for iron regulatory proteins (IRPs), a small (currently 2 member) family of cytoplasmic proteins that responds to cellular signals (see below) by binding to IREs. With information on IRP-2 being fairly limited, I will focus on IRP-1 and refer to it as IRP. When IRP binds to IREs located in the 5' UTR of an mRNA, translation is repressed because the small ribosomal subunit is denied access to the mRNA, provided that the IRE is located within the first ~50 nucleotides from the transcription start site [Gray and Hentze, 1994; Goossen and Hentze, 1992]. IRP bound to the 3' UTR IREs of TfR mRNA stabilises this transcript against degradation, probably by preventing a rate-limiting initial endonucleolytic attack [Binder et al., 1994]. Consequently, IRP binding increases TfR expression while in reduces the expression of those mRNAs that bear a 5' UTR IRE (Fig.1).

What are the signals that determine the IRE-binding activity of IRP and what are the mechanisms by which these signals induce activation? As indicated by the nomenclature, iron was the first signal recognised to control IRP activity. In iron deficient cells, the IRE-binding activity is high, while it is low in iron loaded cells [Rouault et al., 1988]. Since the activation process does not require de novo protein synthesis (it also occurs in cycloheximide-treated cells), activation of IRE binding is achieved by a posttranslational mechanism [Pantopoulos et al., 1995a]. The extensive amino acid sequence homology between IRP and the mitochondrial enzyme aconitase provided the first clue for understanding the mechanism by which cellular iron levels control IRP: aconitase is a protein that requires a cubane 4Fe-4S cluster for the conversion of its substrate, citrate, into isocitrate. The perfect (100%) conservation of the 18 amino acids that form the catalytic center and co-ordinate the Fe-S cluster in mitochondrial aconitase strongly suggested that IRP was also a Fe-S protein [Rouault et al., 1991; Hentze and Argos, 1991]. It was demonstrated subsequently that formation of a 4Fe-4S cluster (in iron loaded cultured cells or in vitro) indeed prevented IRE binding, while the cluster-less apoIRP that predominated in iron deficient cells bound to IREs [Haile et al., 1992; Constable et al., 1992; Kennedy et al., 1992]. Furthermore, 4Fe-4S IRP displays aconitase activity itself and was identified to represent the long known cytoplasmic aconitase [Kennedy et al., 1992; Emery-Goodman et al., 1993]. Thus, iron regulates the cellular IRE binding activity of IRP by posttranslational Fe-S cluster insertion or removal. The aconitase activity of IRP is controlled reciprocally to IRE binding.

Subsequent studies revealed that the conversion of 4Fe-4S IRP into apoIRP is also specifically induced by nitric oxide (NO) and hydrogen peroxide (H_2O_2) (Fig.2).

A new mechanism of posttranslational regulation by signal-dependent Fe-S cluster switching

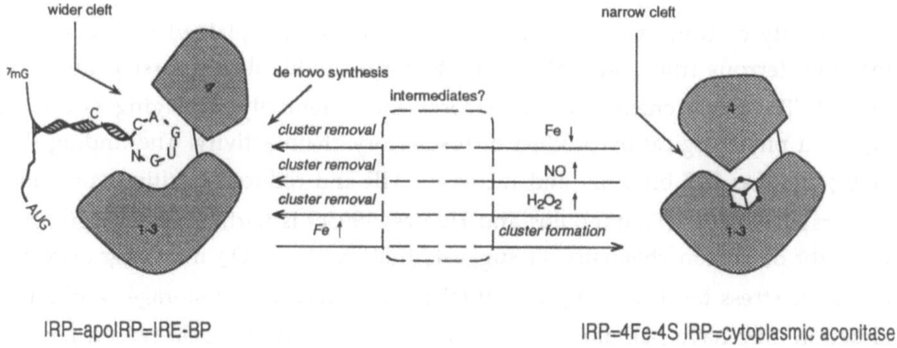

Fig. 2: Regulation of IRP-1 by cellular signals. IRP-1 is depicted as an IRE-binding apoprotein (left) or as a 4Fe-4S cytoplasmic aconitase (right). Nitric oxide, H_2O_2 and iron deficiency are suggested to induce the conversion of 4Fe-4S IRP-1 into apoIRP-1. The conversion may follow different pathways in response to the three signals, and it may involve as yet unidentified intermediates. The IRE-binding site(s) of IRP-1 is not yet fully defined. The possibility of allosteric regulation of IRE-binding is, therefore, still hypothetical.

NO is produced by at least three different isoforms of the enzyme nitric oxide synthase (NOS) involving the conversion of arginine into citrulline. It serves as an inter- and intracellular messenger that controls many physiological responses including blood pressure, platelet aggregation, macrophage-mediated cytotoxicity as well as neuronal processes. When intracellular NO synthesis is induced or cells are exposed to extracellular NO, IRE binding activity increases and the aconitase activity decreases [Drapier et al., 1993; Weiss et al., 1993]. In fibroblasts, ferritin synthesis drops and transferrin receptor levels are elevated [Pantopoulos and Hentze, 1995b]. In the murine macrophage cell lines J774 and RAW 264.7, induction of NO synthesis reduces ferritin synthesis, but no increase in TfR expression can be observed [Drapier et al., 1993; Pantopoulos and Hentze, 1995b; Pantopoulos et al., 1994]. This response may result from an intranuclear mechanism that negatively affects TfR expression. The clinical role of NO-induced activation of IRE binding on iron metabolism or erythropoiesis has not yet been evaluated. The physiological relevance of the regulatory interconnection between iron and NO metabolism remains to be defined more precisely, but is further supported by the finding that cellular iron levels regulate the transcription of the macrophage NOS gene, suggesting the existence of a feedback loop [Weiss et al., 1994].

The toxicity of iron overload can be, at least in part, explained by the Fenton reaction: ferrous iron and H_2O_2 react to form the highly aggressive hydroxyl radical. The same chemistry underscores the potentially damaging effects of H_2O_2, a physiological by-product of respiratory chain activity. The finding that H_2O_2 induces IRE binding (and increased TfR and reduced ferritin expression) with rapid kinetics [Pantopoulos and Hentze, 1995c] is particularly intriguing in the light of Fenton chemistry. It suggests that excess H_2O_2 may aggravate the oxidative stress by increasing iron uptake and reducing iron storage. While this situation may have pathological consequences, the physiological reason for this regulatory loop may have to be sought elsewhere: the mRNA that encodes the mitochondrial enzyme aconitase harbours an IRE in its 5' UTR and could thus be subject to negative regulation by IRP during oxidative stress. This may allow downregulation of the activity of the Krebs' cycle and the respiratory chain during oxidative stress and thus fulfil a protective function.

Further studies of the biology and clinical pathology of the IRE/IRP system may reveal important insights into the regulation of cell growth and metabolic activity by iron, oxidative stress and NO (Fig.3).

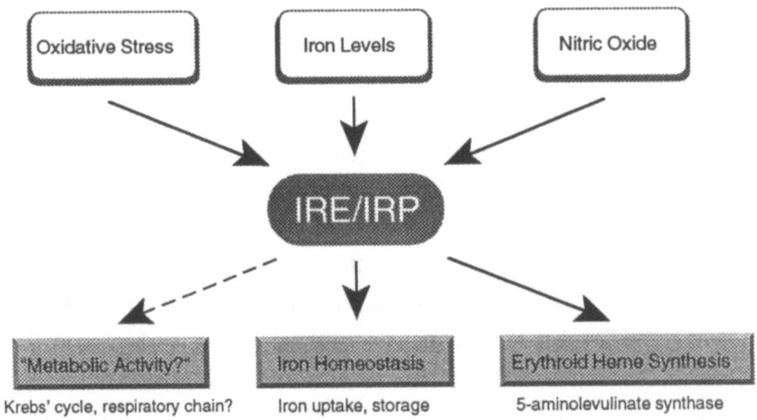

Fig. 3: Synopsis of the biology of the IRE/IRP system. Three signals are depicted to regulate IRP. The IRE-binding activity of IRP is shown to affect iron homeostasis, erythroid heme synthesis and, currently still hypothetically (dashed arrow), the metabolic activity of cells via the Krebs cycle.

REFERENCES

Aziz N and Munro HN (1987) Iron regulates ferritin mRNA translation through a segment of its 5' untranslated region. Proc. Natl. Acad. Sci. USA 84:8478-8482

Binder R, Horowitz JA, Basilion JP, Koeller DM, Klausner RD and Harford JB (1994) Evidence that the pathway of transferrin receptor mRNA degradation involves an endonucleolytic cleavage within the 3' UTR and does not involve poly(A) tail shortening. EMBO J 13:1969-1980

Casey JL, Hentze MW, Koeller DM, Caughman SW, Rouault TA, Klausner RD and Harford JB (1988) Iron-responsive elements: regulatory sequences that control mRNA levels and translation. Science 244:357-359

Constable A, Quick S, Gray NK and Hentze MW (1992) Modulation of the RNAbinding activity of a regulatory protein by iron in vitro: switching between enzymatic and genetic function? Proc. Natl. Acad. Sci. USA 89:4554-4558

Cox TC, Bawden MJ, Martin A and May BK (1991) Human erythroid 5-aminolevulinate synthase: promoter analysis and identification of an iron-responsive element in the mRNA. EMBO J 10:1891-1902

Dandekar T, Stripecke R, Gray NK, Goossen B, Constable A, Johansson HE, and Hentze MW (1991) Identification of a novel iron-responsive element in murine and human erythroid δ-aminolevulinic acid synthase mRNA. EMBO J 10:1903-1909

Drapier JC, Hirling H, Wietzerbin J, Kaldy P and Kühn LC (1993) Biosynthesis of nitric oxide activates iron regulatory factor in macrophages. EMBO J 12:3643-3649

Emery-Goodman A, Hirling H, Scarpellino L, Henderson B and Kühn LC (1993) Iron regulatory factor expressed from recombinant baculovirus: conversion between the RNA-binding apoprotein and Fe-S cluster containing aconitase. Nucl. Acids Res. 21:1457-1461

Goossen B and Hentze MW (1992) Position is the critical determinant for function of iron-responsive elements as translational regulators. Mol. Cell. Biol. 12:1959-1966

Gray NK and Hentze MW (1994) Iron regulatory protein prevents binding of the 43S translation pre-initiation complex to ferritin and eALAS mRNAs. EMBO J 13:3882-3891

Haile DJ, Rouault TA., Harford JB, Kennedy MC, Blondin GA, Beinert H and Klausner RD (1992) Cellular regulation of the iron-responsive element binding protein: Disassembly of the cubane iron-sulfur cluster results in high-affinity RNA binding. Proc. Natl. Acad. Sci. USA 89:11735-11739

Hentze MW, Caughman SW, Rouault TA, Barriocanal JG, Dancis A, Harford JB and Klausner RD (1987) Identification of the iron-responsive element for the translational regulation of human ferritin mRNA. Science 238:1570-1573

Hentze MW and Argos P (1991) Homology between IRE-BP, a regulatory RNA-binding protein, aconitase, and isopropylmalate isomerase. Nucl. Acids Res. 19:1739-1740

Kennedy MC, Mende-Mueller L, Blondin GA and Beinert H (1992) Purification and characterization of cytosolic aconitase from beef liver and its relationship to the iron-responsive element binding protein. Proc. Natl. Acad. Sci. USA 89:11730-11734

Melefors Ö, Goossen B, Johansson HE, Stripecke R, Gray NK and Hentze MW (1993) Translational control of 5-aminolevulinate synthase mRNA by iron-responsive elements in erythroid cells. J. Biol. Chem. 268:5974-5978

Müllner E.W, Neupert B and Kühn LC (1989) A specific mRNA binding factor regulates the iron-dependent stability of cytoplasmic transferrin receptor mRNA. Cell 58:373-382

Pantopoulos K, Weiss G and Hentze MW (1994) Nitric oxide and the post-transcriptional control of cellular iron traffic. Trends Cell Biol. 4:82-86

Pantopoulos K, Gray NK and Hentze MW (1995a) Differential regulation of two related RNA-binding proteins, iron regulatory protein (IRP) and IRP_B. RNA 1:155-163

Pantopoulos K and Hentze MW (1995b) Nitric oxide signaling to iron-regulatory protein (IRP): Direct control of ferritin mRNA translation and transferrin receptor mRNA stability in transfected fibroblasts. Proc. Natl. Acad. Sci. USA 92:1267-1271

Pantopoulos K and Hentze MW (1995c) Rapid responses to oxidative stress mediated by iron regulatory protein. EMBO J 14:2917-2924

Rouault TA, Hentze MW, Caughman SW, Harford JB and Klausner RD (1988) Binding of a cytosolic protein to the iron-responsive element of human ferritin messenger RNA. Science 241:1207-1210

Rouault TA, Stout CD, Kaptain S, Harford JB and Klausner RD (1991) Structural relationship between an iron-regulated RNA-binding protein (IRE-BP) and aconitase: functional implications. Cell 64:881-883

Weiss G, Goossen B, Doppler W, Fuchs D, Pantopoulos K, Werner-Felmayer G, Wachter H and Hentze MW (1993) Translational regulation via iron-responsive elements by the nitric oxide/NO-synthase pathway. EMBO J 12:3651-3657.

Weiss G, Werner-Felmayer G, Werner ER, Grunewald K, Wachter H and Hentze MW (1994) Iron regulates nitric oxide synthase activity by controlling nuclear transcription. J. Exp. Med. 180, 969-976

The Control of Multiplication and Differentiation in Human Myelomonocytic Leukemia Cells

FR Collart, DA Tonetti, and E Huberman
Center for Mechanistic Biology and Biotechnology
Argonne National Laboratory
9700 South Cass Avenue
Argonne, Illinois 60439

Abstract

To decipher the steps in the signal transduction pathway that leads to macrophage differentiation, we used human promyelocytic HL-60 cell variants that are either susceptible or resistant to differentiation induced by phorbol 12-myristate 13-acetate (PMA). The acquisition of a macrophage phenotype in these cells was characterized by cell multiplication arrest, reactivity with maturation specific monoclonal antibodies, and phagocytosis. The PMA-resistant cell variants exhibited (relative to the susceptible cells) discrete deficiencies in the expression of a number of genes that code for certain protein kinase C isoenzymes. By transfecting suitable expression vectors, we were able to circumvent the deficiency and restore the wild-type phenotype in the PMA-resistant cells. These studies thus substantiate the role that specific gene products (deficient in the PMA-resistant cells) play in the signal transduction process that leads to the acquisition of a macrophage phenotype in HL-60 and other related cell types.

We also examined the role of a specific protein complex (PC), which is composed of a 10- and a 14-kDa protein, in regulating cell multiplication during terminal differentiation of myelomonocytic cells. The genes coding for these proteins are expressed during terminal differentiation of HL-60 cells (and related cell types) and in mature peripheral blood granulocytes and monocytes. This expression appears to be associated with terminal differentiation, because incubation of HL-60 cells with an agent or condition that does not induce differentiation but does suppress cell multiplication does not result in the expression of the PC coding genes. We have also found that the purified PC inhibited the multiplication of HL-60 and related cell types. The expression pattern and inhibitory

NATO ASI Series, Vol. H 99
Tumor Biology
Regulation of Cell Growth, Differentiation and Genetics in Cancer
Edited by Asterios S. Tsiftsoglou, Alan C. Sartorelli, David E. Housman and T. Michael Dexter

character of the PC suggest that it may have a role in suppressing cell replication during terminal myelomonocytic differentiation.

Introduction

Human cells produce and respond to growth- and differentiation-inducing factors. Stem cells begin to mature after the interaction of a specific inducer of differentiation with its appropriate cellular receptor. Shortly following this interaction, a series of cellular signals are transmitted from the receptor to the genome, causing the activation and expression of a series of genes, regulatory genes in particular. The products of these early activated genes cause, through positive or negative controls (e.g., through a "trans"-acting process or ligand-receptor interaction), sequential expression of genes that code for the different functions associated with the mature state (including inhibition of cell multiplication). This paper will first deal with the role of a specific protein kinase in the signal transduction processes that lead to differentiation induction in human myeloid leukemia cells and then will describe the role a specific protein complex may play in regulating multiplication during human myelomonocytic cell differentiation.

Part I. Mediation of Human Myeloid Cell Differentiation by Protein Kinase C

Protein kinase C (PKC) is a family of serine-threonine protein kinases comprised of at least 12 isoenzymes: α, βI, βII, γ, δ, ε, ζ, η, θ, ι, λ, and μ (for review see Dekker *et al.* 1995). Among their functions, the PKC isozymes act as cellular transducers of external signals that result in the expression of specific sets of genes and, consequently, the manifestation of physiological traits such as cell multiplication and differentiation. Several investigators have studied the roles that individual PKC isozymes play in chemically induced differentiation by examining, in a number of cell systems, the expression of various PKC genes during this process. In particular, attention has been focused on the human myeloid HL-60 leukemia cell line, because these cells express useful markers of maturation and can be induced with phorbol 12-myristate 13-acetate (PMA) and related agents to acquire a macrophage phenotype (Rovera *et al.* 1979, Huberman *et al.* 1982); with

1,25-dihydroxyvitamin D_3 to acquire a monocytic cell type (McCarthy *et al.* 1983, Murao *et al.* 1983); and with retinoic acid (Breitman *et al.* 1980) or dimethyl sulfoxide to acquire a granulocytic phenotype (Collins *et al.* 1978). In these HL-60 cell studies was demonstrated an accordance between PKC-α and -β gene expression and macrophage differentiation (Nishikawa *et al.* 1990, McSwine-Kennick *et al.* 1991, Solomon *et al.* 1991, Aihara *et al.* 1991, Edashige *et al.* 1992). We have described an HL-60 cell variant, HL-525, which is resistant to PMA-induced macrophage differentiation but not to retinoic acid- or 1,25-dihydroxyvitamin D_3-induced differentiation (Homma *et al.* 1986). This resistant cell variant is also deficient in PKC-β gene expression, the most abundant PKC isozyme in the HL-60 cell line (Tonetti *et al.* 1992). These investigations, however, indicated only an association between the altered expression of certain PKC genes and cell differentiation. An approach that is more direct in determining the roles of individual PKC isozymes involves the use of variant cells that harbor a defect in a particular PKC isozyme. The function of the isozyme of interest can then be substantiated by proving that the cellular behavior associated with the defect is restored to that of the wild type by transfection with an appropriate expression vector. The HL-60 cell variant, HL-525, which is deficient in PKC-β gene expression and is resistant to PMA-induced differentiation, provided us with such a system.

Restoration of PMA-induced Differentiation Markers in PMA-resistant HL-525 Cells after Transfection of PKC-β cDNA. To determine whether the reestablishment of PKC-β gene expression in the PKC-β-deficient HL-525 cells is sufficient to restore susceptibility to PMA-induced differentiation, we transfected these cells with expression plasmids that contain either PKC-βI or -βII cDNA. For comparison, we also transfected the HL-525 cells with expression plasmids that contain only the gene coding for neomycin resistance (*neo*^{R}). Initially, each *neo*^{R} isolate from the various transfections was analyzed for PKC-β gene expression and PMA-induced cell attachment and cell spreading, which are hallmarks of macrophage differentiation in cells from the HL-60 cell line and its clone, HL-205 (Huberman and Callaham 1979; Rovera *et al.* 1979).

Several HL-525/βI *neo*^{R} isolates were obtained which demonstrated an increased level of steady-state PKC-β RNA relative to the untransfected parental HL-525 cells (Fig. 1A) and which exhibited approximately 50% cell attachment and spreading following PMA treatment (Table 1). Similarly, HL-525/βII *neo*^{R} isolates also showed an increased level of PKC-β

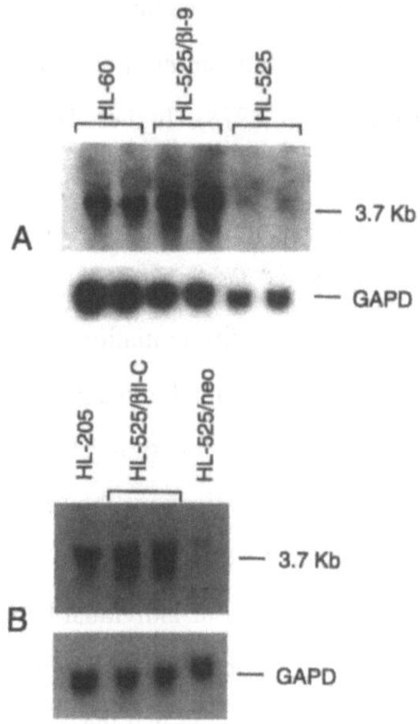

Fig. 1. Northern blot analysis of RNA from HL-205, HL-525/*neo*, HL-525/pMV7 cells and cells from selected HL-525/βI (A) and βII (B) transfectants. Total RNA (20 μg) was hybridized to the human PKC β cDNA probe. Hybridization to the GAPD cDNA probe is shown beneath each lane to assess the quantity of RNA loaded per lane. Each Northern blot represents the earliest analysis following transfection (from Tonetti *et al.* 1994).

RNA relative to the untransfected HL-525 cells (Fig. 1B) and 50% or greater cell attachment and spreading following PMA treatment (Table 1). Cell attachment and spreading were not observed in any of the HL-525/βI or HL-525/βII transfectants that did not exhibit an increased PKC-β RNA level. Also, none of the HL-525/neo^R isolates that were transfected with control plasmids demonstrated increased PKC-β RNA levels or PMA-induced cell attachment and spreading. These results indicate that increasing the steady-state PKC-β RNA levels by productive PKC/βI or PKC/βII cDNA transfection restores PMA-induced cell attachment in HL-525 cells.

Transfection of PKC-β cDNA Restores Susceptibility to PMA-induced Differentiation in PMA-resistant HL-525 Cells. To characterize further the mature phenotype of those

Table 1. Manifestation of differentiation markers in cells treated for three days with 30 nm PMA (%)

Cell type	Inhibition of cell multiplication[a]	Cell Attachment[b]	Cells reacting with OKM1[c]
HL-60	>95	>95	99
HL-205	>95	>95	98
HL-525	22	≤5	≤5
HL-525/pMV7	34	≤5	≤5
HL-525/pRSV-*neo*	37	≤5	≤5
HL-525/βI-7	93	~50	24
HL-525/βI-9	95	~50	49
HL-525/βI-F	95	~50	30
HL-525/βII-I	87	~50	52
HL-525/βII-C	93	~75	73
HL-525/βII-H	80	~75	42

[a] Untreated cells: (2–9) × 10^5 cells/ml, except for the HL-525/βII-H isolate, where there were 5 × 10^4 cells/ml (from Tonetti *et al.* 1994).
[b] Relative to untreated cells, which is < 0.1%.
[c] Relative to untreated cells, which ≤ 5%.

HL-525/βI and HL-525/βII clones exhibiting PMA-induced cellular attachment, we analyzed two additional differentiation markers, namely, inhibition of cell multiplication and reactivity with the OKM1 monoclonal antibody, which are characteristic of mature human monocytes, macrophages, and granulocytes (Foon *et al.* 1982). Treatment of HL-525/β1 and HL-525/βII isolates for three days with 30 nM PMA resulted in 80–90% inhibition of cell multiplication compared with a more than 95% inhibition observed for the HL-60 and HL-205 cell lines and less than 40% for the parental HL-525 variant of the HL-525/*neo* isolates (Table 1). Treatment of these HL-525/βI and HL-525/BII isolates with PMA also resulted in a time- and dose-dependent increase in their cell reactivity with the OKM1 antibody. After a 2- to 3-day treatment with 30 nm PMA, 20–50% of the HL-525/βI cells and 40–70% of the

HL-525/βII cells reacted with the antibody compared, with 5% or less in untreated or PMA-treated parental HL-525 cells or those transfected with control plasmids (Table 1).

In addition, we tested one PKC-βI (HL-525/βI-9) and two PKC-βII (HL-525/βII-I and HL-525/βII-C) transfectants for their ability to phagocytize opsonized fluorescent beads. Although 5% or less of untreated cells or PMA-treated HL-525/*neo* cells engulfed 20 or more beads/cell, 95% or more of PMA-treated HL-205, 75% of PMA-treated HL-525/βI-9 cells, and about 50% of PMA-treated HL-525/βII-I and HL-525/βII-C cells engulfed 20 or more beads/cell (Table 2). These results indicate that these PMA-treated HL-525/βI and HL-525/βII transfectants exhibit a high degree of phagocytosis, which is a primary function of mature macrophages. Thus, productive transfection of the PKC-β-deficient HL-525 cells with expression plasmids that contain either PKC-βI or PKC-βII can restore their susceptibility to PMA-induced macrophage differentiation.

Although all PKC-β transfectants that exhibited markers of differentiation also demonstrated increased levels of PKC-β mRNA, the reverse was not always true. Some transfectants with increased PKC-β mRNA levels did not display markers of differentiation, whereas others having increased RNA levels ceased to display these markers with time. The mechanism of this instability was examined by monitoring over time the PMA-susceptible phenotype in a number of productive transfectants, including an HL-525/βII-C single-cell clone. We determined that in most cases the loss of the phenotype was not due to the outgrowth of an HL-525/*neo* contaminant or a cell with a major deletion in the PKC-β gene,

Table 2. Percentage of phagocytizing cells after treatment for two days with 30 nM PMA[a]

	≥5 beads/cell		≥10 beads/cell	
Cell type	Control	PMA	Control	PMA
HL-205	15	100	3	100
HL-525/βII-I	11	74	0	48
HL-525/βII-C	19	71	4	46
HL-525/*neo*	9	24	0	3

[a] The number of engulfed beads was determined from confocal microscope images of treated and untreated cells (from Tonetti *et al.* 1994).

since a transcript of normal size was produced. Those transfectants that did not display the differentiation markers may therefore either be expressing a defective PKC-β transcript or have a defect in its translation, since the level of PKC protein and activity correlate with the PMA-susceptible phenotype. Therefore, defective gene expression (impairment in mRNA transcription in particular) appears to be a frequent event and may result from the use of nonhuman construct.

The manifestation of the differentiation markers following PMA treatment of the HL-525/βI or βII transfectants is clear. However, the extent of inhibition of cell multiplication, cell attachment, reactivity toward the OKM1 antibody, and phagocytosis is not as profound as seen in the parental HL-60 or HL-205 cells. It appears that reestablishment of PKC-βI or βII gene expression, either at wild-type or overexpressed levels, can only partially restore PMA susceptibility to the HL-525 cell line. This may be the result of the refractory character of HL-525 cells, as was observed by the rapid loss of the PMA-susceptible phenotype, or because the phenotype is controlled by a nonhuman PKC-β isozyme, or because of a deficiency of an additional component that is required for full establishment and maintenance of susceptibility of PMA-induced differentiation.

Despite the unstable nature of the transfectants, we can conclude that introduction of PKC-βI and to a greater extent, PKC-βII, into the PKC-β-deficient HL-525 cells can restore susceptibility to the PMA-induced differentiation in this HL-60 cell variant. We can therefore presume that PKC-β is an essential element in the PMA-induced signal transduction pathway which results in macrophage differentiation in HL-60 cells and, perhaps, other related normal and leukemia cell types.

We propose an outline for such a pathway that is consistent with our findings. Such a pathway would be initiated by the interaction of PMA or a related ligand with the PKC-β isozyme and would subsequently lead to the activation and translocation of the enzyme from the cytosol to the membrane fraction (Aihara *et al.* 1991; Tonetti *et al.* 1992; Hocevar and Fields 1991; Slapak *et al.* 1993). The activated PKC-β isozyme, either directly or indirectly, would phosphorylate a variety of specific proteins, including nuclear proteins (Homma *et al.* 1988; Beckmann *et al.* 1992; Hocevar *et al.* 1993), which in turn would cause the expression of the early response genes (Tonetti *et al.* 1992; Szabo *et al.* 1991; Kharbanda *et al.* 1991; Nguyen *et al.* 1993). These gene products then would elicit the expression of a cascade of genes, including those that code for cellular functions that define the mature phenotype.

Part 2. A Protein Complex Expressed During Terminal Differentiation of Myelomonocytic Cells Is an Inhibitor of Growth

To identify myelomonocytic cell differentiation markers, we developed a set of murine monoclonal antibodies (mAb's) directed against nuclear antigens (Murao *et al.* 1985). One of these antibodies, NM-6, was used to analyze the distribution of its antigen in mature and immature blood cells (Murao *et al.* 1989). The NM-6 antigen was detected in peripheral blood monocytes and granulocytes but not in lymphocytes, and only a limited expression was detected in human myeloid leukemia cell lines (including the HL-60 cells). These results indicated that the NM-6 antigen is a maturation marker in myelomonocytic cells.

To further study its role in differentiation, we purified the NM-6 antigen from a human spleen with the aid of an NM-6 immunoaffinity column. Denaturing gel electrophoresis indicated that the antigen is a protein complex (PC) composed of two proteins having molecular masses of 10 and 14 kDa (Murao *et al.* 1989). Using a polyclonal antibody directed against the purified 14-kDa protein, we isolated and sequenced a cDNA clone from a λgt11 human monocyte cDNA library (Murao *et al.* 1989). The DNA sequence of this clone was the same as that of a cDNA clone that codes for the protein MRP-14 (Dorin *et al.* 1987; Odink *et al.* 1987) except for three nucleotides in the 3′ noncoding region. The amino terminus of the 14-kDa protein was blocked; however, analysis of an internal peptide derived by V8 protease digestion yielded a 25-amino acid segment that corresponded to a coding segment in the human cDNA clone. Amino-terminal sequence analysis indicated that the 10-kDa protein was likely identical to another protein termed MRP-8 (Odink *et al.* 1987). Although the NM-6 mAb recognizes the intact PC, it does not recognize purified MRP-8 or MRP-14 constituent proteins.

We examined the expression and regulation of the MRP8 and MRP14 genes in the human leukemic cell line HL-60 treated with either of two differentiation-inducing agents, mycophenolic acid (MPA) or the biologically active from of vitamin D_3, 1α,25-dihydroxyvitamin D_3 (1,25-$(OH)_2D_3$). The HL-60 cell line is a useful model for the study of monomyelocytic cell differentiation, as treatment with MPA induces features of both mature granulocytes and monocytes (Lucas *et al.* 1983; Collart and Huberman 1990), and 1,25-$(OH)_2D_3$ treatment induces expression of differentiation markers characteristic of monocytic-like cells (Murao *et al.* 1983). These two agents differ in their mechanism of action, in that MPA blocks guanine nucleotide biosynthesis by inhibiting IMP dehydrogenase, whereas 1,25-$(OH)_2D_3$ appears to act predominantly by a classical steroid receptor mechanism.

Inhibition of Cell Multiplication by the PC. Incubation of HL-60 cells with purified PC resulted in a time- and dose-dependent decrease in the rate of HL-60 multiplication. Inhibition of multiplication was observed even after two days of treatment with the PC at a concentration as low as 5 nM. This inhibitory effect was also evident in THP-1 cells, as well as in other cell lines (Table 3).

In contrast to its ability an inhibit cell multiplication, the PC at concentrations of up to 7.5 nM did not induce a mature phenotype in the HL-60 cells (i.e., the treated cells did not exhibit markers associated with myelomonocytic cell differentiation, such as reactivity with OKM1 mAb, reduction of nitroterazolium dye, or staining for nonspecific esterase activity).

Table 3. PC concentration required to produce 50% inhibition of cell multiplication[a] (nM)

Cell Type	Concentration
HL-60 promyelocytic leukemia	8
THP-1 monocytic leukemia	5
CEM T-lymphocytic	8
PN-3 malignant melanoma	5
U-373 glioblastoma	10
MCF-7 mammary adenocarcinoma	15
Y-79 retinoblastoma	15
IMR-70 lung fibroblast (nontransformed)	15

[a] Cells were seeded into 24-well tissue culture plates at 3×10^5 cells in 1 ml of medium and treated with the PC immediately after plating. After four days, the number of cells was determined by hemocytometer counting (from Murao *et al.* 1990).

Similarly, treatment of CEM T-lymphoblastic leukemia cells with the purified PC reduced cell growth but did not alter reactivity with OKT3, OKT4, or OKT8 mAb's, which detect maturation antigens in this cell line (Murao *et al.* 1990). The results suggest a role for the PC as an inhibitor of cell growth.

Time- and Dose-Dependent Increase in PC Levels During MPA Treatment of HL-60 Cells. During four days in culture, less than 7% of untreated HL-60 cells exhibited reactivity with the NM-6 monoclonal antibody (Fig. 2A). Treatment of HL-60 cells with MPA for up to four days resulted in a time- and dose-dependent increase in the percentage of reacting cells (Fig. 2A). After four days of treatment with 3 μM MPA, more than 85% of the cells reacted with the NM-6 mAb (Fig. 2A). Under these conditions, HL-60 cells express antigenic and enzymatic markers of both monocytic and granulocytic differentiation (15). The MPA-mediated increase in reactivity with the NM-6 mAb was accompanied by an inhibition of HL-60 cell proliferation, which was also dose-dependent (Fig. 2B). As we previously demonstrated, this growth inhibition may be a consequence of the maturation agent-mediated increase in the PC level (10).

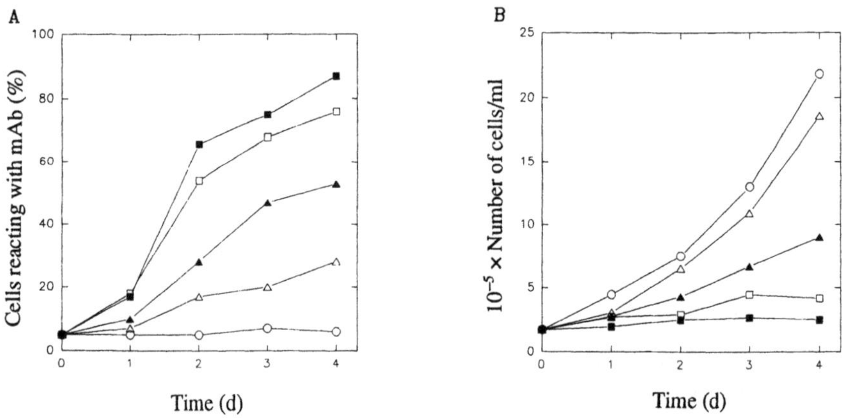

Fig. 2. (A) PC levels and (B) cell growth in untreated and MPA-treated HL-60 cells. PC levels are presented at the percentage of cell reacting with the NM-6 mAb. O, untreated; △, 0.5 μM MPA; ▲, 1.0 μM MPA; □, 2.0 μM MPA; ■, 3.0 μM MPA. Results are the mean of two experiments in which the variation of the data was within 20% (from Warner-Bartnicki *et al.* 1993).

Time- and Dose-Dependent Increase in MRP8 and MRP14 Steady-State mRNA Levels During MPA Treatment of HL-60 Cells. To determine whether changes in the amount of PC complex were due to an increase in PC constituent mRNA steady-state levels, we carried out Northern blot analyses or RNA isolated from untreated and MPA-treated HL-60 cells with the use of cDNA probes specific for MRP8 and MRP14 cDNA probes (Fig. 3). These probes detected the 0.55-kbp MRP8 and 0.75-kbp MRP14 transcripts (Murao *et al.* 1990). For comparison, we included a 1,25-$(OH)_2D_3$ treatment condition which in HL-60 cells can increase the levels of the PC and both MRP8 and MRP14 mRNA (Murao *et al.* 1990). Treatment of the HL-60 cells with MPA or 1,25-$(OH)_2D_3$ resulted in a time- and dose-dependent increase in the steady-state levels of MRP8 and MRP14 mRNA (Fig. 3). Both genes were coordinately expressed, although MRP8 mRNA levels exhibited a two-fold greater increase than MRP14 mRNA levels after MPA treatment (i.e., 65- as compared to 35-fold). Induction with MPA exceeded the 1,25-$(OH)_2D_3$-mediated increases in the steady-state levels of MRP8 and MRP14 mRNA and occurred with faster kinetics. For both maturation agents, transcript induction appeared maximal on either the third or fourth day after treatment.

MRP8 and MRP14 transcripts have been examined in normal blood monocytes (Murao *et al.* 1985), but not in normal blood granulocytes. This is relevant, as some studies

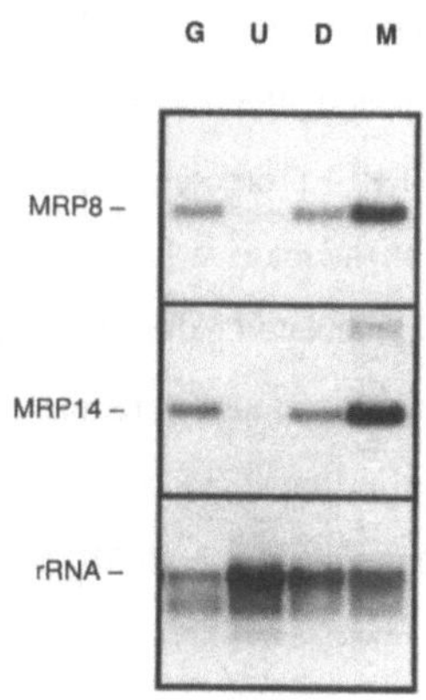

Fig. 3. MRP8 and MRP14 steady-state mRNA levels in normal blood granulocytes and in HL-60 cells. The HL-60 cells were treated or treated for three days with either 0.4 μM 1,25-$(OH)_2D_3$ or 2 μM MPA. G, granulocytes; U, untreated HL-60 cells; D, 1,25$(OH)_2D_3$-treated HL-60 cells; M, MPA-treated HL-60 cells (from Warner-Bartnicki *et al.* 1993).

have reported an additional, larger MRP8 transcript in monocytes (Dorin *et al.* 1987). We therefore examined MRP8 and MRP14 transcripts in normal blood granulocytes and compared them with mRNA obtained from untreated HL-60 cells and from HL-60 cells treated with MPA or 1,25-$(OH)_2D_3$ (Fig. 3). The granulocytes clearly expressed the same size transcripts as did HL-60 cells and did not transcribe additional higher molecular weight MRP8 and MRP14 mRNAs.

Increased Transcription of MRP8 and MRP14 Genes in HL-60 Cells. Two common ways in which mRNA accrual can occur are through an increase in mRNA half-life or through an increase in transcription initiation events at relevant promoters. This former regulatory scheme was investigated for the MRP8 and MRP14 genes by monitoring mRNA half-lives one day and four days after MPA and 1,25-$(OH)_2D_3$ treatment. Both mRNAs exhibited half-lives of about 6–8 h, which did not vary markedly with length or type of treatment (data not shown). The half-lives in untreated HL-60 cells were not measured because of the low steady-state MRP8 and MRP14 mRNA levels.

To determine whether the observed increases in MRP8 and MRP14 mRNA steady-state levels were due to increased transcription initiation at their promoters, we used nuclear run-on transcription assays to assess transcription rates of the MRP8 and MRP14 genes in untreated cells and in HL-60 cells treated for 1, 3, or 4 days with either MPA or 1.25-$(OH)_2D_3$ (data not shown). We standardized MRP8 and MRP14 transcription rates using two common positive controls: β-actin and 28S rRNA. While little or no increase in transcription initiation at MRP8 and MRP14 promoters could be observed at one day after treatment of HL-60 cells with MPA, an increase was observed 3–4 days after treatment (Table 4). In comparison, 1,25-$(OH)_2D_3$-mediated gains in transcription initiation frequency were detected as early as one day after treatment and were maximal 3–4 days after treatment (Table 4). The magnitudes of increase in the MRP8 and MRP14 gene transcription rates were comparable (about an order of magnitude) for both MPA and 1,25-$(OH)_2D_3$ (Table 3).

Possible Mode of Action of the PC. Regardless of these possibilities, the expression of the PC during differentiation and its multiplication-inhibitory effect in different cell types

Table 4. Relative Transcription Rates of the MRP8 and MRP14 Genes (*n*-fold increase)[a]

	MRP8		MRP14	
Agent	Day 3	Day 4	Day 3	Day 4
$1,25\text{-}(OH)_2D_3$[b] (0.4 μM)	6.2	6.6	8.5	7.8
MPA[c] (3.0 μM)	8.5	11.5	13.7	15.1

[a] Transcription rate in untreated cells was set at 1.0. Values are the average of at least two independent experiments.

[b] Transcription rates were normalized to the β-actin transcription rate. Similar to Solomon *et al.* (1991), we observed less that twofold differences between the β-actin and 28S rRNA transcription rates after $1,25\text{-}(OH)_2D_3$ treatment.

[c] Transcription rates were normalized to the 28S rRNA transcription rate (from Warner-Bartnicki *et al.* 1993).

suggest that it plays a role in the physiological functions of myelomonocytic cells in addition to its possible role as a mediator of cell multiplication inhibition during terminal differentiation of these cells. The results of our nuclear run-on transcription assays attest to the transcriptional regulation of MRP8 and MRP14 genes during differentiation induction by MPA and $1,25\text{-}(OH)_2D_3$. While a small increase in the frequency of transcription initiation could be detected as early as one day after treatment, the most pronounced gains occurred three and four days after treatment. These increases in transcription initiation frequency were in agreement with the increments in MRP8 and MRP14 mRNA steady-state levels. Even though the increase in transcription initiation frequency (about an order of magnitude) is not large when compared with the increase in MRP8 and MRP14 transcript levels (as much as 65-fold after MPA treatment), the long half-lives (6–8 h) of both transcripts suggest that relatively small increases in the frequency of transcription initiation can result in substantial accumulation of MRP8 and MRP14 mRNA.

Further studies a) on the involvement of PKC β in the signal transduction processes that result in differentiation induction and b) of the PC in regulating cell replication during terminal differentiation of myelomonocytic cells will provide further insight into these critical cellular events. Moreover, this type of information will lead to a greater understanding of the processes leading to terminal differentiation and the aberrations in these processes that result in malignant transformation.

We find it intriguing that the differentiation-inducing agents MPA and 1,25-$(OH)_2D_3$, which do not have the same mode of action in cells, are capable of precipitating similar substantial increases in the MRP8 and MRP14 proteins during monomyelocytic maturation of HL-60 cells. Whereas 1,25-$(OH)_2D_3$ binds to an intracellular receptor and directly interacts with responsive genes to modulate their transcription, MPA depletes guanine nucleotides necessary for cellular metabolism by inhibiting IMP dehydrogenase, the rate-limiting enzyme in *de novo* guanine nucleotide synthesis. Additionally, the observed elevated MRP8 and MRP14 mRNA and protein levels are consequences of an increase in transcription initiation. This may indicate that the PC complex functions as a common mediator in the maturation response to both agents. Our findings, as well as the abundance of expression of these genes in normal monocytes and granulocytes, substantiate that MRP8 and MRP14 genes play a role in the terminal cell differentiation of human promyelocytes.

Acknowledgment

This work was supported by the U.S. Department of Energy, Office of Health and Environmental Research, under Contract No. W-31-109-ENG-38.

References

Aihara H, Asaoka Y, Yoshida K, Nishizuka Y (1991) Sustained activation of protein kinase C is essential to HL-60 cell differentiation to macrophage. Proc Natl Acad Sci USA 88:11062–11066

Beckmann R, Buchner K, Jungblut PR, Eckerskorn C, Weise C, Hilbert R, Hucho F (1992) Nuclear substrates of protein kinase C. Eur J Biochem 210:45–51

Breitman TR, Selonick SE, Collins SJ (1980) Induction of differentiation of the human promyelocytic leukemia cell line (HL-60) by retinoic acid. Proc Natl Acad Sci USA 77:2936–2940

Collart FR, Huberman E (1990) Expression of IMP dehydrogenase in differentiating HL-60 cells. Blood 75:570–576

Collins SJ, Ruscetti FW, Gallagher RE, Gallo RC (1978) Terminal differentiation of human promyelocytic leukemia cells induced by dimethyl sulfoxide and other polar compounds. Proc Natl Acad Sci USA 75:2458–2462

Dekker LV, Palmer RH, Parker PJ (1995) The protein-kinase-C-related gene families. Current Opinion in Structural Biology 5:396–402

Dorin J, Novak M, Hill RE, Brock DJH, Secher DS (1987) A clue to the basic defect in cystic fibrosis from cloning the CF antigen gene. Nature 326:614–617

Edashige K, Sato EF, Akimuru K, Kasai M, Utsumi K (1992) Differentiation of HL-60 cells by phorbol ester is correlated with up-regulation of protein kinase C-α. Biochem Biophys 299:200–205

Foon KA, Schroff RW, Gale RP (1982) Surface markers on leukemia and lymphoma cells: recent advances. Blood 60:1-19

Hocevar BA, Fields AP (1991) Selective translocation of βII protein kinase C to the nucleus of human promyelocytic (H160) leukemia cells. J Biol Chem 266:28–33

Hocevar BA, Burns DJ, Fields AP (1993) Identification of protein kinase C (PKC) phosphorylation sites on human lamin-B potential roles of PKC in nuclear lamina structural dynamics. J Biol Chem 268:7545–7552

Homma Y, Gemmell MA, Huberman E (1988) Protein kinase C activities with different characteristics, including substrate specificity, from two human HL-60 leukemia cell variants. Cancer Res 48:2744–2748

Homma Y, Henning-Chubb CB, Huberman E (1986) Translocation of protein kinase C in human leukemia cells susceptible or resistant to differentiation induced by phorbol 12-myristate 13-acetate. Proc Natl Acad Sci USA 83:7316–7319

Huberman E, Braslawsky GR, Callaham MF, Fujiki H (1982) Induction of differentiation of human promyelocytic leukemia (HL-60) cells by teleocidin and phorbol 12-myristate 13-acetate. Carcinogenesis 3:111–114

Huberman E, Callaham, MF (1979) Induction of terminal differentiation in human promyelocytic leukemia cells by tumor-promoting agents. Proc Natl Acad Sci USA 76:1293–1297

Kharbanda S, Nakamura T, Stone R, Hass R, Bernstein S, Datta R, Sukhatme VP, Kufe D (1991) Expression of the early growth response-1 and response-2 zinc finger genes during induction of monocytic differentiation. J Clin Invest 88:571–577

Lucas DL, Webster HK, Wright DG (1983) Purine metabolism in myeloid precursor cells during maturation: studies with the HL-60 cell line. J Clin Invest 72:1889–1900

McCarthy DM, San Miguel JF, Freake HC, Green PM, Zola H, Catovsky D, Goldman JM (1983) 1,25-Dihydroxyvitamin D_3 inhibits proliferation of human promyelocytic leukemia (HL-60) cells and induces monocyte-macrophage differentiation in HL60 and normal human bone marrow. Leukemia Res 7:51–55

McSwine-Kennick RL, McKeegan EM, Johnson MD, Morin MJ (1991) Phorbol-diester-induced alterations in the expression of protein kinase C isozymes and their messenger RNAs: analysis in wild-type and phorbol-diester-resistant HL-60 cell clones. J Biol Chem 266:5135–5143

Murao S, Collart F, Huberman E (1990) A protein complex expressed during terminal differentiation of monomyelocytic cells is an inhibitor of cell growth. Cell Growth & Differ 1:447–454

Murao S, Collart FR, Huberman E (1989) A protein containing the cystic fibrosis antigen is an inhibitor of protein kinases. J Biol Chem 264:8356–8360

Murao S, Clevenger CV, Epstein AL, Huberman E (1985) Expression of maturation-specific nuclear antigens in differentiating human myeloid leukemia cells. Cancer Res 45:791–795

Murao S, Anderson NL, Callaham MF, Gemmell MA, Huberman E (1983) Control of macrophage cell differentiation in human promyelocytic HL-60 leukemia cells by

1,25-dihydroxyvitamin D_3 and phorbol 12-myristate 13-acetate. Cancer Res 43:4989–4996

Nguyen HQ, Hoffman-Liebermann B, Liebermann DA (1993) The zinc finger transcription factor Egr-1 is essential for and restricts differentiation along the macrophage lineage. Cell 72:197–209

Nishikawa M, Komada F, Uemura Y, Hidaka H, Shirakawa S (1990) Decreased expression of type-II protein kinase C in HL-60 variant cells resistant to induction of cell differentiation by phorbol diester. Cancer Res 50:621–626

Odink K, Cerletti N, Bruggen J, Clerc RG, Tarcsay L, Zwadlo G, Gerhards G, Schlegal R, Sorg C (1987) Two calcium-binding proteins in infiltrate macrophages of rheumatoid arthritis. Nature 330:80–82

Rovera G, Santoli D, Damski C (1979) Human promyelocytic leukemia cells in culture differentiate into macrophage-like cells when treated with a phorbol diester. Proc Natl Acad Sci USA 76:2779–2783

Slapak CA, Kharbanda S, Saleem A, Kufe DW (1993) Defective translocation of protein kinase C in multidrug-resistant HL-60 cells confers a reversible loss of phorbol-ester-induced monocytic differentiation. J Biol Chem 268:12267–12273

Solomon DH, O'Driscoll K, Sosne G, Weinstein IG, Cayre IE (1991) 1α,25-Dihydroxyvitamin-D_3-induced regulation of protein kinase C gene expression during HL-60 cell differentiation. Cell Growth & Differ 2:187–194

Szabo E, Preis LH, Brown PH, Birrer MJ (1991) The role of *jun* and *fos* gene family members in 12-*O*-tetradecanoylphorbol-13-acetate-induced hematopoietic differentiation. Cell Growth & Differ 2:475–482

Tonetti DA, Horio M, Collart FR, Huberman E (1992) Protein kinase β gene expression is associated with susceptibility of human promyelocytic leukemia cells to phorbol-ester-induced differentiation. Cell Growth & Differ 3:739–745

Tonetti DA, Henning-Chub C, Yamanishi T, Huberman E (1994) Protein kinase C-β is required for macrophage differentiation of human HL-60 leukemia cells. J Biol Chem 23230–23235

Warner-Bartnicki AL, Murao S, Collart FR, Huberman E (1993) Regulated expression of the MRP8 and MRP14 genes in human promyelocytic leukemic HL60 cells treated with the differentiation-inducing agents mycophenolic acid and 1α,25-dihydroxyvitamin D_3. Exp Cell Res 204:241–246

THE ROLE OF ONCOGENES IN THE INTEGRATED CONTROL OF CELL PROLIFERATION AND CELL DEATH

Barbara S. Ink and Gerard I. Evan
Imperial Cancer Research Fund
44 Lincoln's Inn Fields
London WC2A 3PX
United Kingdom

KEYWORDS/ABSTRACT: apoptosis/ c-myc/ cytokines/ IGF-1/ bcl-2/ growth factors/ Bak/ vaccinia virus/ tumor cell/ CHOhr

The c-myc oncogene can act as an inducer of apoptosis in serum-deprived fibroblasts. Cytokines, IGF-1 and PDGF, were identified as the components of serum which are able to delay *c-myc* induced apoptosis. The ability of the cytokines to protect from apoptosis is separate from their ability to function as growth factors.

A variety of apoptotic and anti-apoptotic genes from the cell and virus have been identified. The newest member of the *bcl-2* family, Bak, also induces apoptosis, while a new cowpox virus gene, *CHOhr*, delays virus-induced apoptosis in nonpermissive cells.

Summary

The *c-myc* proto-oncogene encodes an essential component of the cell's proliferative machinery and deregulation of expression of *myc* has been implicated in most tumours. C-Myc can also act as a potent inducer of apoptosis in cells deprived of serum or when arrested with cytostatic drugs at different stages of the cell cycle. Expression levels of myc correlate with the level of apoptosis and deletion mapping shows that regions of c-Myc required for apoptosis overlap with regions necessary for co-transformation, auto regulation, inhibition of differentiation, transcriptional activation and sequence-specific DNA binding. In addition, the heterologous partner of *myc*, Max, is required for *c-myc*-induced apoptosis. All of this strongly implies that *c-myc* drives apoptosis through a transcriptional mechanism: presumably by modulation of target genes.

NATO ASI Series, Vol. H 99
Tumor Biology
Regulation of Cell Growth, Differentiation and Genetics in Cancer
Edited by Asterios S. Tsiftsoglou, Alan C. Sartorelli, David E. Housman and T. Michael Dexter

Two models can be proposed to explain the induction of apoptosis by *c-myc*. One , the "Conflict of Signals" model proposes that death is due to a conflict in growth signals which is generated by the inappropriate or unscheduled expression of *c-myc* under conditions that would normally promote growth arrest. In this model, induction of apoptosis is a pathological consequence of *c-myc* deregulation rather than a normal function. It has importance only for models of carcinogenic progression in which *myc* expression has been disrupted. The other model, the "Dual Signal" model, proposes that induction of apoptosis is a normal obligate function of *c-myc* which is modulated by specific survival factors. Thus, every cell that enters the cycle implements the apoptotic program which must be continuously suppressed by signals from the immediate cellular environment for the proliferating cell to survive.

Much work in the field of apoptosis has focused on identifying both apoptotic and anti-apoptotic genes. These include both cellular and virus genes which function in a variety of systems. The newest member of the *bcl-2* family, *bak*, was recently cloned and shown to antagonise Bcl-2 protection of *c-myc* induced apoptosis and induced apoptosis in RAT-1 cells when placed under an inducible promoter.

In addition to the *bcl-2* family members of genes involved in apoptosis, DNA viruses encode novel anti-apoptotic proteins which show no homology to the *bcl-2* family members or to other virus anti-apoptotic proteins. Viruses utilise these anti-apoptotic genes because an infected cell will often show an inhibition of cell death in lytic and latent infections and during virus transformation. The virus system has been utilised to identify a new anti-apoptotic gene, *CHOhr,* which delays vaccinia virus induced apoptosis in Chinese hamster ovary cells.

Induction of apoptosis by c-myc in serum-deprived fibroblasts

Cell death can occur by one of two mechanisms: necrosis or apoptosis. Necrosis involves swelling of the cell and damage to the cellular organelles. Eventually due to the injury placed upon the cell, the cell will lyse releasing the contents into the extracelluar environment. The debris is taken up by phagocytosis and there is inflammation of the tissues (Kerr et al., 1972; Wyllie et al., 1980). In contrast, apoptosis involves shrinkage of the cell and the organelles remain undamaged. At later stages the DNA becomes fragmented and the cell breaks into membrane-bounded apoptotic bodies. There is also phagocytosis of the fragments, however, there is no inflammation (Kerr et al., 1972; Wyllie et al., 1980).

Apoptosis can be induced by a number of different physiological stimuli. Some of the more commonly studied include UV irradiation and cytostatic drugs, removal of growth factors, and the interaction of Fas ligand with many cells during the immune response. The response of the cell to the apoptotic stimuli depends upon the extracellular and intracellular environment. The control of apoptotsis depends not only on genes identified in the apoptotic pathway and the extracellular environment, but also is regulated by genes which are involved in the control of cell proliferation and differentiation.

The expression of *c-myc* is elevated or deregulated in virtually all tested tumours, suggesting that myc activation is important for carcinogensis (Spencer & Groudine, 1991). *c-myc*, which is one of a family of related *myc* genes present within the mammalian genome (DePinho, 1991), encodes a short-lived sequence-specific DNA binding protein. The *c-myc* protein, c-Myc, is most probably a transcription factor. It posses an N-terminal domain with the ability to activate transcription (Kato et al., 1990; Amati et al., 1992; Kretzner et al., 1992) and a C-terminal DNA-binding/dimerization domain comprised of a basic helix-loop-helix leucine zipper domain, similar to that present in many other transcription factors of the bHLHZ class, which mediates dimerization with its partner Max (reviewed in Evan & Littlewood 1993). However, few target genes for *c-myc* have been identified and their interaction is not well characterized.

c-myc is an immediate early response gene which is induced following mitogen stimulation and is expressed throughout the cell cycle. Deregulated expression of *c-myc* keeps the cells in a continuously proliferating state. Regions of the c-myc protein required for cellular proliferation are identical to those required for *c-myc* to act as transcription factor (Evan et al., 1992). Thus, a likely role for *c-myc* is the regulation of target genes that control entry into and exit from the cell cycle.

In vitro deregulation of *c-myc* expression appears sufficient to generate cells that no longer respond to signals which would normally trigger their growth arrest. Thus, the deregulation of a single gene, *c-myc*, would appear sufficient to convert a normal cell into tumour cell. However, this conclusion is difficult to reconcile with evidence that neoplastic conversion is over a long period of time and requires multiple mutations, which is consistent with the extreme rarity of cancer. Therefore, even though cells with deregulated *c-myc* expression exhibit continuous proliferation *in vitro*, this alone cannot be all that is required for neoplastic transformation.

Examination of the growth rate of such cells under conditions in the presence of limiting growth factors indicates why these cells are not fully neoplastic. When normal fibroblasts, are grown in the absence of growth factors (low serum), *c-myc* expression is rapidly downregulated and the cells arrest in G_1 (Dean et al., 1986;

Waters et al. 1991). Although, fibroblasts with deregulated *c-myc* expression continue to cycle in the absence of growth factors, cell numbers do not generally increase because of substantial cell death. Cell death due to over expression of *c-myc* has all the characteristic features of apoptosis; it is rapid (20-40 min), accompanied by cell surface blebbing, cell shrinkage and fragmentation and cell DNA is cleaved into fragments on nucleosome length (Evan et al., 1992).

Site-directed mutagenesis was used to determine what regions of c-Myc were important to induce apoptosis in serum-deprived fibroblasts. Those regions required for c-Myc to function in cotransformation and those required to trigger apoptosis in low serum were identical (Evan et al., 1992; Stone et al., 1987). Those regions included the N-terminal transactivation domain, the sequence-specific DNA binding basic regions and the intact C-terminal helix-loop-helix-leucine zipper dimerization domain. The exact same regions are also necessary for c-Myc to function as a transcription factor. Thus, the ability of c-Myc to induce apoptosis is genetically inseparable from its ability to promote cell growth, and both functions probably involve the specific modulation of c-Myc target genes. All known transcription factors of the bHLHZ class require dimerization with a partner in order to bind DNA and exert their action.

c-Myc Induced Apoptosis and Carcinogenesis

The ability of *c-myc* to induce apoptosis is consistent with a large amount of data indicating that genetic lesions that suppress apoptosis can synergize with c-Myc oncogenically. One example is the anti-apoptotic oncogene *bcl-2*. Bcl-2 synergizes with c-Myc to promote development of lymphomas by suppressing c-myc induced apoptosis (Vaux et al., 1988). Expression of Bcl-2 also blocks c-Myc induced apoptosis in fibroblasts under conditions of low serum (Bissonnette et al. 1992; Fanidi et al. 1992; Wagner et al. 1993) or in response to DNA damage (Fanidi et al., 1992), however *c-myc* is still able to induce proliferation. In general, deregulation of *c-myc* is found in virtually all tumour cells, thus indicating a role for c-myc activation during carcinogenesis. However, the deregulation of *c-myc* as a potent trigger of apoptosis supports the concept that anti-apoptotic lesions are likely to be common components of carcinogenesis.

Although it is now clear that deregulated expression of c-Myc is a potent trigger of apoptosis, is normal c-Myc expression also involved in promoting apoptosis? Two different models can be invoked to explain the induction of apoptosis by c-Myc following serum-deprivation. The first proposes that cell death is due to a conflict in signals between the proliferative action of c-Myc and the growth inhibitory effect of growth factor deprivation. In this model, induction of apoptosis is a "conflict of signals" due to an imbalance between growth arrest induced by serum deprivation and growth promoting function of c-Myc. An alternative model for c-Myc induced apoptosis is to propose that *c-myc* expression can signal both cell growth and apoptosis i.e. that proliferation and apoptosis are obligatory coupled. In this "Dual Signal" model, successful cell proliferation requires two independent signals, one to trigger mitogenesis and the other to suppress the concomitant apoptotic programme. According to this model, cells expressing c-Myc die in low serum because they are deprived of serum factors required to suppress the c-Myc induced apoptotic programme. Therefore, induction of apoptosis by c-Myc is an obligate and normal physiological aspect of c-Myc function.

The idea that c-Myc induces apoptosis by a transcriptional mechanism led to the argument that in the "conflict" model c-Myc only induces its apoptotic transcriptional programme as a result of a conflict of signals. To test this, cells were treated with either cycloheximide or actinomycin D to block the apoptotic programme. Both cycloheximide or actinomycin D can block specific types of apoptosis, such as dexamethasone-induced apoptosis in thymocytes, suggesting that *de novo* protein synthesis is required for apoptosis. However, not only was there no inhibition of c-Myc induced apoptosis by each drug, but instead apoptosis occurred more rapidly when either drug was added. There was also a correlation between levels of c-myc expression and the rate of apoptosis after the drug was added, i.e. the higher the level of *c-myc*, the more rapid apoptosis occurred. It appears that in these cells the c-Myc induced apoptotic transcriptional programme is already present although it is inactive.

Therefore, *c-myc* induces apoptosis by either activating or suppressing the apoptotic programme. Most likely this programme is induced upon expression of *c-myc*, however, it is inactive in the presence of high serum whereas, in low serum the program is activated.

Cytokines that Block c-Myc-Induced Apoptosis

The induction of apoptosis by *c-myc* in low serum could be due either to nutrient deprivation or the loss of specific cytokines which are needed to suppress apoptosis. Cell death in low serum does not result from nutrient deprivation as replacement of serum with a cytokine free serum substitute which provides the same essential nutrients as serum does not suppress c-Myc induced apoptosis in rat fibroblasts. To determine which cytokines were important for suppression of apoptosis, various cytokines were tested for their abilities to suppress c-Myc induced apoptosis in serum deprived cells. Addition of any one of the cytokines IGF-1, IGF-II, insulin, PDGF AB or PDGF BB significantly suppressed apoptosis in the absence of any other exogenous cytokines or nutrients. However, EGF basic FGF, acidic FGF, Interleukin-1, TGFa, TGFb and bombesin all failed to inhibit apoptosis. The possibility the lack of receptor expression accounted for these observations was tested by examining the expression of the immediate early nuclear proteins, c-Fos and Egr-1/Zif268/NGR1A following the addition of each cytokine. All cytokines tested were found to induce the transient expression of these proteins. In addition, EGF, bFGF and bombesin are all potent mitogens for fibroblasts whereas IGF-I is only poorly mitogenic for fibroblasts indicating that the ability to suppress c-Myc induced apoptosis among tested cytokines is not linked to mitogenicity.

In the "Conflict of Signals" model, the ability of c-myc to promote growth in low serum is counteracted upon by the absence of growth signals from the serum. In this case, the cytokines are necessary to allow the cell to complete a growth cycle, otherwise the cell enters into an aborted cell cycle which results in apoptosis. Cytokines such as IGF-1 and PDGF would provide the growth signals which are needed for successful growth of the cell and prevent *c-myc* induced apoptosis. In contrast, the "Dual Signal" model proposes that IGFs and PDGF function in the apoptotic pathway independent of the stage of the cell cycle. To distinguish between these two models two different questions were asked. First, do cytokines suppress *c-myc* induced cell death under conditions when growth factors are not needed for proliferation. Second, do cytokines inhibit *c-myc* induced apoptosis under conditions where they cannot function as a mitogenic signal. To answer these questions, the effects of IGF-1 were examined in the post-commitment S/G2 stage of the cell cycle when the cells no longer require IGF-1 for completion of the cell cycle.

Expression of *c-myc* in cells growth arrested in post-commitment parts of the cell cycle by either thymidine (S) or etoposide (S/G2) undergo apoptosis (Evan et al. 1992; Fanidi et al. 1992; Harrington et al. 1994). Neither IGF-I nor PDGF had any effect on cells arrested in the S or S/G2 phase of the cell cycle. Therefore the effect of IGF-1 to delay c-Myc-induced apoptosis under conditions where the cells are growth arrested in either S or S/G2 was examined. Both IGF-I and PDGF delayed c-Myc-induced apoptosis in fibroblasts exposed to thymidine or etoposide. Thus, the ability of the cytokines to suppress apoptosis is separate from their ability to promote cell proliferation.

In the cell cycle it is only during the precommitment G1 phase in which growth factors are required (Zetterberg et al. 1985; Pardee et al. 1989). In contrast, the post-commitment S/G2/M fibroblasts are independent of growth factors to complete their growth cycle. We therefore asked whether IGF-I could suppress c-Myc induced apoptosis in these post-commitment fibroblasts, a time in the cell cycle when the cytokine has no identifiable role in promoting cell proliferation. Apoptosis induced by c-Myc in serum-deprived fibroblasts arrested in the post-commitment S/G2 phase is inhibited by both IGF-1 and PDGF. Thus, cytokines suppress apoptosis under conditions where they have no effect on cell cycle progression

The anti-apoptotic effects of IGF-I are evident even in the absence of RNA or protein synthesis, as was shown in cells treated with actinomycin D or cycloheximide. Thus, IGF-I does not require de novo gene expression or protein synthesis to delay apoptosis, as opposed to the mitogenic function of IGF-1 which clearly does require *de novo* expression of a variety of immediate-early growth response genes. The action of PDGF appears to be different from IGF-1 in requiring new gene expression as there is no anti-apoptotic effect in cells treated with cycloheximide or actinomycin D.

Induction of Apoptosis by Bak

A variety of genes have been identified which encode proteins showing homology to Bcl-2. Members of the *bcl-2* family either positively or negatively regulate apoptosis (Boise et al. 1993; Hengartner et al. 1994 Oltvai et al. 1993; Reed et al. 1994). The newest member of the *bcl-2* family, *bak,* was cloned using degenerate oligonucleotide primers corresponding to the two conserved domains, BH1 and BH2 (Chittenden et al. 1995). Bak shows the strongest homology to Bcl-2 and Bcl-X_L.

To determine what effect Bak had on cell survival, *bak* was expressed in RAT-1 cells which expressed both *bcl-2* and *c-myc*. Under conditions of serum starvation, Bcl-2 will protect against cell death when c-Myc is activated (Fanidi et al. 1992). However, coexpression of Bak antagonised the anti-apoptotic effect of Bcl-2.

Transfectants which expressed high levels of Bak were obtained by utilizing an inducible system of the GAL4-oestrogen receptor-VP16 fusion which is activated by oestrogen and transactivates GAL-4 responsive promoters. A RAT-1 cell line was constructed in which the *bak* gene was placed under the control of a synthetic promoter containing GAL4 DNA binding sites. The activation of expression of *bak* in serum starved RAT-1 fibroblasts induced rapid cell death which showed morphological features of apoptosis (Chittenden et al. 1995). Cell death induced by Bak was delayed in the presence of 10% serum, indicating that cytokines can protect against bak induced cell death.

Bak most likely induces apoptosis in the same cell death pathway where the other *bcl-2* family members function. Bak may either simplr antagonize the protective effect of Bcl-2 or Bcl-X_L, or Bak may act directly in activating the cell death pathway or function as a component of the pathway.

Apoptosis and Virus Infection

During a virus infection, the cell and virus compete with each other for their survival. The organism and its component cells attempt to suppress the replication of the virus and prevent its spread within the organism, whilst, the virus wants either to become a resident of the cell, as is the case for persistent infections, or to produce a high number of progeny virions and lyse the cell, as in a lytic infection. Viruses utilise a variety of methods to capture the cellular machinery for their own replication. To combat a virus infection, host cells try to limit virus replication by inhibiting protein synthesis and cell proliferation through the production of interferons. More recently, it has been proposed that cells may also limit the spread of the virus by undergoing suicidal cell death before the virus has completed the replication cycle. Thus, the cell prevents the virus from producing progeny virions that will infect neighbouring cells. Evidence to support this idea comes from recent findings that many viruses contain genes whose function appears to be to suppress cell death during a virus infection (Birnbaum et al. 1985; Clem et al. 1991; Crook et al. 1993; Gregory et al. 1991; Henderson et al. 1991; Henderson et al. 1993; Rao et al. 1992).

Cell death during virus infection can occur either by necrosis or apoptosis. Death by necrosis leads to eventual release of the cellular contents into the extracelluar environment. Most likely this would be an efficient method of spreading the virus because free virus is directly released into the environment (Clouston et al. 1985). Necrosis also frequently triggers an inflammatory reaction and causes tissue damage (Kerr et al. 1972; Wyllie et al. 1980). In contrast to necrotic death, death by apoptosis produces no inflammatory reaction or tissue damage (Kerr et al. 1972; Wyllie et al.

1980). Mature progeny virions will be retained in the membrane-bounded bodies that are rapidly phagocytosed by macrophages. Thus, apoptotic death tends to inhibit release of the virus into the environment (Clouston et al. 1985). Therefore, death of the cell by apoptosis probably leads to less damage to the organism during a virus infection than does death by cell necrosis. In examining the virulence of a virus one important question is how the host cell dies after infection.

Virulence of a virus is determined by a variety of factors, including the host range genes encoded by the virus. Host range genes are necessary for optimal infection and determine the target cell specificity. If host cell apoptosis prior to virus replication acts as a significant limitation to virus infection, one possible function of these host range genes may be to suppress host cell apoptosis.

To understand better how host range genes function during an infection, we examined vaccinia virus infection in Chinese hamster ovary (CHO) cells. Although vaccinia virus, a prototype of the Orthopoxvirus family, usually produces a lytic infection in permissive cell lines, in CHO cells vaccinia virus infection is abortive and triggers a complete shut down of host protein synthesis (Drillien et al. 1978; Hruby et al. 1980; Njayou et al. 1982). The complete shutdown in protein synthesis shows similarity to certain types of cell death by apoptosis. Therefore two questions were asked: first does vaccinia virus induce apoptosis in nonpermissive CHO cells, and second, if so, what is needed to overcome the virus induced apoptosis?

Delay of Virus-Induced Apoptosis by *CHOhr* from Cowpox Virus

To determine if infection of CHO cells by vaccinia virus occurs by apoptosis, we examined infected cells by time-lapse video microscopy. Cell death of CHO cells during infection with vaccinia virus showed all the characteristics of apoptosis: rounding up and blebbing of the cells, shrinkage and eventually death. In addition, fragmentation and shrinkage of the nucleus and the presence of fragmented DNA at 24 hours after infection all gave further confirmation that CHO cells underwent apoptosis when non productively infected with vaccinia virus.

The host range gene, *CHOhr*, from cowpox virus had previously been shown to allow vaccinia virus to productively infect CHO cells (Spehner et al. 1988). *CHOhr* encodes a 77kDa protein which contains an ankyrin-like repeat and shows no significant homology to other proteins (Spehner et al. 1988; Shlchekunov et al. 1993). Therefore two recombinant viruses of the Copenhagen strain of vaccinia virus were used to determine the effect of the *CHOhr* gene on virus infection in CHO cells. Each recombinant virus contained the *CHOhr* gene recombined into the virus genome and had been shown to multiply efficiently in CHO cells (Drillien, per. comm.).

A time course of protein synthesis was used to determine the timing and level of protein synthesis in an abortive and productive infection. At two hours after infection, protein synthesis was similar in CHO cells when infected with either vaccinia virus or a recombinant vaccinia virus, IV-41, containing *CHOhr*. However, whereas by 4 hrs. after infection protein synthesis was not detected in cells infected with vaccinia virus, cells infected with the recombinant virus IV-41 showed a characteristic pattern of synthesis of late virus protein s which continued at least until 10 hrs. after infection. It appears that *CHOhr* is needed to overcome vaccinia virus-induced block to expression of late virus genes at least until 10 hours after infection.

Host cell apoptosis appears to be a limiting factor in productive infection by vaccinia virus in CHO cells, thus, this suggests that the *CHOhr* gene acts either by directly suppressing apoptosis or by overcoming the block to protein synthesis, so allowing the infection to proceed.

The rapid shut off in protein synthesis by 4 hours after infection in CHO cells with vaccinia virus suggests that cell death may occur at this time. However, the translation of late mRNAs until 10 hrs. after infection in CHO cells with the recombinant virus containing *CHOhr* suggests that cell death may begin at this time. Time-lapse videomicroscopy showed that the beginning of cell death coincided temporally with the shut-off in protein synthesis . In the case of vaccinia virus infection, protein synthesis is shut down by 4 hrs. post infection and by 6 hrs. apoptosis begins to occur. On the other hand, CHO cells infected with the recombinant virus IV-41 allowed for the complete replication of the virus in which late proteins were translated at least up to 10 hours after infection.

It appears that CHO cells respond to a vaccinia virus infection in two ways: shut down of protein synthesis and death of the cell by apoptosis. These two phenomena may be casually related. For vaccinia virus to successfully propagate, the virus either delays or prevents apoptosis of host cells and overcomes the block in protein synthesis. Indeed, the effect of the *CHOhr* gene is to both delay apoptosis and to restore protein synthesis of virus late mRNAs during vaccinia infection of CHO cells, so allowing assembly of progeny virions and virus propagation. This lends support to the proposal that early activation of host cell suicide pathways is an important factor in limiting the propagation of many viruses.

The Effect of the Adenovirus *E1B 19K* on Vaccinia Virus-Induced Apoptosis

The use of anti-apoptotic genes by viruses which are necessary for a productive infection raised the question of whether delaying virus-induced apoptosis was all that was needed for a productive infection. To test this, CHO cells were tconstructed which expressed the adenovirus *E1B 19K* gene. The *E1B 19K* gene is similar to *CHOhr* in that mutant adenoviruses which carry a mutant *E1B 19K* gene show a decrease in protein synthesis, virus yield, and no protection against virus-induced apoptosis, (Pilder et al. 1984; White et al. 1984). *E1B 19K* is not functionally interchangeable with *CHOhr* since it was unable to overcome the block in protein synthesis and allow for a productive virus infection. It appears that although virus-induced apoptosis can be suppressed regardless of the shut down in protein synthesis, it is not all that is required for a productive infection. In addition, the correlation between shut down in protein synthesis and apoptosis does not necessarily mean the two events are dependent on each other.

References

Amati, B., S. Dalton, M. Brooks, T. Littlewood, G. Evan, and H. Land (1992) Transcriptional activation by c-Myc oncoprotein in yeast requires interaction with Max. Nature 359:423-426.

Birnbaum, M. J., R. J. Clem, and L. K. Miller (1994) An apoptosis-inhibiting gene from a nuclear polyhedrosis virus encoding a polypeptide with cys/his/ sequence motifs. J. Virol. 68:2521-2528.

Bissonnette, R., F. Echeverri, A. Mahboubi, and D. Green (1992) Apoptotic cell death induced by *c-myc* is inhibited by *bcl-2*. Nature 359:552-554.

Boise, L. H. (1993) *bcl-X*, a *bcl-2* related gene that functions as a dominant regulator of apoptotic cell death. Cell 74:597-608.

Bursch, W., L. Kleine, and M. Tenniswood (1990) The biochemistry of cell death by apoptosis. Biochem. Cell Biol. 68:1071-1074.

Chittenden, T., E. A. Harrington, R. O'Connor, C. Flemington, R. J. Lutz, G. I. Evan, and B. Guild (1995) Induction of apoptosis by the Bcl-2 homologue Bak. Nature 374:733-736.

Cleary, M. L., and J. Sklar (1985) Nucleotide sequence of a t(14:18) chromosomal breakpoint in follicular lymphoma and demonstration of a breakpoint region near a transcriptionally active locus on chromosome 18. Proc. Natl. Acad. Sci. USA 82:7439-7443.

Clem, R. J., M. Fechheimer, and L. K. Miller (1991) Prevention of apoptosis by a baculovirus gene during infection of insect cells. Science 254:1388-1390.

Clouston, W. M., and J. F. R. Kerr (1985) Apoptosis, lymphocytotoxicity and the containment of viral infections. Med. Hypotheses 18:399-404.

Crook, N. E., R. J. Clem, and L. K. Miller (1993) An apoptosis-inhibiting gene with a zinc finger-like motif. J. Virol. 67:2168-2174.

Davis, L. H. and V. Bennett (1990) Mapping the binding sites of human erythrocyte ankyrin for the anion exchanger and spectrin. J. Biol. Chem. 265:10589-10596.

Dean, M., R. A. Levine, W. Ran, M. S. Kindy,, G.E. Sonenshein, and J. Campisi (1986) Regulation of *c-myc* transcription and mRNA abundance by serum growth factors and cell contact. J. Biol. Chem. 261:9161-9166.

DePinho, R. A., N. Schreiber-Agus, and F. W. Alt (1991) *myc* family oncogenes in the development of normal and neoplastic cells. Adv. Cancer Res. 57:1-46.

Drillien, R., D. Spehner, and A. Kirn (1978) Host range restriction of vaccinia virus in Chinese hamster ovary cells: relationship to shutoff of protein synthesis. J. Virol. 28:843-850.

Evan, G. and T. Littlewood (1993) The role of *c-myc* in cell growth. Curr. Opin. Genet. Dev. 3:44-49.

Evan, G., A. Wyllie, C. Gilbert, T. Littlewood, H. Land, et al. (1992) Induction of apoptosis in fibroblasts by c-myc protein. Cell 63:119-125.

Fanidi, A., E. Harrington, and G. Evan (1992) Cooperative interaction between *c-myc* and *bcl-2* proto-oncogenes. Nature 359:554-556.

Gregory, C. D., C. Dive, S. Henderson, C. A. Smith, G. T. Williams, J. Gordon, and A. B. Rickinson (1991) Activation of Epstein-Barr virus latent genes protects human B cells from death by apoptosis. Nature 349:612-614.

Harrington, E. A., M. R. Bennett, A. Fanidi, and G. Evan (1994) c-Myc-induced apoptosis in fibroblasts is inhibited by specific cytokines. Embo J. 13:3286-3295.

Henderson, S. A., D. Huen, M. Rowe, C. Dawson, G. Johnson, and A. B. Rickinson 1993) Epstein-Barr virus-coded BHRF1 protein, a viral homologue of Bcl-2, protects human B cells from programmed cell death. Proc. Natl. Acad. Sci. USA 90:8479-8483.

Henderson, S., M. Rowe, C. Gregory, D. Croom-Carter, F. Wang, R. Longnecker, E. Kieff, and A. Rickinson (1991) Induction of *bcl-2* expression by Epstein-Barr virus latent membrane protein 1 protects infected B cells from programmed cell death. Cell 48:334-336.

Hengartner, M. O., and H. R. Horvitz (1994) C. elegans survival gene *ced-9* encodes a functional homolog of the mammalian proto-oncogene *bcl-2*. Cell 76:665-676.

Hershberger, P. A., J. A. Dickson, and P. D. Friesen (1992) Site-specific mutagenesis of the 35-kilodalton protein gene encoded by Autographa californica nuclear polyhedrosis virus: cell line-specific effects on virus replication. J. Virol. 66:5525-5533.

Hruby, D. E., D. L. Lynn, R. Condit, and J. R. Kates (1980) Cellular differences in the molecular mechanisms of vaccinia virus host range restriction. J. Gen. Virol. 47:485-488.

Kato, G.J., J. Barrett, G. M. Villa, and C.N. Dang (1990) An amino-terminal c-myc domain required for neoplastic transformation activates transcription. Mol. Cell. Biol. 10:5914-5920.

Kerr, J, F. R., A. H. Wyllie, and A. R. Currie (1972) Apoptosis: a basic biological phenomenon with wide ranging implications in tissue kinetics. Brit. J. Cancer 26:239-257.

Kretzner, L., E. Blackwood, and R. Eisenmann (1992) Myc and Max possess distinct transcriptional activities. Nature 359:426-429.

Njayou, M. R., R. Drillien, and A. Kirn (1982) Characteristics of the inhibition of protein synthesis by vaccinia virus in nonessential Chinese hamster ovary cells. Ann. Virol. 133E:393-402.
Oltvai, Z., C. Milliman, and S. Korsmeyer (1993) Bcl-2 heterodimerizes in vivo with a conserved homolog, Bax, that accelerates programmed cell death. Cell 74:609-619.
Pardee, A. B. (1989 G1 events and regulation of cell proliferation. Science 246:603-608.
Pilder, S., J. Logan, and T. Shenk (1984) Deletion of the gene encoding the adenovirus 5 early region 1B-21,000-molecular weight polypeptide leads to degradation of viral and cellular DNA. J. Virol. 52:664-671.
Rao, L., M. Debbas, P. Sabbatini, D. Hockenberry, S. Korsmeyer, and E. White (1992) The adenovirus E1A proteins induce apoptosis, which is inhibited by the E1B 19-kDa and Bcl-2 proteins. Proc. Natl. Acad. Sci. USA 89:7742-7746.
Reed, J. C. (1994) Bcl-2 and the regulation of programmed cell death. J. Cell Biol. (1994) 124:1-6.
Shchelkunov, S. N., V. M. Blinov, and L. S. Sandakhchiev (1993) Ankyrin-like proteins of variola and vaccinia viruses. FEBS 319:163-165.
Spehner, D., S. Gillard, R. Drillien, and A. Kirn.(1988) A cowpox virus gene required for multiplication in Chinese hamster ovary cells. J. Virol. 62:1297-1304.
Spencer, C. A. and M. Groudine (1991) Control of *c-myc* regulation in normal and neoplastic cells. Adv. Cancer Res. 56:1-48.
Stone, J., T. de Lange, G. Ramsay, F. Jakobvits, J. M. Bishop, H. Varmus, and W. Lee (1987) Definition of regions in human c-myc that are involved in transformation and nuclear localization. Mol. Cell. Biol. 7:1697-1709.
Vaux, D. L., S. Cory, and J. M. Adams (1988) *Bcl-2* gene promotes haemopoietic cell survival and cooperates with *c-myc* to immortalize pre-B cells. Nature 335:440-442.
Wagner, A. J., M. B. Small, and N. Hay (1993) *Myc*-mediated apoptosis is blocked by ectopic expression of *bcl-2*. Mol. Cell. Biol. 13:2432- 2440.
Waters, C., T. Littlewood, D. Hancock, J. Moore, and G. Evan (1991) c-myc protein expression in untransformed fibroblasts. Oncogene 6:101-109.
White, E., T. Grodzicker, and B. Stillman (1984) Mutations in the gene encoding the adenovirus early region 1B 19,000-molecular-weight tumor antigen cause the degradation of chromosomal DNA. J. Virol. 52:410-419.
Wyllie, A. H. (1987) Apoptosis: cell death under homeostatic control. Arch. Toxicol. Suppl. 11:3-10.
Wyllie, A. H., J. F. R. Kerr, and A. R. Currie (1980) Cell death: the significance of apoptosis. Int. Rev. Cytol. 68:251-305.
Zetterberg, A. and O. Larson (1985) Kinetic analysis of regulatory events in G1 leading to prolifewration of quiescence of Swiss 3T3 cells. Proc. Natl. Acad. Sci. 82:5365-5369.

Molecular Mechanisms Controlling Susceptibitliy to Tumor Necrosis Factor Induced Cell Death

Juha Klefstrom, Eero Saksela and Kari Alitalo
Molecular/Cancer Biology Laboratory and
Department of Pathology
Haartman Institute
P.O. Box 21
00014 University of Helsinki
Finland

SUMMARY

The receptors of the multifunctional cytokine Tumor Necrosis Factor-α (TNF) belong to a family of structurally related proteins called the TNFR/NGFR-family. Two receptors of this family, TNFR and Fas can trigger cell death in responsive cells. In contrast to Fas, activated TNFR exerts a variety of biologic functions other than death induction. Virus-infected cells, tumor cells and transformed cells are often sensitive to the death-inducing activity of TNF, yet the molecular mechanisms underlying TNF-sensitivity are unclear. We have addressed the question whether normally growing cells may acquire a TNF-sensitive phenotype upon activation of growth-deregulating oncogenes. We have shown that the oncogenic activation of the nuclear phosphoprotein c-Myc renders TNF-resistant fibroblasts sensitive to TNF-mediated death. Moreover, an increased constitutive expression of c-Myc increases the cytotoxic action of TNF. These results suggest that the deregulation of c-Myc, which is common in human tumors and tumor cell lines is one reason why such cells are TNF sensitive. Deregulation of growth by viral or endogenous oncoproteins renders cells more prone to apoptosis in comparison to their normally growing counterparts. The regions of c-Myc required for cell transformation and induction of apoptosis upon growth-factor deprivation are also necessary for the induction of TNF-sensitivity. In addition, the Bcl2 and MnSOD proteins can inhibit the c-Myc dependent apoptosis of growth factor-deprived cells and of TNF-stimulated cells. Deregulation of cell growth may preset the apoptotic machinery and thus render cells susceptible for the death-inducing effects of TNF.

NATO ASI Series, Vol. H 99
Tumor Biology
Regulation of Cell Growth, Differentiation and Genetics in Cancer
Edited by Asterios S. Tsiftsoglou, Alan C. Sartorelli, David E. Housman and T. Michael Dexter

1. CYTOKINE INDUCTION OF CELL DEATH

The receptors of tumor necrosis factor-α (TNF) belong to a family of structurally related proteins called the TNFR/NGFR-family (reviewed in (Bazan, 1993)). A distinct feature of this receptor family is that several of its members have an important role in the control of programmed cell death or apoptosis. Triggering of cell death is commonly associated with two ligands of this receptor family, TNF and the Fas ligand (FasL), which are expressed as both membrane bound and soluble forms (Larrick and Wright, 1990; Nagata and Golstein, 1995). TNF has two receptors, of 55 kd and 70 kd molecular weight. The extracellular domains of the 55 kd TNF receptor and Fas are similar with other members of the TNFR/NGFR-family, but only these two receptors within the family share an unique homology domain in their intracellular parts. This domain, consisting of less than 100 amino acid residues is required for the induction of cell death by the two receptors, hence it is called the "death domain" (Cleveland and Ihle, 1995). The TNF and FasL receptors apparently function through interaction with an overlapping set of cytoplasmic proteins, at least some of which also contain a death domain. The overexpression of an isolated death domain or certain death domain containing intracellular proteins is sufficient to induce apoptosis (Boldin et al., 1995; Chinnaiyan et al., 1995; Hsu et al., 1995; Stanger et al., 1995). Despite of the similarities in the structure and function of these two receptors at the molecular level, the TNF receptor mediates a multitude of other biologic responses in addition to the induction of cell death, whereas the induction of cell death appears to be the main function of Fas.

There is ample evidence that Fas mediates apoptosis both *in vivo* and *in vitro*. Cell death occurs in several cultured cell lines as a response to stimulation of Fas by FasL or by an activating antibody (Itoh et al., 1991; Suda et al., 1993; Trauth et al., 1989). In mice, intraperitoneal administration of the Fas-activating antibody causes a lethal induction of apoptosis in the liver (Ogasawara et al., 1993). Mutations impairing Fas expression (*lpr*-phenotype) or the receptor binding properties of FasL (*gld*-phenotype) cause an impaired function of the immune system in affected mice. These phenotypes result from an abnormal accumulation of lymphoid cells and an autoimmune disease, apparently because of a failure to eliminate activated T-cells by apoptosis (reviewed in (Nagata and Golstein, 1995)). Only T-cells activated by antigen stimulation are susceptible to Fas-induced apoptosis, since resting T-cells express only low levels of Fas in contrast to the activated T-cells (Nagata and Golstein, 1995). This suggests that the physiologic function of the Fas signal is to trigger cell death and that cells can acquire susceptibility to Fas-mediated death by upregulating Fas expression on the cell surface.

In comparison with Fas, TNF receptor signaling involves an additional level of complexity. TNF-receptor stimulation is required for a variety of physiological functions, which often lead to other responses than cell death. TNF is produced mainly by macrophages and lymphocytes and high systemic levels of TNF have been observed in inflammatory and immune responses and in several diseases (reviewed in (Jäättelä, 1991; Old, 1988)). TNF is considered to play a central regulatory role in inflammation and immune responses. For example, TNF activates the vascular endothelium making it more adhesive for leukocytes, enabling them to migrate through the endothelium and enter inflamed tissues (Mackay et al., 1993). In addition, TNF directly stimulates leukocytes to secrete other cytokines with immunomodulatory functions (Vassalli, 1993). The multifunctional nature of TNF is reflected by the fact that TNF may stimulate or inhibit cell growth and promote chemotaxis, cell differentiation and numerous other biological responses depending on the cell type ((Cornelius et al., 1990; Gordon et al., 1992) and references therein). The repertoire of TNF-responding cells is wide since the receptors for TNF are ubiquitously expressed in different cell and tissues types (Lewis et al., 1991; Vilcek and Lee, 1991).

The identification of TNF was based on its ability to exert direct cytotoxic effects against a variety of tumor cells but not against normal primary cells in culture (Old, 1988). The molecular mechanism which distinguishes cells which are killed by TNF (TNF-sensitive cells) from cells which are not killed by this cytokine (TNF-resistant cells) is unclear. Unlike for Fas, there is no good correlation between the number of TNF-receptors and the TNF-sensitivity of cells. As an example, pg/ml concentrations of TNF are sufficient to effectively kill WEHI-164 fibrosarcoma cells, whereas ng/ml concentrations do not kill pre-adipocytic 3T3L1 cells expressing at least sevenfold more receptors (Klefstrom et al., 1993; Smith and Baglioni, 1992). The ubiquitous expression of TNF-receptors, the high systemic levels of TNF and the lack of correlation between the amount of receptors and TNF-sensitivity indicate that the cell death inducing activity of TNF is not determined at the level of ligand-receptor interaction. An important question then is, whether the intracellular changes that occur during the cell transformation process are responsible for the conversion of the TNF-resistant phenotype of normal cells to the TNF-sensitive phenotype of tumor cells. Cellular susceptibility to TNF is increased upon infection with several viruses. In fact, a single viral transforming protein with a cell cycle deregulating capacity can trigger cellular sensitivity to TNF (see below). Thus, TNF-sensitivity of tumor cells may be associated with oncogenic activation of cellular genes capable of causing a deregulation of the cell cycle.

2. INDUCTION OF TNF-SENSITIVITY BY CELL CYCLE DEREGULATING PROTEINS

Viral induction of TNF-sensitivity

Infection of cells by certain viruses increases their TNF-sensitivity. Viral infection may enhance the cytotoxic effects of TNF or even render completely TNF-resistant cells sensitive to the death inducing activity of TNF. Examples of viruses which make cells TNF-sensitive or enhance the cytotoxic effects of TNF are vesicular stomatitis virus, cytomegalovirus and newcastle disease virus (Rubin, 1992). Infection of fibroblasts by several adenovirus mutants also induces TNF-sensitivity, which is a consequence of the expression of either of the two adenoviral E1A proteins 289R or 243R (Chen et al., 1987). However, TNF-sensitivity is not increased in cells infected with wild-type adenoviruses, indicating the existence of viral proteins counteracting the cytotoxic effects of TNF. Indeed, the E1A induced TNF-sensitivity is abrogated by proteins expressed from the E3 region, present in the wild-type adenoviral genome, but missing from the mutants inducing TNF-sensitivity (Duerksen-Hughes et al., 1991). At least in certain cells, the E1A induced sensitivity to TNF can also be prevented by the simultaneous expression of the adenoviral E1B19K protein (White et al., 1992).

The E1A-proteins have functional similarities with the cellular Myc protein. Both proteins modulate transcription and stimulate cell cycle progression (Shenk and Flint, 1991) independently of extracellular stimuli - both E1A and c-Myc can force quiescent cells to enter the S-phase of the cell cycle (Eilers et al., 1991; Stavel et al., 1985). In addition, both E1A and c-Myc are able to induce cell transformation and, in certain circumstances, apoptosis (Evan et al., 1992; Lowe and Ruley, 1993; Ralston and Bishop, 1983). We were therefore interested in the possibility that the TNF-sensitivity of tumor cells would be associated with the oncogenic activation of c-Myc.

Deregulation of c-Myc induces cellular susceptibility to TNF-induced death

Oncogenic activation of c-Myc is common in human cancers and it is considered to result from constitutive protein overexpression due to genetic alterations, such as chromosomal translocation or gene amplification (Alitalo et al., 1992; Grignani et al., 1990). In cells having a normal growth control the expression of c-Myc is regulated: c-Myc is induced upon mitogenic stimulation and is down-regulated following growth factor withdrawal whereupon cells arrest in G0/G1 (Waters et al., 1991). Constitutive, deregulated c-Myc expression disrupts cellular growth control. It prevents growth arrest upon growth factor withdrawal, maintaining cellular DNA-synthesis (Eilers et al., 1991). The number of cells, however, does not increase, since deregulated c-Myc in such conditions causes cells to gradually undergo apoptosis (Evan et al., 1992).

We have used Rat1A rat fibroblast cells containing a hormone-inducible MycER chimera (Rat1A-MycER) to assay whether activation of c-Myc affects cellular TNF-sensitivity. In these cells, the activity of c-Myc can be induced by estrogen, resulting in cell transformation in the presence of serum growth factors or apoptosis in the absence of serum (Eilers et al., 1989; Evan et al., 1992). In the absence of hormone parental Rat1A cells or the Rat1A-MycER cells are not killed by TNF. In Rat1A-MycER cells grown in the presence of serum, the addition of TNF with a simultaeneous activation of c-Myc resulted in extensive cell death. Nucleosomal-length DNA fragmentation and immunohistochemical staining for fragmented DNA (TUNEL-staining; (Gavrieli et al., 1992)) indicated that the form of cell death was apoptotic. These experiments showed that c-Myc is able to convert the TNF-resistant cellular phenotype to the TNF-sensitive one characteristic of many tumor cells.

Ectopic c-Myc expression is known to induce apoptosis in cells which are growth arrested in any phase of the cell cycle (Evan et al., 1992; Fanidi et al., 1992). We therefore tested if TNF-mediated growth arrest of cells expressing active c-Myc was responsible for the induction of apoptosis. TNF-mediated inhibition of cell growth was not observed, neither did TNF alter the proportion of cells in the different phases of the cell cycle as analyzed by flow cytometry. We also investigated whether c-Myc overexpression increases the cytotoxic effects of TNF in sensitive cells. The cytotoxic activity of TNF was compared in cultures of moderately TNF-sensitive NIH3T3 cells and their c-Myc transfected derivatives. The cells were treated with human TNF, which recognizes the mouse 55 kd TNF receptor only, or with mouse TNF, which recognizes also the 70 kd TNF-receptor (Lewis et al., 1991). In both cases, exposure to TNF resulted in more extensive cell death in c-Myc overexpressing cultures than in cultures of the parental cells. This indicates that the 55 kd TNF-receptor mediated the increased cytotoxic activity of TNF. The 55 kd receptor which contains the Fas-homologous "death-domain" has also been reported to mediate a cytotoxic signal in cells spontaneously highly sensitive to TNF (Tartaglia et al., 1993).

The regions of c-Myc required for cell transformation and apoptosis are also necessary for the induction of TNF-sensitivity

c-Myc belongs to a family of basic region-helix-loop-helix-leuzine zipper (bHLHZip) transcription factors. It has both positive and negative effects on transcription. In a heterodimer with Max, another bHLHZip protein, c-Myc binds specific DNA sequences and activates transcription from promoters containing the target sequence (for reviews, see (Blackwood and Eisenman, 1991; Västrik et al., 1994)). Interaction with Max and the ability to activate transcription is necessary for c-Myc induced cell cycle progression, transformation and, in growth factor-deprived conditions, apoptosis (Amati et al., 1993). In contrast, interaction with Max is not an absolute requirement for c-Myc induced transcriptional repression - overexpression of c-Myc can repress transcription from basal promoter elements without the need for Max (Roy et al., 1993). The deletion of the acidic N-terminus (amino acid residues 1-143) of c-Myc abolishes both the transcriptional activation and repression functions of the protein (Li et al., 1994). The constitutive expression of c-Myc has been reported to repress Cyclin D1 mRNA in a Max-independent manner and the promoter region of the Cyclin D1 gene contains elements required for the transcriptional repression by c-Myc (Philipp et al., 1994). Interestingly, a distinct domain (amino acid residues 92-106) within the N-terminus of c-Myc is required for the transcriptional repression of Cyclin D1, whereas it is not necessary for transformation.

We sought to determine which regions of c-Myc are important for the induction of TNF-sensitivity. Mutant forms of c-Myc with different deletions in their N-terminus were introduced into NIH3T3 cells and the TNF-sensitivities of the resulting cell clones were compared with the cells overexpressing wild-type c-Myc. In these assays the transforming c-Myc mutant deficient in Cyclin D1 repression increased TNF-sensitivity as well as wild type c-Myc. On the other hand, the non-transforming c-Myc mutant capable of repressing Cyclin D1 did not increase TNF-sensitivity. An N-terminal deletion mutant, incapable of both transformation and Cyclin D1 repression, did not either increase TNF-sensitivity.

In cells expressing the hormone-inducible MycER protein, addition of estrogen allows MycER to bind Max. Therefore, cellular responses such as transformation or apoptosis, mediated through MycER-Max complexes are hormone-inducible. By contrast, transcriptional repression of Cyclin D1 is a hormone-independent property of MycER, indicating that Max is not required for transcriptional repression (Philipp et al., 1994). In our experiments with Rat1A cells having the MycER chimera the induction of TNF-sensitivity was hormone-dependent. Taken together, our data thus suggest that the activation of transcription through Myc-Max complexes is the critical step in the induction of TNF-sensitivity. The data make it unlikely that c-Myc would sensitize cells to TNF by repressing the biosynthesis of proteins which are involved in the protection of cells from TNF cytotoxicity. In contrast, similar functions of c-Myc as required for cell transformation and apoptosis are also necessary for the induction of TNF-sensitivity

c-Myc induced susceptibility to TNF-mediated cell death is abrogated by Bcl2 and inhibited by MnSOD

TNF kills a variety of transformed and tumor cells. However, many tumor cell lines expressing TNF receptors do not display a detectable death-response even after long exposures to TNF. An estimate of the efficiency of TNF in eliminating tumor cells is provided by the study of Sugarman et al. (1985). In their study of 22 tumor cell lines only seven (about 25%) lost viability during exposure to TNF. Given that growth deregulation may sensitize cells to TNF, why are most tumor cells TNF-resistant? The possiblity that protective proteins are present is exemplified by the adenovirus case. The reason for why adenovirus infection does not induce susceptibility to TNF, whereas the expression of the adenoviral E1A protein does, appears to lie in the expression of the viral E3 or E1B19K proteins which inhibit the E1A-induced TNF-sensitivity. Similarly, in lymphomas c-Myc is often overexpressed concomitantly with Bcl2, which prevents apoptosis (Fanidi et al., 1992). Accordingly, both a productive viral infection and the development of a malignant tumor may require factors which abrogate cellular susceptibility to apoptosis. Thus, the presence of factors preventing apoptosis may help to explain why many types of tumor cells are TNF-resistant.

We have tested whether c-Myc induced TNF-sensitivity is inhibited by two mitochondrially expressed proteins, Bcl2 and the free oxygen radical scavenging enzyme manganese superoxide dismutase (MnSOD). Bcl2 abrogates c-Myc mediated apoptosis upon growth-factor deprivation, as mentioned above, whereas the MnSOD-induced resistance to cell death is more specifically associated with TNF-cytotoxicity (Wong et al., 1989). When Bcl2 and MnSOD were expressed in Rat1A-MycER cells, the c-Myc induced TNF-sensitivity was almost completely abrogated by the Bcl2, whereas the expression of MnSOD in the same conditions inhibited about 50% of cell deaths. When c-Myc activation-induced apoptosis in low-serum was assayed using the same cells, MnSOD was a much weaker inhibitor than Bcl2.

3. CONCLUDING REMARKS

The steady-state cell number in multicellular organisms is maintained by a balance between cell proliferation and induction of cell death. Cytokines play an important role in the maintenance of this homeostasis - by stimulation of cell growth or by specific elimination of unwanted cells. The TNF- and Fas receptors apparently use overlapping sets of signal transducers to induce cell death but in contrast to the FasL, TNF interaction with its receptor triggers a diverse set of responses in cells. The execution of death in TNF-stimulated cells appears to require intracellular conditions which are not present in cultured primary cells or in cells with a normal growth control. The death-inducing activity of TNF often targets cells with a deregulated growth control acquired through viral infection or oncogenic insults of cell cycle regulating genes. These cells are also prone to apoptotic death - cell cycle inhibition in such cells causes apoptosis whereas their normally growing counterparts stop cycling. However, TNF does not cause death by blocking the cell cycle. The regions of c-Myc required for cell transformation and apoptosis are also necessary for the induction of TNF-sensitivity. In addition, the same proteins, such as Bcl2 and MnSOD, can counteract c-Myc dependent apoptosis of the growth factor-deprived cells and of the TNF-stimulated cells. It is possible, that c-Myc activation or a deregulated growth control as such enhances the expression of proteins which have a role in the execution of apoptosis, making cells more vulnerable to TNF. The identification of such possible proteins remains a challenge for further work.

In summary, we suggest that while Fas eliminates cells in physiologic conditions, TNFR eliminates cells which are already prone to apoptosis due to a deregulated growth-pattern. TNF is derived from activated macrophages and systemic levels strongly increase during inflammation and in pathological conditions, e.g. in several cancers. TNF may have a "cleaning-effect" by enhancing the apoptotic program imminent in damaged host cells and also simultaneously killing apoptosis-prone virus-infected or cancer cells.

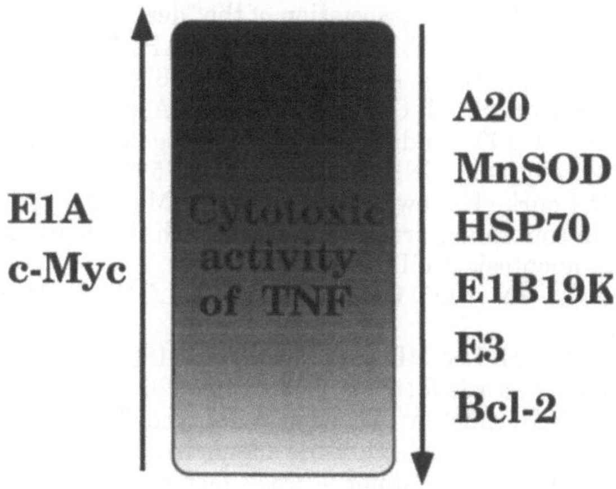

FIG 1. Proteins modulating TNF-cytotoxicity. Ectopic expression of the A20 (Opipari et al., 1992), MnSOD (Wong et al., 1989), HSP70 (Jäättelä et al., 1992), and E1B19K proteins (White et al., 1992) protects spontaneously TNF-sensitive cells. E1A and c-Myc induce TNF-sensitivity in cells which are normally resistant for TNF, and this is counteracted by the adenoviral E1B19K and E3 proteins (White et al., 1992) and by Bcl2 (Klefstrom et al., 1994), respectively.

This review is partially based on the authors' unpublished data. We thank Imre Västrik for critical comments on the manuscript.

4. REFERENCES

Alitalo K, Mäkelä TP, Saksela K, Koskinen P, and Hirvonen H In Book: Kellems R (eds) (1992) Gene amplification in Mammalian Cells: Techniques and Applications. Marcel Dekker, Inc. New York

Amati B, Littlewood TD, Evan GI, and Land H (1993) The c-Myc protein induces cell cycle progression and apoptosis through dimerization with Max. EMBO J 12: 5083-5087

Bazan JF (1993) Emerging families of cytokines and receptors. Curr Biol 3: 603-606

Blackwood EM, and Eisenman RN (1991) Max: A helix-loop-helix zipper protein that forms a sequence-specific DNA-binding complex with myc. Science 251: 1211-1217

Boldin MP, Mett IL, Varfolomeev EE, Chumakov I, Shemer-Avni Y, Camonis JH, and Wallach D (1995) Self-association of the "death domains" of the p55 tumor necrosis factor (TNF) receptor and Fas/APO1 prompts signaling for TNF and Fas/APO1 effects. J Biol Chem 270: 387-391

Chen MJ, Holskin B, Strickler J, Gorniak J, Clark MA, Johnson PJ, Mitcho M, and Shalloway D (1987) Induction by E1A oncogene expression of cellular susceptibility to lysis by TNF. Nature 330: 581-583

Chinnaiyan AM, O'Rourke K, Tewari M, and Dixit VM (1995) FADD, a novel death domain-containing protein, interacts with the death domain of Fas and initiates apoptosis. Cell 81: 505-512

Cleveland JL, and Ihle JN (1995) Contenders in FasL/TNF death signaling. Cell 81: 479-482

Cornelius MJ, Marlowe M, Lee MD, and Pekala PH (1990) The growth factor-like effects of tumor necrosis factor-α. J Biol Chem 265: 20506-20516

Duerksen-Hughes PJ, Hermiston TW, Wold WSM, and Gooding LR (1991) The amino-terminal portion of CD1 of the adenovirus E1A proteins is required to induce susceptibility to tumor necrosis factor cytolysis in adenovirus-infected mouse cells. J Virol 65: 1236-1244

Eilers M, Picard D, Yamamoto KR, and Bishop JM (1989) Chimaeras of Myc oncoprotein and steroid receptors cause hormone-dependent transformation of cells. Nature 340: 66-68

Eilers M, Schirm S, and Bishop JM (1991) The MYC protein activates transcription of the α-prothymosin gene. EMBO J 10: 133-141

Evan GI, Wyllie AH, Gilbert CS, Littlewood TD, Land H, Waters CM, Penn LZ, and Hancock DC (1992) Induction of apoptosis in fibroblasts by c-myc protein. Cell 69: 119-128

Fanidi A, Harrington EA, and Evan GI (1992) Cooperative interaction between c-myc and bcl-2 proto-oncogenes. Nature 359: 554-556

Gavrieli Y, Sherman Y, and Ben-Sasson SA (1992) Identification of programmed cell death in situ via specific labeling of nuclear DNA fragmentation. J Cell Biol 119: 493-501

Gordon HM, Kucera G, Salvo R, and Boss JM (1992) Tumor necrosis factor induces genes involved in inflammation, cellular and tissue repair, and metabolism in murine fibroblasts. J Immunol 148: 4021-4027

Grignani F, Lombardi L, Inghirami G, Sternas L, Cechova K, and Dalla-Favera R (1990) Negative autoregulation of c-myc gene expression is inactivated in transformed cells. EMBO J 9: 3913-3922

Hsu H, Xiong J, and Goeddel DV (1995) The TNF receptor 1-associated protein TRADD signals cell death and NF-kB activation. Cell 81: 495-504

Itoh N, Yonehara S, Ishii A, Yonehara M, Mizushima SI, Sameshima M, Hase A, Seto Y, and Nagata S (1991) The polypeptide encoded by the cDNA for human cell surface antigen Fas can mediate apoptosis. Cell 66: 233-243

Jäättelä M, Wissing D, Bauer PA, and Li GC (1992) Major heat shock protein hsp70 protects tumor cells from tumor necrosis factor cytotoxicity. EMBO J 11: 3507-3512

Jäättelä M (1991) Biology of disease. Biologic activities and mechanisms of action of tumor necrosis factor-α/Cachectin. Lab Invest 64: 724-742

Klefstrom J, Västrik I, Saksela E, Valle J, Eilers M, and Alitalo K (1994) c-Myc induces cellular susceptibility to the cytotoxic action of TNF-α. EMBO J 13: 5442-5450

Klefstrom J, Koskinen PJ, Saksela E, Jäättelä M, Bravo R, and Alitalo K (1993) A subset of immediate early mRNAs induced by tumor necrosis factor-α during cellular cytotoxic and non-cytotoxic responses. Int J Cancer 55: 655-659

Larrick JW, and Wright SC (1990) Cytotoxic mechanism of tumor necrosis factor-α. FASEB J 4: 3215-3223

Lewis M, Tartaglia LA, Lee A, Bennett GL, Rice GC, Wong GH, Chen EY, and Goeddel DV (1991) Cloning and expression of cDNAs for two distinct murine tumor necrosis factor receptors demonstrate one receptor is species specific. Proc Natl Acad Sci U S A 88: 2830-2834

Li LH, Nerlov C, Prendergast G, MacGregor D, and Ziff EB (1994) c-Myc represses transcription in vivo by a novel mechanism dependent on the initiator element and Myc box II. EMBO J 13: 4070-4079

Lowe SW, and Ruley HE (1993) Stabilization of the p53 tumor suppressor is induced by adenovirus 5 E1A and accompanies apoptosis. Genes & Dev 7: 535-545

Mackay F, Loetscher H, Stueber D, Gehr G, and Lesslauer W (1993) Tumor Necrosis Factor α (TNF-α)-induced cell adhesion to human endothelial cells is under dominant control of one TNF receptor type, TNF-R55. J Exp Med 177: 1277-1286

Nagata S, and Golstein P (1995) The Fas death factor. Science 267: 1449-1455

Ogasawara J, Watanabe-Fukunaga R, Adachi M, Matsuzawa A, Kasugai T, Kitamura Y, Itoh N, Suda T, and Nagata S (1993) Lethal effect of the anti-Fas antibody in mice. Nature 364: 806-809

Old LJ (1988) Tumor necrosis factor. Sci Am 258: 41-49

Opipari AW, Hu HM, Yabkowitz R, and Dixit VM (1992) The A20 zinc finger protein protects cells from tumor necrosis factor cytotoxicity. J Biol Chem 267: 12424-12427

Philipp A, Schneider A, Västrik I, Finke K, Xiong Y, Beach D, Alitalo K, and Eilers M (1994) Repression of cyclin D1: a novel function of Myc. Mol Cell Biol 14: 4032-4043

Ralston R, and Bishop JM (1983) The protein products of the *myc* and *myb* oncogenes and adenovirus E1a are structurally related. Nature 306: 803-806

Roy AL, Carruthers C, Gutjahr T, and Roeder RG (1993) Direct role for Myc in transcription initiation mediated by interactions with TFII-I. Nature 365: 359-361

Rubin BY In Book: Aggarwal BB and Vilcek J (eds) (1992) Tumor Necrosis Factors. Marcel Dekker Inc. New York

Shenk T, and Flint J (1991) Transcriptional and transforming activities of the adenovirus E1A proteins. Adv Cancer Res 57: 47-85

Smith RA, and Baglioni C In Book: Aggarwal BB and Vilcek J (eds) (1992) Tumor Necrosis Factors. Marcel Dekker Inc. New York
Stanger BZ, Leder P, Lee TH, Kim E, and Seed B (1995) RIP: A novel protein containing a death domain that interacts with Fas/APO-1 (CD95) in yeast and causes cell death. Cell 81: 513-523
Stavel S, Argos P, and Philipson L (1985) The release of growth arrest by microinjection of adenovirus E1A DNA. EMBO J 4: 2329-2336
Suda T, Takahashi T, Golstein P, and Nagata S (1993) Molecular cloning and expression of the Fas ligand, a novel member of the tumor necrosis factor family. Cell 75: 1169-1178
Sugarman BJ, Aggarwal BB, Hass PE, Figari IS, Palladino MA, and Shepard HM (1985) Recombinant human tumor necrosis factor-α: Effects on proliferation of normal and transformed cells in vitro. Science 230: 943-945
Tartaglia LA, Ayres TM, Wong GHW, and Goeddel DV (1993) A novel domain within the 55 kd TNF receptor signals cell death. Cell 74: 845-853
Trauth BC, Klas C, Peters AMJ, Matzku S, Möller P, Falk W, Debatin KM, and Krammer PH (1989) Monoclonal antibody-mediated tumor regression by induction of apoptosis. Science 245: 301-305
Vassalli P (1993) Tumor Necrosis Factor: Knock-out but not knocked out. Current Biology 3: 607-610
Västrik I, Mäkela TP, Koskinen PJ, Klefstrom J, and Alitalo K (1994) Myc protein, partners and antagonists. Crit Rev Oncogen 5: 59-68
Vilcek J, and Lee TH (1991) Tumor necrosis factor. New insights into the molecular mechanisms of its multiple actions. J Biol Chem 266: 7313-7316
Waters CM, Littlewood TD, Hancock DC, Moore JP, and Evan GI (1991) c-myc protein expression in untransformed fibroblasts. Oncogene 6: 797-805
White E, Sabbatini P, Debbas M, Wold WSM, Kusher DI, and Gooding LR (1992) The 19-kilodalton adenovirus E1B transforming protein inhibits programmed cell death and prevents cytolysis by tumor necrosis factor α. Mol Cell Biol 12: 2570-2580
Wong GH, Elwell JH, Oberley LW, and Goeddel DV (1989) Manganous superoxide dismutase is essential for cellular resistance to cytotoxicity of tumor necrosis factor. Cell 58: 923-931

Cell-Cell Interactions in Osteoblastic Metastasis Caused by Advanced Prostate Cancer

Michael Koutsilieris
Molecular Endocrinology Laboratory and Clinic of Hormones and Cancer Therapies
CHUL Research Center
Centre Hospitalier de L' Universite Laval (CHUL)
Ste-Foy (Quebec), G1V 4G2
Canada

Introduction

Among solid tumors metastasizing to the skeleton prostate adenocarcinoma and breast cancer are those producing an osteoblastic reaction. The incidence of pure osteoblastic metastases is 95% for metastatic prostate cancer and 8% for metastatic breast cancer (Sharpe and McDonald, 1942). Pure lytic lesions in metastatic prostate cancer is not a common finding. Consequently, metastatic prostate adenocarcinoma is a unique cancer with regard to its ability to promote new bone formation. Among other important events in the process of bone metastasis, cell-cell interaction between metastatic cancer cells and host tissue regulate not only the establishment of the metastatic tumor but also the development of the local osteoblastic reaction (Galasko, 1982). Skeletal metastases in prostate cancer patients are frequently the only sites of disease progression after long-lasting remission achieved by androgen depletion therapies and practically determine the outcome of prostate cancer patients (Koutsilieris and Tolis, 1985).

In 1891, Von Recklinghausen proposed that any osseous new growth in response to carcinomatous cells is partly the result of either chronic venous congestion (circulatory disturbance hypothesis) or reactive irritation at the edges of metastatic deposits (foreign body or defense reaction hypothesis). Other investigators proposed that the osteoblastic reaction produced by metastatic cancer cells can be related to osteogenic substance(s) secreted by the metastatic prostate cancer cells (Sharpe and McDonald, 1942; Galasko, 1982 and references therein). The latter should be considered the first claim for the presence of prostate-derived growth factors for osteoblasts. Growth factor involvement in the pathogenesis of osteoblastic reaction has been thoroughly investigated over the last 10 years

NATO ASI Series, Vol. H 99
Tumor Biology
Regulation of Cell Growth, Differentiation and Genetics in Cancer
Edited by Asterios S. Tsiftsoglou, Alan C. Sartorelli, David E. Housman and T. Michael Dexter

(Jacobs et al, 1979; Jacobs and Lawson, 1980; Koutsilieris et al, 1986; Koutsilieris, 1987; Koutsilieris et al, 1987a and 1987b).

Prostate-Derived Growth Factors and Osteoblastic Reaction

Direct evidence that human prostate cancer tissues contain mitogens for osteoblasts has emerged when human prostate cancer, benign prostatic hyperplasia, normal adult, and normal prepubertal prostate tissues were extracted and assessed for mitogenic activity on several cell bioassay systems (Jacobs et aL, 1979; Jacobs and Lawson, 1980; Koutsilieris et al, 1986; Koutsilieris, 1987; Koutsilieris et al, 1987a and 1987b).

Prostate extracts contain several mitogens for mesenchymal cells (fibroblasts and osteoblast-like cells). Because tryptic digestion and exposure at 100 ^{o}C inhibited this mitogen(s), the nature of this material is proteinaceous. Furthermore, prostate-derived mitogen(s) for osteoblast-like cells was acid-stable which retained and eluted from cartridges of the C18 silica. Therefore, reverse-phase high performance liquid chromatography (r-HPLC) was employed to purify this material (Koutsilieris, 1987). Indeed, r-HPLC purified several mitogens with broad cell type specificity from prostate cancer and benign prostatic hyperplasia extracts. Among them was material selectively mitogenic for osteoblasts (Koutsilieris et al, 1987a and 1987b). To date, r-HPLC has enabled purification of several peptides with molecular weights varying from 6 kDa to 24 kDa, all exerting selective mitogenic activity for osteoblasts. Therefore, it is conceivable that all selective mitogens for osteoblast could be cleavage products of one molecule (Koutsilieris, 1987). The final chemical characterization of this selective mitogen for osteoblasts that would permit structural analysis has not yet been accomplished directly from human prostate tissues.

In addition, there is evidence now documenting that extracts of benign prostatic hyperplasia, prostate cancer, and adult prostate contain this selective mitogen(s) for osteoblasts but not prepubertal human prostate (Koutsilieris, 1987). This finding indicates an androgen-dependent regulation of expression. Because other than the prostate none of the normal and malignant extracts tested so far contained such selective mitogen for osteoblasts, prostate tissue is thought to possess either the exclusive production or the exclusive processing of this material to its active form (Koutsilieris, 1987).

Therefore, metastasis of prostate cancer in bone is a unique anatomical situation where prostate cancer cells are very close to osteoblasts, thus by producing prostate-derived growth factor(s) directly stimulate osteoblasts to produce woven bone.

In Vivo Models for the Analysis of the Osteoblastic Reaction

Spontaneous skeletal metastases do not occur in any of the available animal models for prostate cancer, despite they are quite common in humans. This is probably due to less developed anastomoses between the pelvic vein and vertebral vein system in rat than it is in man. Therefore, there exist nonspontaneous *in vivo* models that allow the study of osteoblastic metastases (Koutsilieris, 1993).

The Pollard tumors are prostate tumors that developed spontaneously in 10% of aged, germ-free, and conventional Lobund-Wistar (L-W) rats. Four PA cell lines have been isolated from four different Pollard tumors (PA-I, -II, -III, and -IV), all transplantable only to L-W rats. These PA tumor cell lines have had a predictable metastatic spread (Pollard and Luckert, 1985; Pollard et al, 1988). The PA-III cells demonstrated the capacity, when deposited over either the calvarium or the scapula of L-W rats, to produce (a) local tumors on bones and (b) the osteoblastic reaction at the site of transplantation onto rat skeleton. This was not seen with PA-II cells; and PA-I or PA-IV cells produced less frequently osteoblastic lesions (Pollard and Luckert, 1985). The development of PA-III cell-induced tumors on rat bone required disruption of local periosteum by the inoculating needle. Otherwise, periosteum acted as a barrier for the establishment of PA-III tumors and the blastic reaction. Although there is evidence now indicating that approximately 30% of these PA tumors are of seminal vesicle origin (Hoover et al, 1990) the ability of transplanted PA-III cells to produce an osteoblastic reaction was exploited as a pertinent *in vivo* model to study the paracrine interactions leading to an osteoblastic reaction (Koutsilieris, 1992).

The R3327 prostate tumor model developed originally in the Copenhagen rat produced different R3327 tumor variants with different characteristics regarding hormone sensitivity, histology and organ specific metastatic potential (Isaacs et al, 1981). The R3327-MATLyLu tumor variant is a rapidly growing, anaplastic, androgen-independent variant (Geldof and Rao, 1990a and 1990b). The R3327-MATLyLu tumor cells were injected intravenously in male Copenhagen rats that had transient surgical clamping of the lower caval vein. This method produced microscopic visible metastases in the lumbar spine within 4 days. These metastatic bone tumors were accompanied by an osteoblastic reaction (Geldof and

Rao, 1990a and 1990b), and treatment with bisphosphonates suppressed metastatic potential and delayed the development of the hind leg paralysis in these animals (Sun et al, 1992).

In addition, the MatLyLu cells, when inoculated also into the left ventricle of the heart in Copenhagen rats, adhered preferably to bone marrow stromal endothelial cells rather than to other bone-derived cells, including nonendothelial bone marrow cells, fibroblasts, osteoblasts, and hepatic endothelial cells. These results suggested that the adhesion of prostate cancer cell preferably to bone marrow endothelium may play a role in the incidence of bone metastases in prostate cancer (Haq et al, 1992). The functional properties of metastatic prostate cancer cells with respect to cell interactions with bone cells has been studied in various *in vitro* cell culture systems. From all these various cell lines, the PA-III rat and PC-3 human metastatic prostate cancer cells are used most frequently.

The Role of Urokinase-Type Plasminogen Activator (uPA), Insulin-Like Growth Factors (IGFs), and IGF-Binding Proteins (IGFBPs) in the Pathogenesis of Osteoblastic Reaction

The ability of PA-III cells to produce the osteoblastic reaction into rat bone was reported previously in this review. The development of PA-III cell-induced tumor into the bone required disruption of the local periosteum (Pollard and Luckert, 1985), suggesting that development of the PA-III tumor required stimuli originated from rat bone. From the histological point of view, subsequent bone reaction was similar to the osteoblastic metastases in advanced prostate cancer patients. Therefore, PA-III rat prostate cancer cells and UMR 106 osteoblast-like rat osteosarcoma cells have been used as an homologous *in vitro* cell culture system to study the cell interactions leading to the osteoblastic reaction (Koutsilieris, 1988; Polychronakos et al, 1991; Koutsilieris and Polychronakos, 1992; Koutsilieris et al, 1993).

There is now data confirming bone matrix, after the liver as the richest source of IGFs (Canali et al, 1988). Hence studies have examined the possible implication of IGFs in the Pollard model for bone metastases (Koutsilieris, 1988 and 1993). These studies documented the following (a) that cell-conditioned media (CM) from UMR 106 osteoblast-like cells stimulated the proliferation of PA-III cells via an IGF-dependent mechanism (anti-IGF-I monoclonal antibody neutralized the stimulation of UMR 106 cell CM in PA-III cells); (b) that PA-III cell-CM stimulated the proliferation of UMR 106 osteoblast-like cells via an IGF-dependent mechanism (anti-IGF-I monoclonal antibody blocked the stimulation of PA-III cell-CM in UMR 106 cells), (c) that IGF-I but not IGF-II receptor was present on the membrane of PA-III cells, and (d) that IGF-I receptor mediated the mitogenic effects of IGF-I, IGF-II, and insulin in PA-III cells (Polychronakos et al, 1991; Koutsilieris and Polychronakos, 1992). Therefore, osteoblast-derived IGFs were the probable paracrine stimulus for the initial growth of PA-III cells leading to the establishment of PA-III cell-induced tumor into rat bone.

Because evidence exists that IGF-binding proteins (IGFBPs), which have an affinity for IGFs higher than IGF-I receptor (IGF.R) and IGF-II.R, regulate the bioavailability of IGFs in tissues such as bone (Binoux et al, 1991), it was conceivable that prostate cancer cells (PA-III cells) could modulate the proliferation of osteoblast-like cells (UMR 106 cells) via hydrolysis of IGFBPs. Indeed, PA-III cell-CM hydrolyzed the IGFBPs of UMR 106 cell-CM as assessed by ligant blotting assays (Koutsilieris and Polychronakos, 1992). These data indicated that protease(s) from prostate cancer cells (PA-III cells) may modulate the interactions with osteoblasts (Koutsilieris and Polychronakos, 1992).

Purification and sequence analysis of this putative protease(s) in PA-III cell CM revealed complete homology with amino acid sequence of rat urokinase-type plasminogen activator. Rat uPA hydrolyzed IGFBPs in UMR 106 cell-CM and stimulated the proliferation of UMR 106 cells (Koutsilieris et al, 1993). Therefore, (a) osteoblasts such as UMR 106 cells in producing IGFs enable the development of PA-III cell-induced tumor in rat bone after disruption of local periosteum, and (b) PA-III cells stimulated osteoblasts such as UMR 106 cells via an IGF-dependent mechanism by involving the uPA/plasmin-induced hydrolysis of IGFBPs, thus the increasing bioavailability of IGFs, locally. This mechanism could explain the IGFs role in (a) the development and evolution of PA-III cell-induced tumor, and (b) the subsequent osteoblastic reaction (Koutsilieris and Polychronakos, 1992; Koutsilieris et al, 1993; Koutsilieris, 1995).

Other investigators have used the PC-3 human prostate cancer cell line established from a hormone-indepedent osteoblastic in a prostate cancer patient for the study of local interactions with bone cells (Simpson et al, 1985; Perkel et al, 1990; Rabbani et al, 1990). Indeed, PC-3 cell mRNA, when microinjected into xenopus oocytes, directed the synthesis of substances mitogenic to osteoblasts (Simpson et al, 1985). Similar, if not identical, mitogens to those described in human prostate tissues (selective mitogen(s) for osteoblasts) were reported in PC-3 cell-CM (Perkel et al, 1990). One of these substance(s) in PC-3 cell-CM was documented to share identical NH_2-terminal sequence with urokinase-type plasminogen activator (uPA). Initially, this material was reported to stimulate osteoblasts such as human Saos-2 osteoblast-derived human osteosarcoma cells by an action that was independent from uPA's catalytic activity (Rabbani et al, 1990).

Recent data indicate that human uPA stimulates the proliferation of osteoblast-like cells such as MG-63 osteoblast-derived human osteosarcoma cells via an IGF-dependent mechanism that involves hydrolysis of IGFBPs (Campbell et al, 1991; Koutsilieris and Polychronakos, 1992; Koutsilieris, 1995). Therefore, different cell culture systems (homologous cultures of rat PA-III/UMR 106 cells and human PC-3/Saos-2 or MG-63 cells) have indicated the pivotal implication of uPA in the osteoblastic reaction.

Herein we must mention that earlier clinical studies have documented elevated uPA plasma levels in stage D2 prostate cancer patients (Hienert et al, 1988a and 1988b). The plasma levels of uPA correlated well with the number of bone metastases in prostate cancer patients (Hienert et al, 1988b). Recently, the Mat LyLu rat prostate cancer cells, stably transfected with the cDNA of uPA, thus expressing 10-fold higher uPA, was documented to produce hindlimb paralysis (bone metastases) earlier than the original MatLyLu cells in male Copenhagen rats (Achabarou et al, 1993). In addition, the uPA/plasmin system was reported to activate the latent form of transforming growth factor beta 1 (TGFβ1) which is a very important growth factor for the osteoblast biology (Gehron-Robey et al, 1987; Lyons et al, 1988; Pfeilschifter et al, 1990). Therefore, uPA should play a pivotal role in the physiology of bone and pathophysiology of bone metastasis.

The Role of Plasma Membrane and Extracellular Matrix Proteins in Osteoblastic Metastases

The ability of a cell to move is, for the most part, mediated through the cytoplasmic actin microfilament system and its modulation by a group of associated proteins (Way and Weeds, 1990). Depending on the cell type and state of differentiation, actin-binding proteins (ABPs) permit specific rearrangements of the three-dimensional state of the microfilaments in the peripheral cytoplasm, which in turn allows the cell to become (Janmey et al, 1990; Finidori et al, 1992). Since differences exist with respect to the motility of tumor cells, it was speculated that such differences may be due to corresponding differences in the expression and functional deployment of specific ABPs. In this respect, it is interesting to note that several cell lines of human melanoma that are deficient in ABP had a significantly decreased translocation capacity, as evaluated in a chemotactic chamber (Cunningham et al, 1989). Certainly more extensive research efforts in this direction should prove fruitful.

The interface between the cytoskeleton and the cell membrane constitutes a point of control of cell shape and cell attachment to adjacent cells and matrix proteins. The proteins found in this particular domain of the cell are implicated in cellular motility and cell responsiveness to growth factors and other external stimuli (Cunningham, 1992; Luna and Hift, 1992 and references therein). Cytoplasmic actin filaments terminate at the plasma membrane in specialized regions referred to as *focal adhesions*. These supramolecular assemblies are concentrated in those areas of the membrane that are involved in adhesion to the substratum and are associated with the transmembrane adhesion proteins, the *integrins*. The number of focal adhesions is inversely related to the rate of cellular translocation and focal adhesions appear to be regulated by phosphorylation, which may also be indicative of a signalling role for focal adhesions (Burridge et al, 1992; Zachary and Rozengert, 1992; Koutsilieris and Rogers, 1994, and references therein).

The plasma membrane plays a key role in the establishment of structural and functional interactions between (a) extracellular matrix and cytoskeleton, (b) among various cell types, and (c) control of the bioavailability of certain growth factors such as bFGF, IGFs and TGFβ1. Cell surface glycocoproteins such as integrins, laminin receptor and matrix molecules such as fibronectin, type I collagen, vitronectin, laminin, fibrin, Willebrand factor and throbospondin play a crucial role for cell motility and cell function (Janmey et al, 1990).

In the plasma membrane of osteoblasts uPA.R-uPA complexes are located very close to IGFBPs-IGFs complexes positioned by RDS sequences of the integrin-like adhesion molecules or by the integrin-bound or fibronectin-bound proteoglycans, such as syndecans (Ellis et al, 1992; Koutsilieris and Rogers, 1994; Koutsilieris, 1995). In addition, osteoblasts produce the highest levels of IGFs and TGFβ1 after liver and platelets, respectively (Gehron-Robey et al, 1987; Canalis et al, 1988; Binoux et al, 1991). Colocalization of these important regulators onto osteoblast membrane-mediated reservoir of growth factors enables osteoblasts to control tightly extracellular matrix deposition (Koutsilieris, 1993; Koutsilieris and Rogers, 1994). Activation of this plasma membrane-dependent growth factor reservoir by proteinolytic substances such as the uPa/plasmin system and other extracellular matrix enzymes regulates bone matrix formation. This interplay of local regulators, including the activation of TGFβ1, bFGF and IGFs increase osteoblast production of matrix proteins and osteoblast growth but also inhibit uPA production and increase plasminogen activator inhibitor (PAI) expression, thus inactivating the uPA/plasmin system which in its turn decrease bioavailability of IGFs and TGFβ1. This system could be part of the mechanism regulating the deposition of bone matrix (Koutsilieris, 1995).

It is tempting to postulate that metastatic prostate cancer cells, producing unregulated amounts of uPA, probably deregulate this mechanism of bone matrix deposition. Therefore, the local increase in uPA levels would increase the local bioavailability of these growth factors, thus stimulating the growth (a) of metastatic prostate cancer cells (paracrine effects of IGFs on metastatic prostate cancer cells), and (b) of osteoblasts (autocrine actions of activated osteoblast-derived TGFβ1 and IGFs). This vicious cycle of cell-cell interactions into bone martix may result in the formation of the osteoblastic reaction presented as woven bone at the sites of prostate cancer metastasis (Koutsilieris, 1995; Koutsilieris and Dimopoulos, 1995).

Three Dimensional Type I Collagen Matrix System for the Study of Paracrine Interactions in Osteoblastic Metastases

Recently, a three-dimensional type I collagen gel has supported the function of osteoblast-like cells in culture. This type I collagen matrix system was modified to allow coculturing of osteoblast-like cells and metastatic prostate cancer cells (Koutsilieris et al, 1994). Indeed, the inoculation of PC-3 prostate cancer cells into this type I collagen gel containing human MG-63 osteoblast-like cells produced the osteoblastic reaction. This finding confirmed the specific cell interactions between PC-3 human prostate cancer cells and MG-63 osteoblast-like cells leading to the osteoblastic metastasis. This system, unique in its ability to study human cell interactions in a three-dimensional structure, is expected to provide further evidence for the specific cell interactions involved in the mechanism of the osteoblastic reaction (Koutsilieris et al, 1994; Koutsilieris, 1995; Koutsilieris and Dimopoulos, 1995).

Concluding Remarks

There is no doubt that at this point in time our understanding of the pathophysiology of osteoblastic metastases has progressed significantly. Nonetheless, it is also evident that, from a clinical point of view, this has not translated to a better therapeutic approach for advanced stage prostate cancer patients. Our belief is that elucidation and understanding of the rules governing metastasis is the only route to better therapeutic interventions. This progress should be directly related to research on metastatic tumor cell biology, cell-cell interactions and tumor cell-cytoskeleton interactions. It is conceivable that our appreciation of the molecular aspects of prostate cancer metastasis may determine the pharmaceutic agents that could modify the fetal evolution of prostate cancer metastases in bones.

Acknowledgements

Michael Koutsilieris is a scholar (Chercheur-Boursier) of the Medical Research Council of Quebec (Fonds de la Recherche en Sante du Quebec; FRSQ) supported by the Medical Research Council of Canada (MRC of Canada) and Cancer Research Society of Montreal, Inc (CRS, Montreal Inc.).

References

Achbarou A, Kaiser S, Tremblay D, Goltzman D, Rabbani SA (1993) Overexpression of urokinase by prostate cancer cells in vivo results in increased skeletal metastasis. J Bone Miner Res (Suppl 1) 8:S133.

Binoux M, Roghani M, Hossenlopp P, Hardouin S, Goumelen M (1991) Molecular forms of the human IGF binding proteins-physiological implications. Acta Endocrinol 124:241-245.

Burridge K, Petch LA, Romer LH (1992) Signals from focal adhesions. Curr Biol 10:537-539.

Campbell PG, Novak JF (1991) Insulin-like growth factor binding protein (IGFBP) inhibits IGF action on human osteosarcoma cells. J Cell Physiol N149:293-300.

Canalis E, McCarthy M, Centrella M (1988) Isolation and characterization of insulin-like growth factor I (somatomedin C) from cultures of fetal rat calvariae. Endocrinology 122:22-27.

Cunningham CC, Byers HR, Hartiwig JH, Stossel TP, Kwiatkowski DJ (1989) Morphologic properties of the three actin-binding protein (ABP-nonmuscle filamin)-deficient human malignant melanoma cell lines are consistent with reduced cortical actin gelation J Cell Biol 109:277-285.

Cunningham CC (1992) Actin structural proteins in cell motility. Cancer Metastasis Rev 11:69-77.

Ellis V, Pyke C, Eriksen J, Solberg H, Dano K (1992) The urokinase receptor: involvement in cell surface proteolysis and cancer invasion. Ann N Y Acad Sci 667:13-31.

Finidori J, Friederich E, Kwiatkowski DJ, Louvard D (1992) In vivo analysis of functional domains from villin and gelsolin. J Cell Biol 116:1145-1155.

Galasko CSB (19982) Mechanisms of lytic and blastic metastatic disease of bone. Clin Orthop Rel Res 12:20-27.

Gehron-Robey P, Young MF, Flanders KC, Roche NS, Kondaiah P, Reddi AH, Termine JD, Sporn MB, Roberts AB (1987) Osteoblasts synthesize and respond to transforming growth factor-type b (TGF-b) in vitro. J Cell Biol 105:457-464.

Geldof AA, Rao BR (1990a) Factors in prostate cancer metastasis. Anticancer Res 10:1303-1306.

Geldof AA, Rao BR (1990b) Prostatic tumor (R3327) skeletal metastasis. Prostate 16:279-290.

Haq M, Goltzman D, Tremblay G, Brodt P. Rat prostate adenocarcinoma cells disseminated to bone and adhere preferentially to bone marrow-derived endothelial cells. Cancer Res 52:4613-4619, 1992.

Hienert G, Kirchheimer JC, Pfluger H, Binder BR (1988a) Urokinase-type plasminogen activator as a marker for the formation of distant metastases in prostatic carcinomas. J Urol 140:1466-1469.

Hienert G, Kirchheimer JC, Christ G, Pfluger H, Binder BR (1988b) Plasma urokinase-type plasminogen activator corralated to bone scintigraphy in prostatic carcinoma. Eur Urol 15:256-258.

Hoover DM, Best KL, McKenney BK, Tamura, RN, Neubauer BL (1990) Experimental induction of neoplasia in accessory sex organs of male Lobund-Wistar rats. Cancer Res 50:142-146.

Isaacs JT, Yu GW, Goffey DS (1981) The characterization of a newly identified prostatic adenocarcinoma system: the MATLyLu tumor. Invest Urol 19:20-23.

Jacobs SC, Pinka D, Lawson RK (1979) Prostatic osteoblast factor. Invest Urol 17:195-198.

Jacobs SC, Lawson RK (1980) Mitogenic factor in human prostate extracts. Urology 16:488-491.

Janmey PA, Hvidt S, Lamb J, Stossel TP (1990) Resemblance of actin-binding protein/actin gels to covalently crosslinked networks. Nature 345:89-92.

Koutsilieris M, Tolis G (1985) Long term follow-up of patients with advanced prostatic carcinoma treated with either buserelin (HOE 766) or orchiectomy: classification of variables associated with disease outcome. Prostate 7:31-39.

Koutsilieris M, Rabbani SA, Goltzman D (1986) Selective mitogens can be extracted from prostatic tissue. Prostate 9:109-115.

Koutsilieris M (1987) Characterization of growth factors for osteoblast from human prostatic tissue: implications for the pathogenesis of osteoblastic metastases. Ph.D Thesis, McGill University.

Koutsilieris M, Rabbani SA, Goltzman D (1987a) Effects of prostatic mitogens on rat bone cells and fibroblasts. J Endocrinol 115:447-454.

Koutsilieris M, Rabbani SA, Bennett HPJ, Goltzman D (1987b) Characteristics of prostate-derived growth factors for cells of the osteoblast phenotype. J Clin Invest 80:941-946.

Koutsilieris M (1988) Prostate-derived growth factors for bone cells: implications for bone physiology and pathophysiology. In Vivo 2: 377-384.

Koutsilieris M (1992) PA-III rat prostate adenocarcinoma cells. In Vivo 6:199-204.

Koutsilieris M, Polychronakos C (1992) Proteinolytic activity against IGF-binding proteins involved in the paracrine interactions between prostate adenocarcinoma cells and osteoblasts. Anticancer Res 12:905-910.

Koutsilieris M (1993) Osteoblastic metastasis in advanced prostate cancer. Anticancer Res 13:443-450.

Koutsilieris M, Frenette G, Lazure C, Lehoux JG, Govindan MV, Polychronakos C (1993) Urokinase-type plasminogen activator: a paracrine factor regulating the bioavailability of IGFs in PA-III cell-induced osteoblastic metastases. Anticancer Res 13:481-486.

Koutsilieris M, Rogers P (1994) The role of plasma membrane in the pathogenesis of osteoblastic metastasis. In: Bkaily G (ed). Membrane Pathophysiology,pp 375-386.

Koutsilieris M, Sourla A, Pelletier G, Doillon C (1994) A three-dimensional type I collagen gel system for the study of osteoblastic metastases produced by metastatic prostate cancer cells. J Bone Miner Res 11:1823-1832.

Koutsilieris M (1995) Skeletal metastases in advanced prostate cancer: cell biology and therapy. Cr Rev Hema/Onco 18:51-64.

Kourtsilieris M, Dimopoulos M-A (1995) The molecular concept of prostate cancer. Cancer J, in press.

Lyons RM, Keski-Oja J, Moses HL (1988) Proteolytic activation of latent transforming growth factor-b from fibroblast-conditioned medium. J Cell Biol 106:1659.

Luna EJ, Hitt AL (1992) Cytosceletal-plasma membrane interactions. Science 258:955-963.

Perkel VS, Mohan S, Herring SJ, Baylink DJ, Linkhart TA (1990) Human prostatic cancer cells, PC3, elaborate mitogenic activity which selectively stimulated human bone cells. Cancer Res 50:6902-6906.

Pfeilschifter J, Erdmann J, Schmidt W, Naumann A, Minne HW, Ziegler R (1990) Differential regulation of plasminogen activator inhibitor by osteotropic factors in primary cultures of mature osteoblasts and osteoblast precursors. Endocrinology 126:703-706.

Pollard M, Luckert PH (1985) Effects of dichloromethylene diphosphate on the osteolytic and osteoplastic effects of rat prostate adenocarcinoma cells. J Natl Cancer Inst 75:949-954.

Pollard M, Luckert PH, Scheu J (1988) Effects of diphosphonate and X-rays on bone lesions induced in rats by prostate cancer cells. Cancer 61: 2027-2032.

Polychronakos C, Janthly U, Lehoux JG, Koutsilieris M (1991) Mitogenic effects of insulin and insulin-like growth factors on PA-III rat prostate adenocarcinoma cells: characterization of the receptors involved. Prostate 19:313-321.

Rabbani S, Desjardins J, Bell AW, Danville D, Mazar A, Henkin J, Goltzman D (1990) An amino-terminal fragment of urokinase isolated from a human prostate cancer cell line (PC-3) is mitogenic for osteoblast-like cells. Biochem Biophys Res Commun 173:1058-1064.

Sharpe WS, McDonald JR (1942) Reaction of bone to metastasis from carcinoma of the breast and prostate. Arch Pathol 17:312-325.

Simpson E, Harrod J, Eilon G, Jacobs JW, Mundy GR (1985) Identification of a messenger ribonucleic acid fraction in human prostatic cancer cells coding for a novel osteoblast-stimulating factor. Endocrinology 117:1615-1619.

Sun YC, Geldof AA, Newling DW, Rao BR (1992) Progression delay of prostate tumor skeletal metastasis effects by bisphosphonates. J Urol 148:1270-1273.

Way M, Weeds A (1990) Actin-binding proteins. Cytosceletal ups and downs. Nature (News and Views) 344:292-293.

Zachary I, Rozengurt E (1992) Focal adhesion kinase (p125FAK): a point of convergence in the action of neuropeptides, integrins, and oncogenes. Cell 71:891-894.

APOPTOSIS IN COLONIC EPITHELIUM: A MESSAGE FROM THE CRYPT

Sergio A. Lamprecht, Sarit Lifshitz, Sylvie Polak-Charcon* and Bertha Schwartz

Gastroenterology Department, Soroka Medical Center and Clinical Biochemistry Unit, Ben-Gurion University of the Negev, Beer Sheva, and *Institute of Pathology, Sheba Medical Center, Israel

Introduction

Colonic epithelium represents a unique system for the analysis of cellular and molecular events controlling cell proliferation, differentiation and death. The crypt of the large intestine is the anatomical residence of cells in a continuous process of self-renewal. Colonic cells exhibit mitotic activity in the lower crypt zones and acquire the differentiated phenotype during migration to the upper crypt regions (Williamson, 1978; Lipkin, 1987; Hall and Watt, 1992; Gordon et al., 1992).

Loss of effete cells from the colonic lumenal surface ensues periodically concomitantly with replenishment by a new wave of differentiated colonocytes. The constancy of cell number along the colonic crypt axis is obviously under stringent control to assure the accurate balance between the rate of new cell production and the rate of cell death. The precise molecular basis of the counting process responsible for compiling the cellular census remains elusive.

Apoptosis is an integral part of the intricate processes controlling overall tissue homeostasis. The physiological deletion of distinct cells by apoptotic death is a well recognized process during metazoan morphogenesis, tissue development and remodeling (Garcia-Martinez et al., 1993; Concouvanis et al., 1995). The execution of the death program is associated with stereotyped morphological changes and is remarkably similar among cell types of disparate lineages. These include loss of intercellular contacts, cell shrinkage, compaction of chromatin abutting beneath the inner surface of the nuclear membrane, cellular budding with formation of apoptotic bodies and rapid phagocytosis by

NATO ASI Series, Vol. H 99
Tumor Biology
Regulation of Cell Growth, Differentiation and
Genetics in Cancer
Edited by Asterios S. Tsiftsoglou, Alan C. Sartorelli,
David E. Housman and T. Michael Dexter

neighboring cells. DNA strands fragmentation at internucleosomal linker sites is observed in many cells, leading to the typical DNA ladder in agarose gel electrophoresis (Wyllie, 1992; 1993; Wyllie and Duvall, 1992; Schwartzman and Cidlowski, 1993; Martin, 1993; Martin et al., 1994; Hockenbery, 1995; Majno and Joris, 1995; Steller, 1995).

Apoptosis in colonic cells: facts and speculations

Apoptosis has been observed within the murine and human colonic crypt axis. Potten, (1977), Potten et al. (1990, 1992) and Merrit et al. (1994) reported a low spontaneous rate of apoptosis in the proliferative compartment of the colonic crypt that increased after a genotoxic insult in a p53-dependent manner. Gravieli et al. (1992) noted that the in situ labeling of DNA strand breaks is apparent only in the uppermost nuclei at the edge of the human and rodent colonic crypt. Recently, Hall and colleagues (1994) have described in a careful study the occurrence of apoptotic bodies predominantly at the tip of colon crypts. These authors propose a central role for apoptosis in the regulation of cell number in intestinal epithelium.

The restricted zonal distribution of apoptotic cells along the colonic crypt continuum in whole tissue preparations might signify that the initiation and execution of the death program in colonic cells is strictly contextual, depending on the apoptosis-permissive crypt microenvironment. We investigated this possibility by monitoring programmed cell death in whole rat colonic tissue specimens, in isolated crypts (Ecke et al. , 1995) and in colonic cells deprived of crypt ambient. Colonic cell populations were sequentially collected from the crypt axis at precise stages of proliferation and differentiation using an established non-enzymatic isolation procedure (Schwartz et al., 1991). Apoptotic death was assessed by agarose gel electrophoresis analysis of oligonucleosomal DNA fragmentation (Sellins and Cohen, 1987) and by the terminal deoxynucleotidyl transferase (TdT)-mediated dUTP nick end labeling (TUNEL) procedure (Gavrieli et al., 1992; Ben-Sasson et al., 1995). Typical morphological features were verified by transmission electron microscopy (Kerr et al., 1995).

Consistent with the abovementioned reports, apoptotic cells in whole tissue colonic specimens were found extensively in the upper zone of the crypt continuum. Isolated colonic crypts showed an identical zonal distribution of apoptosis. Surprisingly, however, we have observed rapid and extensive apoptosis in *all* isolated colonic cells irrespective of their actual position along the crypt axis at harvest time. The percentage of DNA fragmentation was identical in all cell populations. The present results, previously reported in meeting communications (Lifshitz et al., 1994a,b; 1995a,b) have recently been confirmed by Bedi et al. (1995) in human normal heterogeneous colorectal cell populations.

The apoptotic process in rat colonocytes is not triggered by the separation procedure based on Ca^{2+} chelation since it has also been reproduced in murine colonocytes isolated using an alternative enzymatic method (Lifshitz et al. 1995a). In this context is noteworthy that Bates et al. (1994) have recently shown that the removal of calcium from culture medium disrupts the structural organization of the LIM 1863 colon carcinoma cell line resulting in suspension of single cells which, if prevented from reaggregating, undergo rapid apoptosis. One may therefore surmise that colonic cells possess an intrinsic program for their own demise which is ineluctably activated in all cell populations once they are deprived of the crypt ambient irrespective of their hierarchical status within the crypt continuum. In contrast, colonic cells moving in tight cohorts along the crypt axis would be allowed entry into the apoptotic pathway only within precise windows of space and time.

The precise molecular mechanisms inducing cell death within the colonic crypt continuum remain to be elucidated. A prevailing view, sustained by vast experimental evidence, holds that cells would die by default unless kept alive by specific signals originating from neighboring cells (Raff, 1992; Raff et al., 1993; Collins et al., 1994). This model seems to us very attractive for explaining the collective and indiscriminate suicide of colonic cells deprived of crypt habitat. Putative survival factors present in colonic epithelium include TGF-α, EGF, gastrin, IGF and bcl-2 (Burgess and Sizeland, 1990; Hockenbery, et al., 1992; Avery et al., 1993.; Podolski, 1994; Lamprecht and Hanby, unpublished observation).

The survival model outlined above implies a central role in cell-to-cell communication in preventing unscheduled and autonomous apoptosis in the colonic crypt. An intriguing observation is consonant with this view. As outlined in the preceding page, the sequential stripping of colonic mucosa affords the harvest of single cell suspensions. Occasionally, however, cells in clusters are collected. We have consistently noticed that, in contrast to the isolated colonocytes, cells in the aggregates did not exhibit signs of apoptotic distress. This finding strengthens the notion that cell-to-cell communication impedes the apoptotic process.

A key role of adhesion molecules as components of the survival system is supported by several observations (Meredith et al., 1993; Ruoslahti and Reed, 1994). Frish and Francis (1994) have presented evidence that cross-talk between extracellular matrix and epithelial cells is critical to prevent apoptosis: disruption of communication results in cell death. The authors have aptly named this phenomenon "anoikis", a Greek word for homelessness. Bates et al. (1994) have provided convincing results showing that anti-integrin antibodies are able to trigger apoptosis in isolated colonic cancer cells through their capacity to inhibit close cellular contacts. Loss of integrin-mediated cell matrix communication has been shown to induce apoptosis in cultured cells (Montgomery et al., 1994; Boudreau et al., 1995).

E-cadherin, a member of Ca^{2+} dependent cell adhesion proteins (Grunwald, 1993) , is involved in homophylic interactions between epithelial cells and is essential for the induction and maintenance of normal epithelial integrity (Takeichi, 1991).

Herminston and Gordon (1995) have recently provided cogent evidence for an essential role of E-cadherin as a survival molecule in preventing cell death in mouse enterocytes. Transfection of a dominant N-cadherin mutant disrupts E-cadherin function in villus-associated enterocytes of chimeric transgenic mice resulting in a loss of the differentiated polarized phenotype and in precocious entry of the cells into the death program.

The apoptotic process in colonic epithelial cells may therefore be regulated by permissive and inhibiting instructive interactions and signaling pathways between epithelial cells, the extracellular matrix and mesenchymal cells. This indeed seems to be the overall strategy for the control of the ordered growth and bipolar migration of colonic cells along the crypt continuum (Podolsky, 1993; Gordon and Hermiston, 1994). Our scant understanding of this complex informative network may underlie the persistent failure to maintain normal colonic cells in primary culture.

The apoptotic process is very rapid. Frank signs of apoptotic distress are exhibited by colonic cells within fifteen minutes from collection time. It is difficult to reconcile the recognized rapidity of the cascade of apoptotic events with the elaborate, multi-step cellular machinery set in motion by the expression of putative death genes (Steller, 1995). The possibility must therefore be considered that cells constitutively possess the lethal molecular signals for their own demise awaiting only the appropriate contextual cues to rapidly engender apoptosis. Alternatively, survival factors may be rendered redundant in cells poised to die. The plasticity of these responses may rely on post-translational modifications of critical cell components. A salient case in point is presented by the proteolytic cleavage of pro-ICE-like cysteine proteases to the active molecular form involved in the effector pathway of apoptosis (Jacobson and Evan, 1994; Kumar, 1995; Nicholson et al., 1995; Whyte and Evan, 1995). Loss of function induced by phosphorylation of the E-cadherin/cathenins complex (Grunwald, 1993; Takeichi, 1993; Birchmeier and Behrens, 1994; Hink et al., 1994) would provoke the collapse of cell-to-cell communication networks triggering the cell's overt commitment to apoptotic death. The tenable possibility has been considered that regulation of apoptosis occurs in the same cell at multiple tiers of control including transcriptional and post-translational steps (Martin, 1993; Jacobson and Evan, 1994; Kumar, 1995).

Conclusions

It is apparent that the apoptotic process plays a role in the maintenance of cellular census in the intestine. Once a redundant cell has been informed of the termination of its biological purpose, programmed death is activated. Hall et al. (1994) have presented evidence indicating that the cell loss in the normal murine intestine can be largely explained on the basis of apoptosis. Hermiston and Gordon (1995) have proposed that preparation for death is a normal part of the intestinal cell lineage's terminal differentiation program. Indeed, failure of apoptosis may be of key importance for the development and progression of colorectal cancer (Bedi, et al., 1995; Watson, 1995a,b).

In summary our results suggest that the life and death of the colonic cell are stringently controlled by signaling pathways and molecular cues originating within and outside the crypt milieu. Continuous regulation is needed for the entry into and escape from the apoptotic pathway. The indiscriminate self-destruction of colonic cells deprived of crypt context is consistent with the view that cells are constitutively wired with a latent suicide program and express the apoptosis phenotype once reprieving cross-talk with their neighbors becomes disorderly and communication is severed.

References

Avery, A., Paraskeva, C., Hall, P., Flanders, K.C., Sporn, M. and Moorghen, M. (1993) TGFα expression in the human colon: Differential immunostaining along crypt epithelium. Br. J. Cancer 68:137-139.

Bates, R.C., Buret, A., van Helden, D.K., Horton, M.A. and Burns, G.F. (1994) Apoptosis induced by inhibition of intercellular contact. J. Cell Biol. 125:403-415.

Bedi, A., Pasricha, P.J., Akhtar, A.J., Barber, J.P., Bedi, G.C., Giardello, F.M., Zehnbauer, B.A., Hamilton, S.R. and Jones, R.J. (1995) Inhibition of apoptosis during development of colorectal cancer. Cancer Res. 55:1811-1816.

Ben-Sasson, S., Sherman, Y. and Gavrieli, Y. (1995) Identification of dying cells - *In Situ* Staining. Methods in Cell Biology 46:29-39.

Birchmeier, W. and Behrens, J. (1994) Cadherin expression in carcinomas: Role in the formation of cell junctions and the prevention of invasiveness. Biochim. Biophys. Acta 1198:11-26.

Boudreau, N., Sympson, C.J., Werb, Z. and Bissell, M.J. (1995) Suppression of ICE and apoptosis in mammary epithelial cells by extracellular matrix. Science 267:891-893.

Burgess, A.W. and Sizeland, A.M. (1991) Growth factors and the gut. J. Gastroenterol. Hapatol. 1:10-21.

Collins, M.K.L., Perkins, G.R., Rodriguez-Tarduchy, G., Nieto, M.A. and Lopez-Rivas, A. (1994) Growth factors as survival factors: regulation of apoptosis. BioEssays 16:133-138.

Coucouvanis, E.C., Martin, G.R. and Nadeau, J.M. (1995) Genetic approaches for studying programmed cell death during development of the laboratory mouse. Methods in Cell Biology 46:387-440.

Ecke, D., Bleich, M., Lehrmann, E., Hropot, M, Englert, H.C., Lang, H.J., Warth, R., Rohm, W., Schwartz, B., Fraser, G. and Greger, R. (1995) A chromanol type of K+ channel blocker inhibits forskolin but not carbachol-mediated Cl^- secretion in rat and rabbit colon. Cell Physiol. Biochem. 5:204-210.

Frisch, S.M. and Francis, H. (1994) Disruption of epithelial cell - matrix interactions induces apoptosis. J. Cell Biol. 124:619-626.

Garcia-Martinez, V., Macias, D., Ganan, Y., Garcia-Lobo, J.H., Francia, M.V., Fernandez-Teran, M.A. and Hurle, J.M. (1993) Internucleosomal DNA fragmentation and programmed cell death (apoptosis) in the interdigital tissue of the embryonic chick leg bud. J. Cell Sci. 106:201-208.

Gavrieli, Y., Sherman, Y. and Ben-Sasson, S.A. (1992) Identification of programmed cell death in situ via specific labeling of nuclear DNA fragmentation. J. Cell Biol. 119:493-501.

Gordon, J.I., Schmidt, G.H. and Roth K.A. (1992) Studies of intestinal stem cells using normal, chimeric, and transgenic mice. FASEB J. 6:3039-3050.

Gordon, J.I. and Hermiston, M.L. (1994) Differentiation and self-renewal in the mouse gastrointestinal epithelium. Curr. Opin. Cell Biol. 6:795-803.

Grunwald, G.B. (1993) The structural and functional analysis of cadherin calcium-dependent cell adhesion molecules. Curr. Opin. Cell Biol. 5:797-805.

Hall, P.A. and Watt, F.A. (1989) Stem cells: the generation and maintenance of cellular diversity. Development 106:619-633.

Hall, P.A., Coates, P.J., Ansari, B. and Hopwood, D. (1994). Regulation of cell number in the mammalian gastrointestinal tract: the importance of apoptosis. J. Cell Sci. 107:3569-3577.

Hermiston, M.L. and Gordon, J.I. (1995) In vitro analysis of cadherin function in the mouse intestinal epithelium: Essential roles in adhesion maintenance of differentiation, and regulation of programmed cell death. J. Cell Biol. 129:489-506.

Hinck, L., Nathke, I.S., Papkoff, J. and Nelson, W.J. (1994) β-catenin: A common target for the regulation of cell adhesion by Wnt-1 and Src signaling pathways. TIBS 19:538-542.

Hockenbery, D. (1995) Defining apoptosis. Am. J. Pathol. 146:16-19.

Hockenbery, D.M., Zutter, M., Hickey, W., Nahm, M. and Korsmeyer, S.J. (1991). BCL2 protein is topographically restricted in tissues characterized by apoptotic cell death. Proc. Natl. Acad. Sci. USA 88:6961-6965.

Jacobson, H.D. and Evan, G.I. (1994) Breaking the ICE. Curr. Biol. 4:337-340.

Kerr, J.F.R., Gobe, C.G., Winterford, C.M. and Harmon, B.V. (1995) Anatomical methods in cell death. Methods in Cell Biology 41:1-27.

Kumar, S. (1995) ICE-like proteases in apoptosis. Trends Biol. Sci. 20:189-202.

Lifshitz, S., Schwartz, B., Polak-Charcon, S., Mikhailowski, R. and Lamprecht, S.A. (1994a) Apoptosis (programmed cell death) in colonic cells. Genetic and Biochemical Approaches for Studying Cell Death. ASBMB Fall Symposia, Granlibakken, USA, Abst 29.

Lifshitz, S., Schwartz, B., Polak-Charcon, S., Mikhailowski, R., Niv, Y. and Lamprecht, S.A. (1994b) Apoptosis in murine epithelial colon cells. World Congresses of Gastroenterology, Los Angeles, USA, Abst. 1505.

Lifshitz, S., Schwartz, B., Polak-Charcon, S., Benharroch, S. and Lamprecht, S.A. (1995a) Disruption of colonic cell-cell interaction induces apoptosis. American Gastroenterology Association Meeting, San Diego, USA, Gastroenterology 108:Abst. 498.

Lifshitz, S., Schwartz, B., Polak-Charcon, S., Benharroch, D. and Lamprecht, S.A. (1995b) Indiscriminate apoptotic death of colon cells deprived of crypt habitat. Regulation of Cell Growth, Differentiation and Genetics in Cancer. NATO Advanced Study Institute, Porto Carras, Greece. NATO ASI Abstracts' Book.

Lipkin, M. (1987) Proliferation and differentiation of normal and diseased gastrointestinal cells. In: Physiology of the Gastrointestinal Tract. Johnson, L.R. (ed), pp 255-284, Raven Press, New York.

Majno, G. and Joris, I. (1995) Apoptosis, oncosis and necrosis: an overview of cell death. Am. J. Pathol. 146:3-15.

Martin, S.J., Green, D.R. and Cotter, T.G. (1994) Dicing with death: dissecting the components of the apoptosis machinery. Trend. Biochem. Sci. 19:26-30.

Martin, S.J. (1993) Apoptosis: suicide, execution or murder? Trends Cell Biol. 3:141-144.

Meredith, J.E., Fazeli, J.B. and Schwartz, M.A. (1993) The extracellular matrix as a cell survival factor. Mol. Biol. Cell 4:953-961.

Merrit, A.J., Potten, C.S., Kemp, C.J., Hickman, J.A., Balmain, A., Lane, D.P. and Hall, P.A. (1994) The role of p53 in spontaneous and radiation-induced apoptosis in the gastrointestinal tract of normal and p53-deficient mice. Cancer Res. 54:614-617.

Merrit, H.J., Potten, C.S., Watson, A.J.M., Loh, D.Y., Makayama, K.I., Makayama, K. and Hickman, J.A. (1995) Differential expression of bcl-2 in intestinal epithelia. J. Cell Sci. 108:2261-2271.

Montgomery, A.M.P., Reisfeld, R.A. and Cheresh, D.A. (1994) Integrin rescues melanoma cells from apoptosis in three-dimensional dermal collagen . Proc. Natl. Acad. Sci. USA 91:8856-8860.

Nicholson, D.W., Ali, A., Thornberry, N.A., Vaillancourt, J.P., Ding, C.K., Gallant, M., Gareau, Y, Griffin, P.R., Labelle, M., Lazebnik, Y.A., Munday, N.A., Raju, S.M., Smulson, M.E., Yamin, T-T., Yu, V.L. and Miller, D.K. (1995) Identification and inhibition of the ICE/CED-3 protease necessary for mammalian apoptosis. Nature 376:37-43.

Podolski, D.K. (1994) Peptide growth factors in the gastrointestinal tract. In: Physiology of the Gastrointestinal Tract, Johnson, L.R. (ed). pp 129-167. Raven Press, New York.

Podolski, D.K. (1993) Regulation of intestinal epithelial proliferation: A few answers, many questions. Am. J. Physiol. 264:G179-G186.

Potten, C.S. (1977) Extreme sensitivity of some intestinal crypt cells to X and γ irradiation. Nature 169:518-521.

Potten, C.S. (1990) The significance of spontaneous and induced apoptosis in the gastrointestinal tract of mice. Cancer Metastasis Rev. 11:179-195.

Raff, M.C. (1992) Social controls of cell survival and cell death. Nature 356:397-400.

Raff, M.C., Barres, B.B., Burne, J.F., Coles, H.S., Ishizaki, Y. and Jacobson, M.D. (1993) Programmed cell death and the control of cell survival: lessons from the nervous system. Science 262:691-700

Ruoslahti, E. and Reed, J.C. (1994) Anchorage dependence, integrins and apoptosis. Cell 77:477-479.

Schwartz, B., Avivi, C. and Lamprecht, S.A. (1991) Isolation and characterization of normal and neoplastic colonic epithelial cell populations. Gastroenterology 101:697-702.

Schwartzman, R.A. and Cidlowski, J.A. (1993) Apoptosis: the biochemistry and molecular biology of programmed cell death. End. Rev 14:133-151.

Sellins, K.S. and Cohen, J.J. (1987) Gene modification by γ-irradiation leads to DNA fragmentation in lymphocytes. J. Immunol. 139:3199-3206.

Steller, H. (1995) Mechanisms and genes of cellular suicide. Science 167:1445-1449.

Takeichi, M. (1991) Cadherin cell adhesion receptors as a morphogenetic regulator. Science 251:1451-1455.

Takeichi, M. (1993) Cadherins in cancer: implications for invasion and metastasis. Curr. Opinion Cell Biol. 3:806-811.
Watson, A.J.M. (1995a) Manipulation of cell death - The development of novel strategies for the treatment of gastrointestinal disease. Aliment. Pharmacol. Ther. 9:215-226.
Watson, A.J.M. (1995b) Necrosis and apoptosis in the gastrointestinal tract. Gut 37:165-167.
Whyte, M. and Evan, G. (1995) The last cut is the deepest. Nature 376:17-18.
Wyllie, A.H. (1992) Apoptosis and the regulation of cell numbers in normal and neoplastic tissues: An overview. Cancer Metastasis Rev. 11:95-103.
Wyllie, A.H. (1993) Apoptosis. Br. J. Cancer 67:205-208.
Wyllie, A.H. and Duvall, E. (1992) Cell injury and death. In: Oxford Textbook of Pathology, McGee, J.O'P., Isaackson, P.G., Wright, N.A. (eds), pp 141-152, Oxford University Press, New York.
Williamson, R.C.N. (1978) Intestinal adaptation. New England J. Med. 190:1393-1402; 1444-1450.

Characterisation of a Retroviral Insertion Mutagenesis Protocol to Obtain Mutants with Activated Apoptosis Inhibitory Genes

Leverrier Yann
Thomas Joëlle
Mangeney Marianne
Marvel Jacqueline
Immunologie Cellulaire
LBMC ENS Lyon
CNRS UMR49 INRA LA913
46 allée d'Italie
69364 Lyon cedex 07
France

Introduction

Apoptosis or programmed cell death is a cell autonomous suicide pathway characterised by a reduction in cell volume, chromatin condensation and the activation of an endogenous endonuclease that digest DNA into oligonucleosome length fragments. The cellular machinery necessary for the execution of the death program is present in most cell type in an inactivated form (Raff, 1993) This state of inhibition is dependent on survival signals delivered by exogenous molecules like growth factors or adhesion molecules (Bates, 1994; Harrington, 1994). An increasing number of genes involved in the regulation of apoptosis have been described in the past few years, however the mechanisms responsible for the activation or inactivation of the death program remain poorly understood. The Bcl-2 gene family plays a key role in the regulation of the death program. Over-expression of the Bcl-2 family members such as Bcl-2 or Bcl-XL can inhibit apoptosis induced by removal of survival signals or delivery of a death signal (Oltwai, 1994). To understand the regulation of apoptosis, we would like to identify genes which are involved in the inhibition of apoptosis. These genes could be identified on the basis of

NATO ASI Series, Vol. H 99
Tumor Biology
Regulation of Cell Growth, Differentiation and Genetics in Cancer
Edited by Asterios S. Tsiftsoglou, Alan C. Sartorelli, David E. Housman and T. Michael Dexter

homologies with known apoptosis inhibitors or because their product binds to apoptosis regulatory proteins. Alternatively, genes coding for proteins which do not share these properties could be identified by mutagenesis or "random" cDNA expression. We have used retroviral insertion mutagenesis to identify genes involved in the regulation of apoptosis.

It has been shown in a number of systems that retroviral insertion can induce mutations by disrupting or activating genes. Insertional activation could be the product of the integration of viral-promoter enhancer elements contained within the long terminal repeat (LTR) sequences in the vicinity of a silent gene, resulting in the increased transcription of that gene (Stocking, 1993).

We have used the murine interleukin-3 (IL-3) dependent bone marrow derived BAF-3 cell line to identify genes involved in apoptosis inhibition. These cells grow in culture in the presence of IL-3 and die by apoptosis upon IL-3 removal (Rodriguez-Tarduchy, 1990)

In this study, we describe the retroviral insertion mutagenesis protocol and the selection steps which have allowed us to obtain apoptosis resistant mutants.

Material and methods

Cell culture and reagents

BAF-3 cells were maintained in Dulbecco's modified Eagle's medium (DMEM) containing 6% foetal calf serum (FCS, Boehringer Mannheim), 2 mM L-glutamine (Gibco BRL) and 5 % WEHI 3B cell-conditioned medium as a source of IL-3. Cells were grown at a density of $5x10^4$ to $5x10^5$ per ml. To remove IL-3, cells were washed twice in DMEM/FCS 6%.

Cells are cloned in Methocel MC 1,5% (methylcellulose, Fluka) solubilised in DMEM containing 6% FCS and 2 mM L-glutamine.

Propidium iodide (PI) exclusion was used to measure the percentage of viable cells. $5\ 10^3$ to 10^4 cells incubated with PI at a concentration of 5 µg/ml were analysed on a FACScan (Becton-Dickinson) using a FSC/SSC gate to exclude debris.

Retroviral infection of BAF-3 cells

To permit a high rate of infection we have used the amphotropic replication defective recombinant M3Pneo-sup retrovirus that show high levels of expression in murine hematopoietic precursors (kind gift of F. Farzaneh). BAF-3 cells were infected by 48 hours coculture with M3Pneo-sup producing cells in presence of polybren 8 µg/ml (SIGMA).

Results and discussion

The overexpression of apoptosis inhibitory genes only delays the onset of death following growth factor removal. Hence, the duration of IL-3 starvation allowing the survival of mutants BAF-3 cell expressing one of these genes but not of parental cells has to be defined carefully. To define these conditions we took advantage of the BAF-3 cell line over-expressing one prototype of apoptosis inhibitory gene: the survival gene Bcl-2 (BAF/Bcl-2). Fig1a shows that the parental BAF-3 cells start to dye after 8 hours of culture in liquid medium without IL-3 and FCS while the BAF/Bcl-2 are still alive after 24 hours. The BAF/Bcl-2 cells are also more resistant to IL-3 starvation in Methocel MC semi-solid medium compare to BAF-3 cells (fig1b).

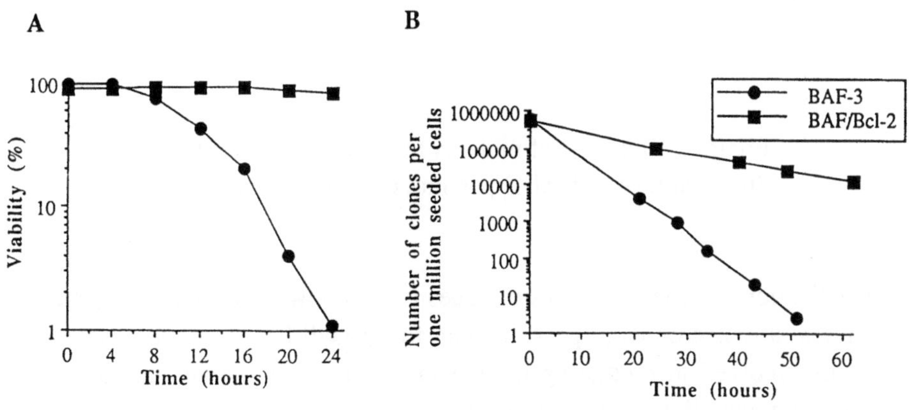

Figure 1: Bcl-2 protects against apoptosis induced by growth factor starvation in liquid (A) and semi-solid medium (B)

After mutagenesis by retroviral infection, the cells were starved of IL-3 and FCS, in liquid medium until more than 90% of the cells were dead. Dead cells were then eliminated by Ficoll density gradient and surviving cells cloned in semi-solid medium in the absence of IL-3 and in the presence of G418. IL-3 was added to the dishes after 30 hours. Individual clones were further expanded in liquid medium 8 days later and survival in absence of IL-3 was monitored for each clone. Fig. 2 shows the result for one mutagenesis experiment. In that experiment, 6 clones out of 57 were found to be strongly resistant to IL-3 starvation i.e. more than 60% of surviving cells 20 hours after growth factor removal.

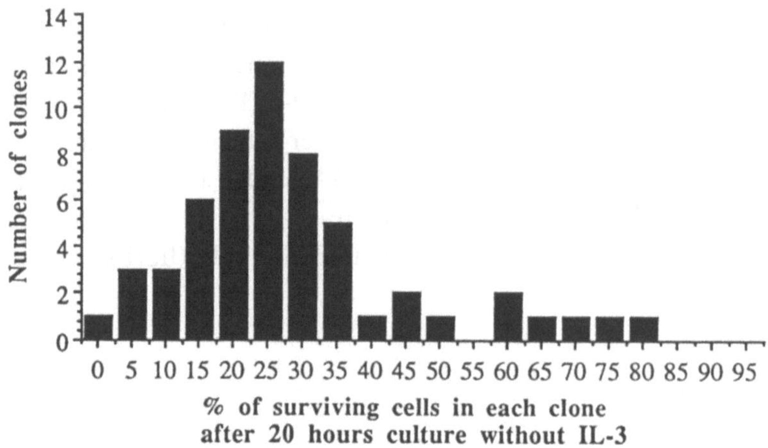

Figure 2: Survival of mutants following IL-3 starvation
(the viability of BAF-3 parental cells is less than 20%)

The activation of at least 3 types of genes could allow these BAF-3 mutants to survive in absence of IL-3: genes coding for growth factors, genes coding for proteins involved in mediating growth factor survival signals and apoptosis inhibitory genes. We have used different drugs known to induce apoptosis to distinguish these different types of mutants. Fig.3 shows that apoptosis induced by etoposide is inhibited by Bcl-2 and by IL-3 while apoptosis induced by staurosporine is only inhibited by Bcl-2 suggesting that only mutants having activated a "survival gene" acquire resistance to staurosporine .

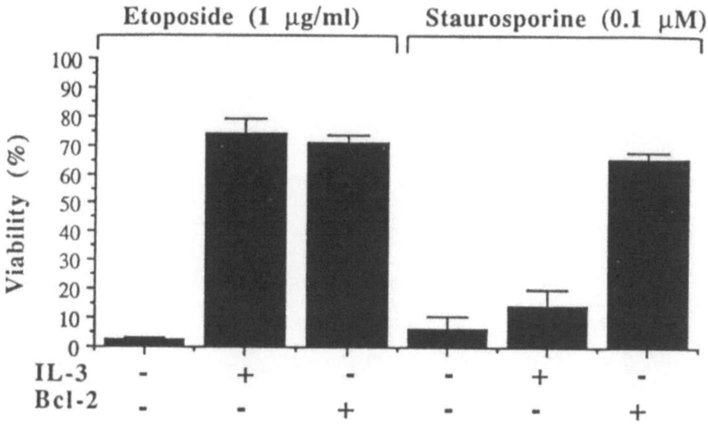

Figure 3: IL-3 and Bcl-2 protect against etoposide but only Bcl-2 protects against staurosporine

Among the clones strongly resistant to IL-3 starvation obtained in 2 mutagenesis experiments, 9 clones were found to be resistant to etoposide and 2 to staurosporine. The later overexpress a survival gene (Bcl-2 in one case and Bcl-X in the other).

References

Bates RC, Buret A, van Helden DF, Horton MA, and Burns GF (1994) J Cell Biol 125: 403-415.

Harrington EA, Bennet MR, Fanidi A, and Evan GI (1994) Embo J 13: 3286-3295.

Oltvai ZN, and Korsemeyer SJ Cell (1994) 79: 189-192

Raff MC, Barres BA, Burne JF, Coles HS, Ishizaki Y, and Jacobson MD (1993). Science 262: 695-700.

Rodriguez-Tarduchy G, Collins MK,and Lopez-Rivas RA (1990) Embo J 9: 2997-3002

Stocking C, Bergholz U, Friel J, Klingler K, Wagener T, Starke C, Kitamura T, Miyajima A, and Ostertag W (1993) Growth Factors 8: 197-209.

Structure-Function Studies of Murine MIP-2, the Homologue of Melanoma Growth Stimulating Activity/gro-α and IL-8

Elias Lolis and L. Fred Jerva
Department of Pharmacology
Yale University School of Medicine
New Haven, CT 06510

Abstract

Members of the chemokine superfamily of proteins, particularly interleukin-8 and melanoma growth stimulating activity/gro-α, are autocrine growth factors for human melanoma cells. The expression of these proteins correlates with the metastatic potential of human melanoma cells in nude mice. An understanding of the molecular mechanism of action of these proteins may provide the basis for development of anti-melanoma compounds.

Two IL-8 receptors have been cloned and characterized. The type A receptor binds only IL-8 with high affinity, whereas the type B receptor binds IL-8, gro-α, neutrophil activating peptide-2 (NAP-2), the KC gene product, and macrophage inflammatory protein-2 (MIP-2) with high affinity. Macrophage inflammatory protein 2 (MIP-2) is believed to be the murine homologue of IL-8 and the three related proteins gro-α, -β, and -γ. MIP-2 has been crystallized and the determination of its three-dimensional structure is in progress. Receptor binding profiles and analysis of sequence and structures are used to determine the receptor binding residues of these chemokines. Mutational analysis of MIP-2 and other chemokines supports the hypothesis that the amino terminus is involved in receptor binding. The information generated from these approaches will be used to model the ligand-receptor complex and understand how these proteins activate the receptor.

NATO ASI Series, Vol. H 99
Tumor Biology
Regulation of Cell Growth, Differentiation and Genetics in Cancer
Edited by Asterios S. Tsiftsoglou, Alan C. Sartorelli, David E. Housman and T. Michael Dexter

Introduction

Proteins that are involved in initiating the immune and inflammatory response have recently been implicated in cancer. These proteins are members of the chemokine superfamily (Figure 1), which is characterized by polypeptides that are 70-100 residues in length and have four conserved cysteines (Oppenheim, et al., 1991). The superfamily has been separated into two families based on the local sequence of the first two cysteines. In the α-chemokine (C-X-C) family, the two cysteines are separated by a single intervening amino acid, and in the β-chemokine (C-C) family, the two cysteines are adjacent to one

CHEMOKINE SUPERFAMILY

α-Chemokine Family

mMIP-2	AVVASELR**C**Q**C**LKT-LPRVDFKNIQSLSVTPPGPH**C**AQTEVIATLKG-GQKV-**C**LDPEAPLVQKIIQKILNKGKAN
hIL-8	...RSAKELR**C**Q**C**IKTYSKPFHPKFIKELRVIESGPH**C**ANTEIIVKLSD-GREL-**C**LDPKENWVQRVVEKFLKRAENS
hgro-α	ASVATELR**C**Q**C**LQT-LQGIHPKNIQSVNKKSPGPH**C**AQTEVIATLKN-GRKA-**C**LNPASPIVKKIIEKMLNSDKSN
hgro-β	APLATELR**C**Q**C**LQT-LQGIHLKNIQSVKVKSPGPH**C**AQTEVIATLKN-GQKA-**C**LNPASPMVKKIIEKMLKNGKSN
hgro-γ	ASVVTELR**C**Q**C**LQT-LQGIHLKNIQSVNVRSPGPH**C**AQTEVIATLKN-GKKV-**C**LNPASPMVQKIIEKILNKGSTN
hagro-α	APVANELR**C**Q**C**LQT-MTGVHLKNIQSLKVTPPGPH**C**TQTEVIATLKN-GQEA-**C**LNPEAPMVQKIVQKMLKSGIRK
mgro-α	APIANELR**C**Q**C**LQT-MAGIHLKNIQSLKVLPSGPH**C**TQTEVIATLKN-GREA-**C**LDPEAPLVQKIVQKMLKGVPK
bPF-4	EGGEDKDLQ**C**V**C**VKT-TSGINPRHISSLEVIGAGTH**C**PSPQLLATKKT-GRKI-**C**LDQQRPLYKKILKKLDGDES
hPF-4	EAEEDGDLQ**C**L**C**VKT-TSQVRPRHITSLEVIKAGPH**C**PTAQLIATLKN-GRKI-**C**LDLQAPLYKKIIKKLLES
rPF-4	...ESDGDLS**C**V**C**VKTSSRGIHLKRITSLEVIKAGPH**C**AVPQLIATLKN-GSKI-**C**LDRQVPLYKKIIKKLLES
muKC	.GAPIANELR**C**Q**C**LQT-MAGIHLKNIQSLKVLPSGPH**C**TQTEVIATLKN-GREA-**C**LDPEAPLVQKIVQKMLKGVPK
hPBP	...DLYAELR**C**M**C**IKT-TSGIHPKNIQSLEVIGKGTH**C**NQVEVIATLKD-GRKI-**C**LDPDAPRIKKIVQKKLAGDESAD
TG	...DLYAELR**C**M**C**IKT-TSGIHPKNIQSLEVIGKGTH**C**NQVEVIATLKD-GRKI-**C**LDPDAPRIKKIVQKKLAGDESAD
NAP-2	AELR**C**M**C**IKT-TSGIHPKNIQSLEVIGKGTH**C**NQVEVIATLKD-GRKI-**C**LDPDAPRIKKIVQKKLAGDESAD
hIP-10	VPLSRTVR**C**T**C**ISISNQPVNPRSLEKLEIIPASQF**C**PRVEIIATMKKKGEKR-**C**LNPESKAIKNLLKAVSKEMSKRSP
C9E3	...KMGNELR**C**Q**C**ISTHSKFIHPKSIQDVKLTPSGPH**C**KNVEIIATLKD-GREV-**C**LDPTAPWVQLIVKALMAKAQLNSDAPL
rCINC	APVANELR**C**Q**C**LQT-VAGIHFKNIQSLKVMPPGPH**C**TQTEVIATLKN-GREA-**C**LDPEAPMVQKIVQKMLKGVPK
rCINC-2	RELR**C**Q**C**LKT-LPRVDFENIQSLTVTPPGPH**C**TQTEVIATLKD-GQEV-**C**LNPQAPRLQKIIQKLLKSPSL

β-Chemokine Family

mMIP-1α	APYGADTPTA**C**-**C**FSY-SRKIPRQFIVDY-FETSSL-**C**SQP-GVIF-LTKRNRQI**C**ADSKETWVQEYITDLELNA
mMIP-1β	APMGSDPPTS**C**-**C**FSYTSRQLHRSFVMDY-YETSSL-**C**SKP-AVVF-LTKRGRQI**C**ANPSEPWVTEYMSDLELN
hMIP-1α	ASLAADTPTA**C**-**C**FSYTSRQIPQNFIADY-FETSSQ-**C**SKP-GVIF-LTKRSRQV**C**ADPSEEWVQKYVSDLELSA
hMIP-1β	APMGSDPPTA**C**-**C**FSYTARKLPRNFVVDY-YETSSL-**C**SQP-AVVF-QTKRSKQV**C**ADPSESWVQEYVYDLELN
hG-26	APMGSDPPTA**C**-**C**FSYTARKLPRNFVVDY-YETSIL-**C**SQP-AVVF-QTKRSKQV**C**ADPSETWVQEYVYDLELN
mSIS	APMGSDPPTS**C**-**C**FSYTSRQLHRSFVMDY-YETSSL-**C**SKP-AVVF-LTKRGRQI**C**ANPSEPWVTEYMSDLELN
hRANTES	ASPYSSDTTP**C**-**C**FAYIARPLPRAHIKEY-FYTSGK-**C**SNP-AVVF-VTRKNRQV**C**ANPEKKWVREYI--NSLEMS
mTY5	APYGADTPTA**C**-**C**FSY-SRKIPRQFIVDY-FETSSL-**C**SQP-GAIF-LTKRNRQI**C**ADSKETWVQEYITDLELNA
mTCA3	KSMLTVSNS**C**-**C**LNTLKKELPLKFIQCYRKMGSS--**C**PDPPAVVFRLNK-GRES**C**ASTNKTWVQNH-L-KKVNPC
hMCP-1	PDAINAPVT**C**-**C**YNFTNRKISVQRLASYRRITSSK-**C**PKE-AVIF-KTIVAKEI**C**ADPKQKWVQDSMDHLDKQTQTP..
mJE	LT**C**-**C**YSFTSKMIPMSRLESYKRITSSR-**C**PKE-AVVF-VTKLKREV**C**ADPKKEWVQTYIKNLDRNQMRS

Figure 1. The amino acid sequences of proteins in the chemokine superfamily. Proteins above the dashed line belong to the α-chemokine family, whereas those below the line are in the β-chemokine family. Conserved cysteines are shown in bold. Legend: b, bovine; c, chicken, ha, hamster; m, mouse; r, rat.

another. The segregation of the superfamily into two families has functional significance. Proteins in the α-chemokine family act almost exclusively on neutrophils. Proteins in the β-chemokine family exert their effects on macrophages, monocytes, T-cells, basophils, and eosinophils, but not neutrophils. Many of the proteins in the α-chemokine family, particularly interleukin-8 (IL-8), melanoma growth stimulating activity/gro-α, and macrophage inflammatory protein-2 (MIP-2), have been found to be growth factors for melanoma cells (Schadendorf, et al., 1993; Richmond, et al., 1988; Sherry, personal communication). The following discussion will review the activities of these proteins and analyze their sequence and three-dimensional structures for the purpose of determining the interactions with cell surface receptors that lead to signal transduction. It is hoped that this information will eventually be used for the design of drugs for melanoma as well as inflammatory disorders.

α-Chemokine Activities

Interleukin-8 is the best characterized protein in the chemokine superfamily. Its function in inflammation is well-established. Its inflammatory activities are initiated by the diffusion of IL-8 from the site of tissue injury to the vasculature where it induces neutrophils to express Mac-1, the CD11b component of the CD11b/CD18 adhesion complex (Detmers, et al., 1990). The expression of Mac-1 on neutrophils provides a ligand for ICAM-1 and ICAM-2 molecules expressed on endothelial cells (Lo, et al., 1989). The interactions between neutrophils and endothelial cells on the surface of a blood vessel prevent the continued circulation of neutrophils in the blood and allow neutrophils to migrate across the endothelial boundary (Huber, et al., 1991). Once out of the blood vessel, neutrophils migrate in response to the IL-8 concentration gradient to the site of infection or tissue injury (Lindley, et al., 1988). There is now sufficient evidence for other members of the chemokine superfamily to invoke a similar general mechanism to explain the recruitment of cells of the immune system to injured sites.

In light of the role of chemokines in initiating the immune and inflammatory responses, it is not a surprise to find that IL-8, gro-α, and MIP-2 among other chemokines are implicated in inflammatory diseases, such as arthritis, psoriasis, and glomerulonephiritis, where the secretion of these chemokines is unregulated (Kasama, et al., 1995; Schroder, et al., 1992; Tang, et al., 1995). Chemokine expression, however, also appears to be involved in the progression to malignant forms of cancer. For example, ultraviolet UV-B radiation increases the aggressiveness of human cutaneous melanoma for growth and metastasis by inducing IL-8 mRNA and protein secretion in cells which normally have negligible expression of IL-8 (Singh, et al., 1995). The constitutive secretion of IL-8 and gro-α (and presumably other chemokines) continues in malignant melanoma cells (Schadendorf, et al., 1993; Bordoni, et al., 1990). Clinically, serum IL-8 is elevated in patients with metastatic melanoma and correlates with tumor load (Scheibenbogen, et al., 1995). The metastatic potential of human melanoma cell lines in nude mice correlates with IL-8 expression (Singh, 1994). From these studies, a hypothesis has been formulated regarding the role of IL-8 and other chemokines in cancer. The induction of these proteins in cutaneous cells results in the continuous presence of a mitogen which supports tumor growth. The correlation with metastasis suggests that chemokines are involved in establishing new tumor sites. Involvement in metastasis may be the result of two *in vitro* activities that have been described for IL-8: (1) it induces the migration of melanoma cells (Wang, et al., 1990); and (2) it is an angiogenic factor (Koch, et al., 1992). The release of IL-8 from sites of inflammation may lead to secondary localization of tumors at these sites. Additionally, constitutive release of an angiogenic factor from malignant melanoma cells results in the formation of blood vessels for transporting nutrients and supporting the growth of the tumor at the new location. Recent clinical data also support a role for IL-8 in metastasis: IL-8 levels are higher in patients with colon cancer who have lung and liver metastases than in those patients that have no apparent metastases (Ueda, et al., 1994).

α-Chemokine Receptors

Although IL-8 is the best characterized protein in the α-chemokine family, it is clear from molecular studies that other proteins in the family will probably have similar effects. Direct binding experiments show that gro-α, β, γ, neutrophil activating peptide-2 (NAP-2), the KC gene product, and MIP-2 all compete with IL-8 for binding to human neutrophils (Oppenheim, et al., 1991; Moser, et al., 1991; Petersen, et al., 1994; Geiser, et al., 1993; Schumacher, et al., 1992). Thus far, two receptors for human IL-8 have been cloned (Holmes, et al., 1991; Murphy and Tiffany, 1991). These proteins are members of the G protein-coupled receptor superfamily. The two receptors are 77% identical in sequence with the N-terminal region being the area that is most dissimilar. Despite the large sequence identity between the two receptors, the binding profiles for ligands are very different. The type A IL-8 receptor is specific for only IL-8 with a dissociation constant between 0.5-5 nM (Holmes, et al., 1991). Other members of the α-chemokine family have no affinity or weak affinity for this receptor. MIP-2, for example, has a dissociation constant for the type A IL-8 receptor of about 100 nM (Figure 2). The type B IL-8 receptor, however, has a much broader binding profile. MIP-2, NAP-2, the KC gene product, and gro-α each display high affinity binding for this

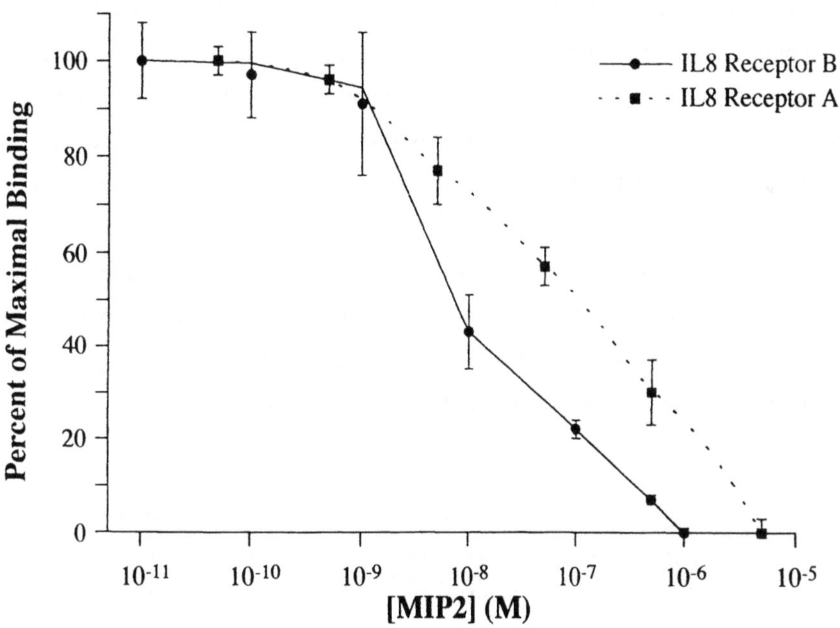

Figure 2. MIP-2 competition studies with COS-7 cells transfected with either the type A or type B IL-8 receptors. Radiolabeled IL-8 was competed with increasing amounts of recombinant MIP-2.

receptor (Lee, et al., 1992; Cerretti, et al., 1993; Bozic, et al., 1995). Furthermore, many of these proteins are able to activate the signal transduction cascade associated with IL-8.

Structural Analysis of α-Chemokines

An analysis of the sequences of the five proteins (IL-8, gro-α, NAP-2, KC, MIP-2) that each bind and activate the type B IL-8 receptor can be used to formulate a hypothesis regarding the receptor binding residues of these proteins. It is likely that the receptor binding determinants for each protein is composed of residues that are identical or highly conserved in all five chemokines. Figure 3 displays the sequence alignment of the five proteins and focuses on the residues that are identical. Nineteen of approximately 70 residues are strictly conserved and are found to be scattered along the entire length of the polypeptide. Among these 19 residues are the four cysteines which define the superfamily and are known to be involved in disulfide bonds. Of the remaining residues, three are either glycine or proline which normally have structural roles in proteins, five constitute residues such as leucine or isoleucine that are normally found in the hydrophobic interior of proteins, one is threonine, and there are six charged residues which normally account for 25% of protein-protein interactions (Janin and Chothia, 1990). Although it is difficult to conclude whether any or all of these residues form part of the receptor binding site, this analysis allows residues that are *likely* to be involved in receptor binding to be identified and targeted for mutational analysis.

```
mMIP-2      AVVASELRCQCLKT-LPRVDFKNIQSLSVTPPGPHCAQTEVIATLKG-GQKV-CLDPEAPLVQKIIQKILNKGKAN
hIL-8    ...RSAKELRCQCIKTYSKPFHPKFIKELRVIESGPHCANTEIIVKLSD-GREL-CLDPKENWVQRVVEKFLKRAENS
hgro-α      ASVATELRCQCLQT-LQGIHPKNIQSVNKKSPGPHCAQTEVIATLKN-GRKA-CLNPASPIVKKIIEKMLNSDKSN
muKC     .GAPIANELRCQCLQT-MAGIHLKNIQSLKVLPSGPHCTQTEVIATLKN-GREA-CLDPEAPLVQKIVQKMLKGVPK
 NAP-2           AELRCMCIKT-TSGIHPKNIQSLEVIGKGTHCNQVEVIATLKD-GRKI-CLDPDAPRIKKIVQKKLAGDESAD
```

Figure 3. Sequence alignment of chemokines that bind to the IL-8 type B receptor with high affinity. Identical amino acids for all five proteins are shown in bold.

The results of the above sequence analysis can be viewed in a structural context as the three-dimensional structures of IL-8 (Baldwin, et al., 1991; Clore, et al., 1990), gro-α (Fairbrother, 1994; Kim, 1994), and NAP-2 (Malkowski, 1995) have been determined. MIP-2 has been crystallized (Lolis, et al., 1992) and the determination of its structure by X-ray crystallography and NMR is in progress. The backbone structure of IL-8 is shown in Figure 4. Each monomer is composed of an extended amino terminus followed by two β-strands and a carboxy terminal α-helix. In IL-8, two monomers associate such that the β-strands form a 4-stranded β-sheet with two antiparallel α-helices above the β-sheet. All other α-chemokines for which high-resolution structures are available also associate. Gro-α and cytokine-induced neutrophil chemoattractant (CINC) (Hanzawa, et al., 1994) form dimers similar to IL-8, and platelet factor-4 (St. Charles, et al., 1989) and NAP-2 (Malkowski, 1995) each associate to form tetramers. Although all of these proteins are dimers or tetramers, it is important to note that the functional form may not correspond to the observed form, as structure determinations are done at higher concentrations than those in the biologically effective range. It is likely that at biologically active nanomolar concentrations, the equilibrium favors the formation of monomers over oligomers. In fact, there is experimental evidence to

(A) **(B)**

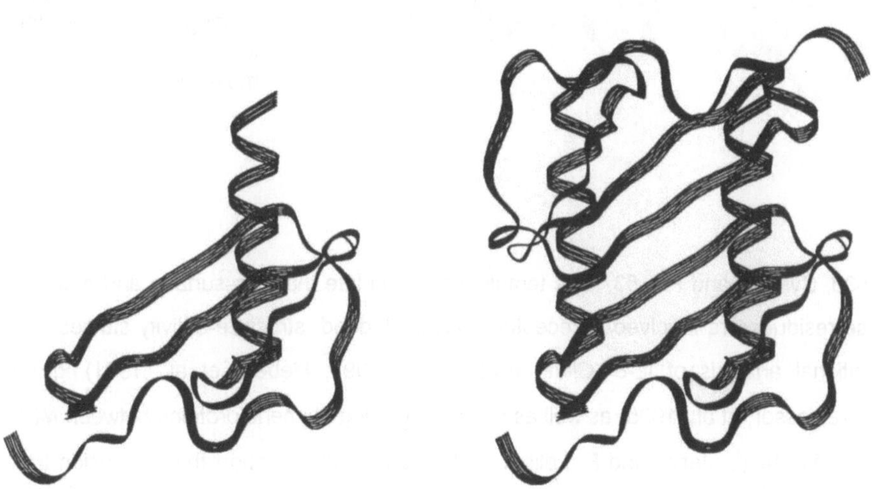

Figure 4. Three-dimensional structure of IL-8 (Clore, et al., 1990). (A) Monomer and (B) dimer.

suggest that the monomeric forms of these proteins are biologically active (Rajarathnam, et al., 1994).

The positions of the identical residues among IL-8, gro-α, MIP-2, the KC gene product, and NAP-2 in the context of the three-dimensional structure is revealing (Figure 5). Although the identical residues occupy positions along the entire length of the polypeptide of the linear sequence, the folding of the protein brings most of these residues together to form a fairly extended surface of about 10 X 20 $Å^2$. This area is comprised of Ala-2, Glu-4, Leu-5, Arg-6, Cys-7, Cys-9, Thr-12, Gly-31, His-33, Cys-34,

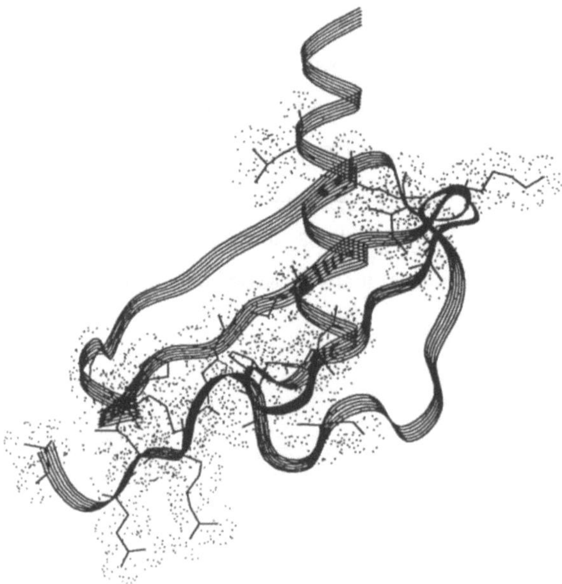

Figure 5. Atoms from the identical residues among IL-8, gro-α, NAP-2, KC, and MIP-2 are highlighted in the context of the backbone structure of IL-8. The exposed surface area generated by these atoms is also shown. The presumed receptor binding site is on the lower left of the molecule.

Glu-38, Cys-50, and Pro-53. It is tempting to speculate that this surface and many of these residues are involved in receptor binding. Indeed, structure-activity studies and mutational analysis of IL-8 (Clark-Lewis, et al., 1991; Hebert, et al., 1991), gro-α (Hesselgesser, et al., 1995) as well as studies involving chimeric proteins between MIP-2 and MIP-1α (L. Jerva and E. Lolis, unpublished results) support the conclusion that the ELR motif at the amino terminus is critical for receptor binding and activation. Whether any of the other identical residues are also involved in receptor binding and activation remains to be determined. It is interesting that other non-identical residues are also important. Conversion of Tyr-28 and Arg-30 in the β-chemokine monocyte chemoattractant protein-1 (MCP-1) to the corresponding IL-8 amino acids (Leu-25 and Val-27) converts this protein to a neutrophil chemoattractant (Beall, et al., 1992). The authors of the study postulate that both of these amino acids are part of the receptor binding groove of IL-8. In a complementary study, Leu-25 of IL-8 was mutated to tyrosine, the amino acid most frequently found at this position for β-chemokines (Lusti-Narasimhan, et al., 1995). This mutant has lowered affinity for the types A and B IL-8 receptors, decreased potency in neutrophil chemotaxis, and exhibits a novel monocyte chemoattractant activity. These studies demonstrate that the receptor binding site of IL-8 extends beyond the residues that are identical.

We have used site-directed mutagenesis and created chimeric proteins between MIP-2 with MIP-1α (unpublished results) to identify specific roles for residues without regard to whether they are identical among all of the type B IL-8 receptor-binding chemokines. In addition to finding that the amino terminus is an important receptor binding determinant, our findings thus far implicate the carboxy terminal α-helix in heparin binding. This observation is consistent with the positively charged cluster of residues involving Lys-64, Arg-60, and Lys-20 which is in close proximity to the helix. What role heparin-binding plays in receptor binding and activation is still under investigation.

While the determination of the receptor binding site of these chemokines is not yet complete, it is sufficiently defined to allow chemists to begin designing compounds to interact with these proteins and prevent receptor binding. This should pave the way for the development of clinically useful drugs for inflammatory disorders and malignant melanoma.

REFERENCES

Baldwin, ET, Weber, IT, St. Charles, R, Xuan, J-C, Apella, E, Yamanda, M, Matsushima, K, Edwards, BFP, Clore, GM, Gronenborn, AM and Wlodower, A (1991) Crystal structure of IL-8: symbiosis of NMR and crystallography. Proc. Natl. Acad. Sci. USA 88:502-506

Beall, CJ, Mahajan, S and Kolattukudy, PE (1992) Conversion of monocyte chemoattractant protein-1 into a neutrophil attractant by substitution of two amino acids. J. Biol. Chem. 267:3455-3459

Bordoni, R, Fine, R, Murray, D and Richmond, A (1990) Characterization of the role of melanoma growth stimulating activity (MGSA) in the growth of normal melanocytes, nevocytes, and malignant melanocytes. J. Cell. Biochem. 44:207-19

Bozic, CR, Kolakowski Jr, LF, Gerard, NP, Garcia-Rodriguez, C, von Uexkull-Guldenband, C, Conklyn, MJ, Breslow, R, Showell, HJ and Gerard, C (1995) Expression and biologic characterization of the murine chemokine KC. J. Immunol. 6049-6057

Cerretti, D, Kozlosky, C, Bos, T, Nelson, N, Gearing, D and Beckmann, M (1993) Molecular characterization of receptors for human Interleukin-8, gro/melanoma growth-stimulatory activity and neutrophil activating peptide-2. Mol. Immunol. 30:359-367

Clark-Lewis, I, Schumacher, C, Baggiolini, M and Moser, B (1991) Structure-activity relationships of interleukin-8 determined using chemically synthesized Analogs. J. Biol. Chem. 266:23128-23134

Clore, MG, Appella, E, Yamada, M, Matsushima, K and Gronenborn, A (1990) Three dimensional structure of interleukin-8 in solution. Biochemistry 24:1689-1696.

Detmers, PA, Lo, SK, Olsen-Egbert, E, Walz, A, Baggiolini, M and Chon, ZA (1990) Neutrophil-activating protein 1/interleukin 8 stimulates the binding activity of the leukocyte adhesion receptor CD11b/CD18 on human neutrophils. J. Exp. Med. 171:1155-1162

Fairbrother, WJ, Reilly, D., Colby, T.J., Hesselgesser, J., Horuk, R (1994) The solution structure of melanoma growth stimulating activity. J. Mol. Biol. 242:252

Geiser, T, Dewald, B, Ehrengruber, MU, Clark-Lewis, I and Baggiolini, M (1993) The interleukin-8-related chemotactic cytokines gro-a, gro-b, and gro-c activate neutrophil and basophil leukocytes. J. Biol. Chem. 1993:15419-15424.

Hanzawa, H, Haruyama, H, Watanabe, K and Tsurufuji, S (1994) The three-dimensional structure of rat cytokine CINC/Gro in solution by mononuclear 3D NMR. FEBS Letters 354:207-212

Hebert, CA, Vitangcol, RV and Baker, JB (1991) Scanning mutagenesis of IL-8 identifies a cluster of residues required for receptor binding. J. Biol. Chem. 266:18989-18994

Hesselgesser, J, Chitnis, CE, Miller, LH, Yansura, DG, Simmons, LC, Fairbrother, WJ, Kotts, C, Wirth, CG-C, Beth L and Horuk, R (1995) A mutant of melanoma growth stimulating activity does Not activate neutrophils but blocks erythrocyte invasion by malaria. J. Biol. Chem. 270:11472-11476

Holmes, WE, Lee, J, Kuang, WJ, Rice, GC and Wood, WI (1991) Structure and functional expression of a human interleukin-8 receptor. Science 253:1278-1280

Huber, A, Todd, S and Weiss, S (1991) Regulation of transendothelial neutrophil migration by endogenous IL-8. Science 254:99-102

Janin, J and Chothia, C (1990) The Structure of protein-protein recognition sites. J. Biol. Chem. 265:16027-16030

Kasama, T, Streiter, R, Lukacs, N, Lincoln, P, Burdick, M and Kunkel, S (1995) Interleukin-10 expression and chemokine regulation during the evolution of murine type II collagen-induced arthritis. J. Clin. Inv. 95:2868-2876

Kim, K-S, Clark-Lewis, I, Sykes, BD (1994) Solution structure of GRO/melanoma growth stimulating activity determined by ^{1}H NMR spectroscopy. J. Biol. Chem. 269:32909-32915

Koch, AE, Polverini, PJ, Kunkel, SL, Harlow, LA, DiPietro, LA, Elner, VM, Elner, SG and Strieter, RM (1992) Interleukin-8 as a macrophage-derived mediator of angiogenesis. Science 258:1798-1802

Lee, J, Horuk, R, Rice, G, Bennett, G, Camerato, T and Wood, W (1992) Characterization of two high affinity human interleukin-8 receptors. J. Biol. Chem 267:16283-16287

Lindley, I, Aschauer, H, Seifert, J-M, Lam, C, Brunowsky, W, Kownatzki, E, Thelen, M, Peveri, P, Dewald, B, Von Tscharner, V, Walz, A and Baggiolini, M (1988) Synthesis and expression in Escherichia coli of the gene encoding monocyte-derived neutrophil-activating factor: biological equivalence between natural and recombinant neutrophil-activating factor. Proc. Natl. Acad. Sci. USA 85:9199-9203

Lo, SK, Van Seventer, GA, Levin, SM and Wright, SD (1989) Two leukocyte receptors (CD11a/CD18 and CD11b/CD18) mediate transient adhesion to endothelium by binding to different ligands. J. Immunol. 143:3325-29.

Lolis, E, Sweet, RM, Cousens, LS, Tekamp-Olson, P, Sherry, B and Cerami, A (1992) Preliminary crystallographic analysis of murine macrophage inflammatory protein 2. J. Mol. Biol. 225:913-915

Lusti-Narasimhan, M, Power, CA, Allet, B, Alouani, S, Bacon, KB, Mermod, J-J, Proudfoot, AEI and Wells, TNC (1995) Mutation of Leu-25 and Val-27 introduces CC chemokine activity into interleukin-8. J. Biol. Chem. 270:2716-21

Malkowski, MG, Wu, JY, Lazar, JB, Johnson, PH, and Edwards, BFP (1995) The crystal structure of recombinant human neutrophil-activating peptide-2 (M6L) at 1.9 A Resolution. J. Biol. Chem. 270:7088-7087

Moser, B, Schumacher, C, von Tscharne, V, Clark-Lewis, I and Baggiolini, M (1991) Neutrophil-activating Peptide 2 and gro/melanoma growth-stimulatory activity interact with neutrophil-activating peptide 1/interleukin-8 receptors on human neutrophils. J. Biol. Chem. 266:10666-10671

Murphy, B and Tiffany, HL (1991) Cloning of complementary DNA encoding a functional interleukin-8 receptor. Science 253:1280-1283

Oppenheim, JJ, Zachariae, COC, Mukaida and Matsushima, K (1991) Properties of the novel proinflammatory supergene "intercrine" cytokine family. Annu. Rev. Immunol. 9:617-48

Petersen, F, Flad, H and Brandt, E (1994) Neutrophil-activating peptides NAP-2 and IL-8 bind to the same sites on neutrophils but interact in different ways. J. Immunol. 152:2467-2478

Rajarathnam, K, Sykes, BD, Kay, CM, Dewald, B, Geiser, T, Baggiolini, M and Clark-Lewis, I (1994) Neutrophil activation by monomeric interleukin-8. Science 90-92

Richmond, A, Balentien, E, Thomas, HG, Flaggs, G, Barton, DE, Spiess, J, Bordoni, R, Francke, U and Derynck, R (1988) Molecular characterization and chromosomal mapping of melanoma growth stimulatory activity, a growth factor structurally related to beta-thromboglobulin. EMBO J. 7:2025-33

Schadendorf, D, Moller, A, Algermissen, B, Worm, M, Sticherling, M and Czarnetzki, BM (1993) IL-8 produced by human malignant melanoma cells in vitro is an essential autocrine growth factor. J. Immunol. 15:2667-75

Scheibenbogen, C, Mohler, T, Haefele, J, Hunstein, W and Keilholz, U (1995) Serum interleukin-8 (IL8) is elevated in patients with metastatic melanoma and correlates with tumor load. Melanoma Res. 5:179-181.

Schroder, JM, Gregory, H, Young, J and Christophers, E (1992) Neutrophil-activating proteins in psoriasis. J. Inv. Dermatol. 98:241-47.

Schumacher, C, Clark-Lewis, I, Baggiolinin, M and Moser, M (1992) High- and low-affinity binding of gro-a and neutrophil-activating peptide 2 to interleukin 8 receptors on human neutrophils. Proc. Natl. Acad. Sci. USA 89:10542-10546

Singh, RK, Gutman, M, Reich, R and Bar-Eli, M (1995) Ultraviolet B irradiation promotes tumorigenic and metastatic properties in primary cutaneous melanoma via induction of interleukin 8. Cancer Res. 55:3669-3674

Singh, RK, Gutman, M, Radinsky, R, Bucana, CD, and Fidler, IJ (1994) Expression of interleukin 8 correlates with the metastatic potential of human melanoma cells in nude mice. Cancer Res. 54:3242-47.

St. Charles, R, Walz, DA and Edwards (1989) BFP The three-dimensional structure of bovine platelet factor 4 at 3.0Å Resolution. J. Biol. Chem. 264:2092-2099

Tang, WW, Yin, SM, Wittwer, AJ and Qi, MY (1995) Chemokine gene expression in anti-glomerular basement membrane antibody glomerulonephritis. Am. J. Physiol. 38:F323-F330

Ueda, T, Shimada, E and Urakawa, T (1994) Serum levels of cytokines in patients with colorectal cancer: possible involvement of interleukin-6 and interleukin-8 in hematogenous metastasis. J. Gestroenterol. 29:423-429.

Wang, JM, Taraboletti, G, Matsushima, K, Van Damme, J and Mantovani, A (1990) Induction of haptotactic migration of melanoma cells by neutrophil activating protein/interleukin-8. Biochem. Biophys. Res. Comm. 169:165-170.

Effect of Retinoic Acid on CD14 Expression and the Intracellular Distribution of Inositol Following LPS-Stimulation in U937 Cells

Nathalie Ménager and Dino Rotondo
Department of Immunology
University of Strathclyde
Glasgow G4 0NR, Scotland, U.K.

ABSTRACT: The effect of bacterial endotoxin (LPS) on the intracellular distribution of [^{3}H]-inositol in human monocytic cells was studied. U937 cells were exposed to retinoic acid (RA) which has been shown to induce differentiation towards a monocytic phenotype. Inositol distribution in normal peripheral blood monocytes (Mø) was also studied as a comparison. Incubation of RA-treated U937 cells and blood Mø with LPS was followed by increased levels of [^{3}H]-inositol phosphates and a decrease in [^{3}H]-inositol incorporation into organic compounds. In parallel, retinoic acid upregulated the expression of the monocytic cell-surface marker CD14 in U937 cells. It is not certain that the onset of responsiveness to LPS in the generation of inositol phosphates and the increased expression of CD14 are related. However, it would appear that RA-treated U937 cells may be a useful model of blood monocytes for *in vitro* experimentation.

INTRODUCTION

It is well accepted that bacterial endotoxin (lipopolysaccharide, LPS) is one of the most potent natural stimulators of blood monocytes (Mø) which play an important role in immunological and inflammatory responses. Tobias *et al.* (1986) have identified an LPS binding protein (LBP) which is believed to interact with LPS to induce cell activation. It is now thought that the major cell surface marker for the monocytic phenotype, CD14, may function as a receptor for the LPS-LBP complex as CD14 binds to these structures (Wright *et al.*, 1990). However, the CD14 / LPS-LBP complex does not cross the plasma membrane and thus, little is known about the triggering of the intracellular signal. The role of phospholipids in cell signalling, has in contrast, been extensively investigated, especially the hydrolysis of phosphatidylinositol 4,5-bisphosphate (PIP_2) to yield diacylglycerol (DAG) and inositol 1,4,5-trisphosphate (IP_3). Thus, studying intracellular changes in [^{3}H]-inositol distribution following LPS-stimulation could help to understand the intracellular signalling pathways responsible for cell activation. There are two basic approaches to studying LPS-induced monocytic cell activation a) using peripheral blood Mø or b) using a monocytic cell line. It has been shown that physiological concentrations of *all-trans* retinoic acid can induce monoblast-like human histiocytic lymphoma cells, such as U937 cells, to differentiate towards a mature monocytic phenotype (Olsson & Breitman, 1982). Here, we investigated the effect of RA on the incorporation of [^{3}H]-inositol into inositol lipids after LPS-stimulation, and in parallel the effect of RA on CD14 expression in U937 cells; compared to peripheral blood Mø.

NATO ASI Series, Vol. H 99
Tumor Biology
Regulation of Cell Growth, Differentiation and
Genetics in Cancer
Edited by Asterios S. Tsiftsoglou, Alan C. Sartorelli,
David E. Housman and T. Michael Dexter

METHODS

Reagents and medium: All reagents were 'AnalaR' except for the following: RPMI 1640, HBSS and foetal calf serum (FCS) were purchased from Gibco (Scotland). Retinoic acid, Ficoll-histopaque 1077 and anti-CD14 were obtained from Sigma (U.K.), and [^{3}H]-inositol was purchased from Amersham (U.K.).

Cell culture: U937 cells (ECACC no 87010802, U.K.) were cultured in RPMI 1640 (supplemented with 10 % FCS) at 37 °C, 5 % CO_2 and 100 % humidity. For all experiments, U937 cells were resuspended at a concentration of 2 x 10^5 cells/ml and incubated with 100 nM RA for 4 days at 37 °C. Peripheral blood Mø were obtained from whole blood (Glasgow and West of Scotland Blood Transfusion Service, Carluke, Scotland) by Histopaque density centrifugation followed by adherence to plastic. Cell viability in both U937 cells and blood Mø was greater than 90 % as determinated by Trypan Blue exclusion. Cell preparations were then prelabelled with 10 µCi/ml [^{3}H]-inositol for 20 hours at 37 °C, 5 % CO_2 and 100 % humidity and, at the end of the labelling period, cells were washed twice with HBSS.

Analysis of inositol lipids production upon LPS-stimulation: Cell aliquots (20 x 10^6/ml) were incubated with either RPMI alone or containing 1 µg/ml LPS for a further 2 hours. Incubations were terminated by the addition of cold HBSS and centrifuged at 600 g (5 mins at 20 °C) followed by removal of the supernatant. The cell pellet was resuspended in chloroform : methanol (2:1 v/v) to yield aqueous and organic phases. Radioactivity from the aqueous phase was measured by scintillation counting after anion-exchange chromatography using triethyl-ammonium bicarbonate buffer as eluant to separate anionic radioactivity (inositol phosphates) (Maslanski & Busa, 1990). Radioactivity in the organic phase was measured by scintillation counting after one dimensional thin-layer chromatography to separate inositol lipids as described by Mitchell *et al.* (1986).

Analysis of CD14 cell-surface expression: CD14 expression in U937 cells was estimated by binding of PE-conjugated anti-CD14. Non-specific binding was estimated using a PE-conjugated mouse isotype-matched antibody. Samples were analysed by flow cytometry.

Statistical analysis of the data: Results are expressed as the means of $n = 3 \pm$ s.d. Students t-test was used and difference between control and LPS-treated batch was considered to be significant for p values < 0.05.

RESULTS

U937 cells treated with RA appeared larger in size as observed by light microscopy, cell growth was reduced by 43 - 48 % (as estimated by cell counting) when compared to untreated U937 cells, but the cell viability was not affected and remained greater than 90 % in all experiments. The total uptake of [^{3}H]-inositol into U937 cells pre-exposed to RA was greater than in untreated cells. The level of [^{3}H]-inositol in untreated cells was in the order of 25 - 30 % that in RA-treated cells.

Effect of retinoic acid on inositol phosphate production in monocytic cells after LPS-stimulation: Production of inositol phosphates upon LPS-stimulation was significantly increased in U937 pre-exposed to RA (from 212 ± 10 cpm to 235 ± 12 cpm, $p < 0.05$) as well as in peripheral blood Mø which were not exposed to RA (from 248 ± 16 cpm to 440 ± 114 cpm, $p < 0.05$). However untreated U937 cells were unable to respond to LPS-stimulation by production of inositol phosphates (Figure 1).

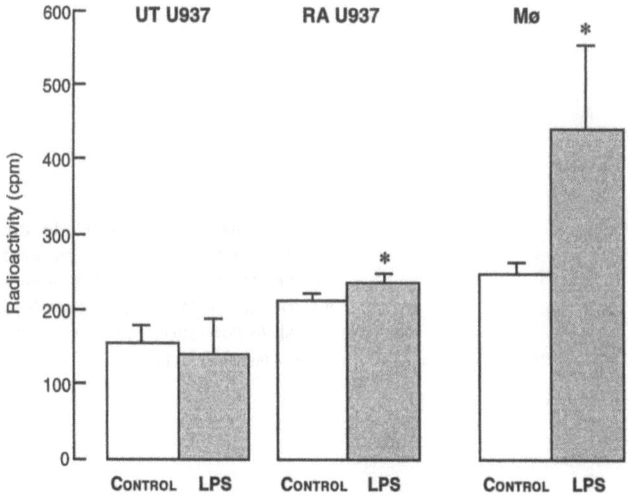

Figure 1: Inositol phosphate production in monocytic cells after incubation with LPS. Aliquots (2 x 10^6 cells) of untreated (UT U937), RA-treated (RA U937) and blood Mø were incubated with RPMI alone or containing 1 µg/ml LPS for 2 hours. Productions of inositol phosphates were determinated by anion-exchange chromatography using triethyl-ammonium bicarbonate buffer as eluant. Values represent means for n = 3 ± s.d., * for $p < 0.05$.

Effect of retinoic acid on the production of inositol lipids in monocytic cells after LPS-stimulation: Incorporation of [^{3}H]-inositol into inositol lipids was significantly decreased in both RA-treated U937 cells (from 1661 ± 37 cpm to 1427 ± 136 cpm, $p < 0.05$) and peripheral blood Mø (from 2595 ± 63 cpm to 2454 ± 54 cpm, $p < 0.05$) after cell-stimulation with LPS. No changes in inositol lipid contents were observed in untreated U937 cells after LPS-stimulation (Figure 2).

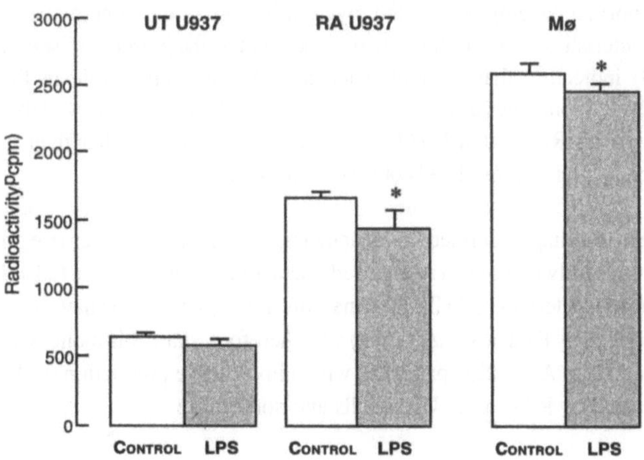

Figure 2: Inositol lipid analysis in monocytic cells after incubation with LPS. Aliquots (2 x 10^6 cells) of untreated (UT U937), RA-treated (RA U937) and blood Mø were incubated with RPMI alone or containing 1µg/ml LPS for 2 hours. Analysis of inositol lipids was determinated by one dimensional thin-layer chromatography. Values represent means for n = 3 ± s.d., * for $p < 0.05$.

Effect of retinoic acid on CD14 expression in U937 cells: In order to indicate cell differentiation in RA-treated U937 cells, CD14 expression, which is associated with maturation to a monocytic cell type, was determinated. Blood Mø expressed high levels of CD14 as expected (100% of cells). In U937 cells, the Relative Fluorescence Intensity increased from 8.3 ± 0.3 in untreated cells to 22.0 ± 1.3 in RA-treated U937 cells (Figure 3).

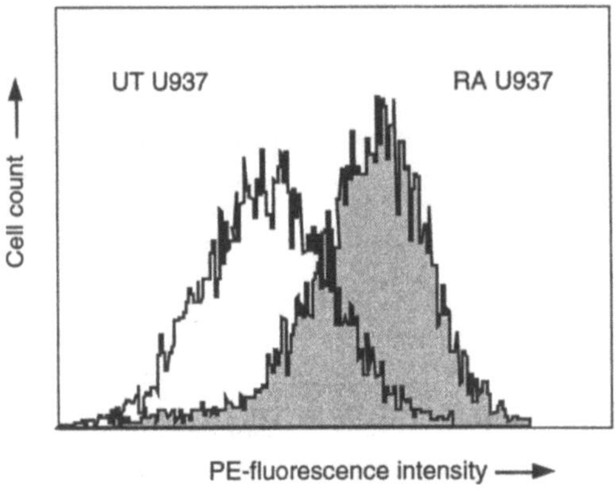

Figure 3: CD14 expression in U937 cells after exposure to RA. U937 cells were pretreated with 100 nM RA for 4 days. CD14 expression was analysed using PE-conjugated monoclonal antibody by flow cytometry. Values represent Relative Fluorescence Intensity.

DISCUSSION

U937 cells had been defined as a monocytic-like cell line by Sundström and Nilsson (1976). In this regard it had been found that U937 cells expressed Fc receptors on their cell surface. Olsson & Breitman (1982) had shown that RA can induce U937 cells to reduce nitroblue tetrazolium (NBT) indicating that RA-treated U937 cells were able to produce superoxide, and thus that RA-treated U937 cells acquired specific monocyte characteristics. In the present experiments, U937 cells were pretreated with physiological concentrations of RA to induce differentiation towards monocytic-like cells. In the normal development of the myeloid lineage in the bone marrow, acquisition of monocytic characteristics has been shown to be dependent on the presence of serum retinoids (Oren *et al.*, 1963) indicating that retinoids such as RA may play a role in the differentiation of myeloid cells to monocytes *in vivo*. In the present study, it was observed that levels of [^{3}H]-inositol taken up by RA-treated U937 cells were considerably higher than in untreated cells and were much similar to those levels observed for blood Mø.

Monocyte activation had been originally defined by Mackaness (1962) as an "increase in bactericidal activity" but more recently it has been recognised that activation in response to LPS results in the release of the cytokine interleukin-1 (IL-1) (Dinarello, 1991). U937 cells have been shown to produce IL-1 as described by Palacios *et al.* (1982) and, therefore, share this monocytic characteristic with normal blood Mø. Activation pathways which involve the production of IL-1 should, therefore, also be shared by RA-treated U937 cells and normal Mø.

Wijelath *et al.* (1988) reported that IL-1 induced inositol phosphate accumulation in murine macrophages. Here, we also studied the accumulation of inositol phosphates and incorporation of [^{3}H]-inositol into the lipid fraction following LPS-stimulation in U937 cells and human Mø. It was found that LPS-stimulation was followed by an increased formation of [^{3}H]-inositol phosphates, and a decrease in [^{3}H]-inositol recovered from the lipid fraction in both RA-treated U937 cells and blood Mø. No changes in radioactivity in inositol phospholipids were observed in untreated U937 cells stimulated with LPS, indicating that only normal Mø and RA-pretreated U937 cells were able to respond to LPS-stimulation with changes in inositol distribution. It is not certain by which mechanisms the cells respond to LPS following exposure to RA. An increase in inositol phosphate formation presumably occurs as a result of the activation of phospholipase C (PLC) (for the hydrolysis of PIP_2 and subsequently the formation of IP_3) or of accessory molecules such as G-proteins which appear to be necessary for PLC activation. It is, therefore, possible that RA may be involved in either a) the induction of a PLC isoform or of an associated regulatory G-protein or b) in the direct or indirect activation of a latent PLC. Alternatively, RA may be involved in the induction of a receptor for LPS, the absence of which would obviously result in a failure to respond. It has been suggested that CD14, the cell surface marker which characterises Mø, is a membrane receptor for LPS-LBP complexes (Wright *et al.*, 1990). It is possible that RA influences U937-cell responsiveness via an effect on the expression of CD14. It was observed that RA increased CD14 expression in U937 cells. In this regard RA, by upregulating CD14 on U937 cells, may facilitate the interaction between LPS/LBP and CD14 and allow an intracellular response to occur in the presence of LPS, including changes in the intracellular distribution of [^{3}H]-inositol. However, in the present study no direct evidence was presented to directly relate the upregulation of CD14 expression and the onset of responsiveness to LPS by changes in inositol phosphates.

In conclusion, these observations indicate that U937 cells, when exposed to RA, are a useful model for studying peripheral blood Mø *in vitro*.

ACKNOWLEDGMENTS: The authors are grateful to Dr. G. Templeton from the Glasgow and West of Scotland Blood Transfusion Service for the supply of human blood, and to the ECACC for their kind gift of U937 cells. N.M. is in receipt of a John Anderson Award from the University of Strathclyde.

REFERENCES

Dinarello CA (1991) Interleukin-1 and interleukin-1 antagonism. Blood 77 : 1627-1652

Mackaness GB (1962) Cellular resistance to infection. J. Exp. Med. 116 : 381-406

Maslanski JA & Busa WB (1990) in Methods in inositide research. pp113-126. Raven press Oxford

Mitchell KT, Ferrell JE & Huestis WH (1986) Separation of phosphoinositides and other phospholipids by two-dimensional thin-layer chromatography. Anal. Biochem. 158 : 447-453

Olsson IL & Breitman TR (1982) Induction of differentiation of the human histiocytic lymphoma cell line U-937 by retinoic acid and cyclic adenosine 3':5'-monophosphate-inducing agents. Cancer Res. 42 : 3924-3927

Oren R, Farnham AE, Saito K, Milofsky E & Karnovsky ML (1963) Metabolic patterns in three types of phagocytizing cells. J. Cell Biol. 17 : 487-502

Palacios R, Ivhed I, Sideras P, Nilsson K, Sugawara I & Fernandez C (1982) Accessory function of human tumor cell lines. Production of interleukin-1 by the human histiocytic lymphoma cell line U-937. Eur. J. Immunol. 12 : 895-899

Tobias PS, Soldau K & Ulevitch RJ (1986) Isolation of a lipopolysaccharide-binding acute phase reactant from rabbit serum. J. Exp. Med. 164 : 777-793

Sundström C & Nilsson K (1976) Establishment and characterization of a human histiocytic lymphoma cell line (U-937). Int. J. Cancer 17 : 565-577

Wijelath ES, Kardasz AM, Drummond R & Watson J (1988) Interleukin-one induced inositol phospholipid breakdown in murine macrophages : possible mechanism of receptor activation. Biochem. Biophys. Res. Commun. 152 : 392-397

Wright SD, Ramos RA, Tobias PS, Ulevitch RJ & Mathison JC (1990) CD14 serves as the cellular receptor for complexes of lipopolysaccharide binding protein. Science 249 : 1431-1433

Regulation of the Proto-oncogene Product c-Jun by Phosphorylation-Mediated Intramolecular Signaling

Athanasios G. Papavassiliou
Differentiation Programme
European Molecular Biology Laboratory
Postfach 10.2209, Meyerhofstrasse 1
D-69012 Heidelberg
Germany

The biological function of the proto-oncogene product c-Jun as a transcription factor (AP-1 constituent) has been reported to be regulated by complex phosphorylation events in different parts of the protein. The first phosphorylation event that was characterized in detail involved a cluster of COOH-terminal residues in direct neighborhood to the DNA-binding domain. It was shown that threonine 231, serine 243, and serine 249 (referred to here as 'COOH-terminal cluster') are phosphorylated in resting cells with low c-Jun activity, whereas in cells that were stimulated by the addition of growth factors or phorbol ester tumor promoters (e.g. TPA, an agonist of protein kinase C (PKC)) or by introduction of activated alleles of Ras this phosphorylation decreases [Boyle *et al.*, 1991; Papavassiliou *et al.*, 1992]. Comparison of the DNA-binding potential of phosphorylated and non-phosphorylated forms of c-Jun revealed that these phosphorylations preclude DNA binding [Boyle *et al.*, 1991; Papavassiliou *et al.*, 1992]. The mechanism of this inhibition is unclear but it seems plausible that the extra negative charges on c-Jun in close proximity to the DNA-binding domain would cause an electrostatic repulsion between the protein molecule and the acidic phosphate groups of the target DNA sequence [Hurley *et al.*, 1990]. It was postulated that dephosphorylation of c-Jun in the COOH-terminal cluster following mitogenic activation of cells was the effect of a signal-induced phosphatase or a down-regulated kinase or both, i.e. a primary event of c-Jun activation [Karin, 1994]. Therefore it was interesting to investigate which kinase mediates the effect in order to unravel the hypothetical signal transduction pathway "from the bottom to the top".

The initial suggestion, that the kinase catalyzing phosphorylation of the c-Jun COOH-terminal cluster was glycogen synthase kinase 3 (GSK3) [Boyle *et al.*, 1991] was abandoned for several reasons, and it is currently believed that casein kinase

NATO ASI Series, Vol. H 99
Tumor Biology
Regulation of Cell Growth, Differentiation and Genetics in Cancer
Edited by Asterios S. Tsiftsoglou, Alan C. Sartorelli, David E. Housman and T. Michael Dexter

II (CKII) phosphorylates c-Jun at these sites [Lin *et al.*, 1992]. The identification of CKII as the Jun COOH-terminal kinase posed a problem with the above scenario, as CKII activity had never been found to be down-regulated by growth factors or Ras; on the contrary, some studies report an augmentation of CKII activity in response to growth stimulation [Carrol *et al.*, 1988]. Similarly, the existence of a regulated phosphatase targeting the c-Jun COOH-terminal region has not been documented. Still more puzzling was the finding that dephosphorylation of the COOH-terminal cluster can occur as a consequence of DNA binding, even in the absence of Ras or PKC activation [Papavassiliou *et al.*, 1992]. Thus, the problem of how the PKC or growth factor signal was translated into a dephosphorylation of the inhibitory sites of c-Jun remained obscure.

A second phosphorylation event leading to c-Jun activation after growth factor or Ras stimulation is a simultaneously with the COOH-terminal dephosphorylations increased phosphorylation of NH_2-terminally located sites, mainly serine 63 and serine 73 (human amino acid sequence) [Smeal *et al.*, 1991; Binétruy *et al.*, 1991]. In transient transfection experiments c-Jun derivatives carrying point mutations at these sites, which cannot serve as phosphorylation substrates, have lost the ability of wild-type c-Jun to mediate an increase of AP-1 transcriptional activity [Smeal *et al.*, 1991; Binétruy *et al.*, 1991]. Furthermore, hybrid transcription factors consisting of the c-Jun NH_2-terminal region and a heterologous DNA-binding domain derived from the growth hormone factor 1 (GHF-1) confer Ras responsiveness to GHF-1 target genes. If a similar fusion protein is constructed that carries the point mutations in serines 63 and 73, inducibility by Ras is lost [Smeal *et al.*, 1991]. These results implied that a Ras-induced phosphorylation in the NH_2-terminal region of c-Jun activates the protein by enhancing its transactivation potential.

This notion was soon substantiated by the finding that the Ras-dependent mitogen-activated protein kinases (MAPKs) of the ERK-, JNK-, and SAPK-classes specifically phosphorylate c-Jun on serines 63 and 73 and on threonine 91 and/or 93 in vitro, albeit with varying efficacy [Pulverer *et al.*, 1991; Hibi *et al.*, 1993; Kyriakis *et al.*, 1994; Dérijard *et al.*, 1994; Papavassiliou *et al.*, 1995]. These kinases perfectly fit the "job description" for an in vivo c-Jun NH_2-terminal kinase: They are inducible by growth factors and other stimulators of AP-1 activity, they operate downstream of Ras and Raf, and they can be found in the nucleus. Contrary to the situation in the c-Jun COOH-terminus, the connection between stimulus (growth factor) and transcription factor phosphorylation appears quite clear, with c-Jun being at the nuclear end of one of the best described intracellular signal transduction pathways [Egan and Weinberg, 1993]. The remaining unresolved issue is how exactly does NH_2-terminal phosphorylation switch on c-Jun activity? The mechanism appears to be complex, but certain aspects of it are now beginning to be understood, as evidence emerges that c-Jun phosphorylation events are interdependent and hierarchical [Papavassiliou *et al.*, 1995]. This idea was first formulated in our laboratory from the analysis of different point mutations in the NH_2-terminal phosphorylation sites serine 63, serine 73, and threonine 91/93. These amino acids were exchanged either for alanine or for the negatively charged aspartic acid residue, in an attempt to mimic the unphosphorylated state of c-Jun in a resting cell, or the phosphorylated state in a growth-stimulated cell, respectively [Papavassiliou *et al.*, 1995]. When expressed in tissue culture, such mutants behave in a strikingly different way than wild-type c-Jun with respect to the phosphorylation of the COOH-terminal cluster which determines the DNA-binding activity of the protein. Whereas wild-type c-Jun is COOH-terminally hyperphosphorylated in uninduced cells and hypophosphorylated in stimulated cells, both mutants around position 90 are no longer responsive. The alanine substitution is constitutively hyperphosphorylated and the aspartic acid substitution is constitutively hypophosphorylated [Papavassiliou *et al.*, 1995]. Consequently, the DNA-binding activity of the alanine substitution mutant is permanently low, and that of the aspartic acid substitution mutant is permanently high [Papavassiliou *et al.*, 1995]. Importantly, the binding activity of the alanine substitution mutant can be rescued by in vitro dephosphorylation of the protein [Papavassiliou *et al.*, 1995]. In summary, the phosphorylation state of the NH_2-terminal MAPK substrate sites, as mimicked by the alanine or aspartic acid substitution, governs the DNA-binding potential of c-Jun by influencing the phosphorylation state of the COOH-terminally located residues. One can deduce from this interpretation that MAPK-catalyzed phosphorylation of serines 63 and 73 and of threonine 91 and/or 93 is the primary event on c-Jun in response to a signal, and that the removal of the inhibitory phosphate groups from the COOH-terminal region is a consequence of this event. In other words, the dephosphorylation of these inhibitory COOH-terminal residues is triggered by a change in the substrate, c-Jun, rather than a modulation of the cognate enzymatic activities. This modification in c-Jun's properties as substrate for a constitutively acting protein kinase/phosphatase system might be caused by a conformational change of the molecule as a result of MAPK phosphorylation, which alters the accessibility of the COOH-terminal phosphoacceptor sites for the aforementioned enzymes. Indeed, phosphorylation of the NH_2-terminal sites considerably affects the electrophoretic mobility of c-Jun, which might be a reflection of such a structural alteration [Pulverer *et al.*, 1991; Radler-Pohl *et al.*, 1993; Papavassiliou *et al.*, 1995].

The data presented above can be rationalized in an appealingly simple model for the post-translational control of c-Jun in response to extracellular signals: MAPK phosphorylation at the NH_2-terminus causes dephosphorylation of the COOH-terminal cluster, which in turn causes binding to and activation of Jun target genes (Figure 1). This model integrates prior observations on what were believed to be two separate mechanisms of c-Jun regulation and provides an illustrative example of intramolecular signaling.

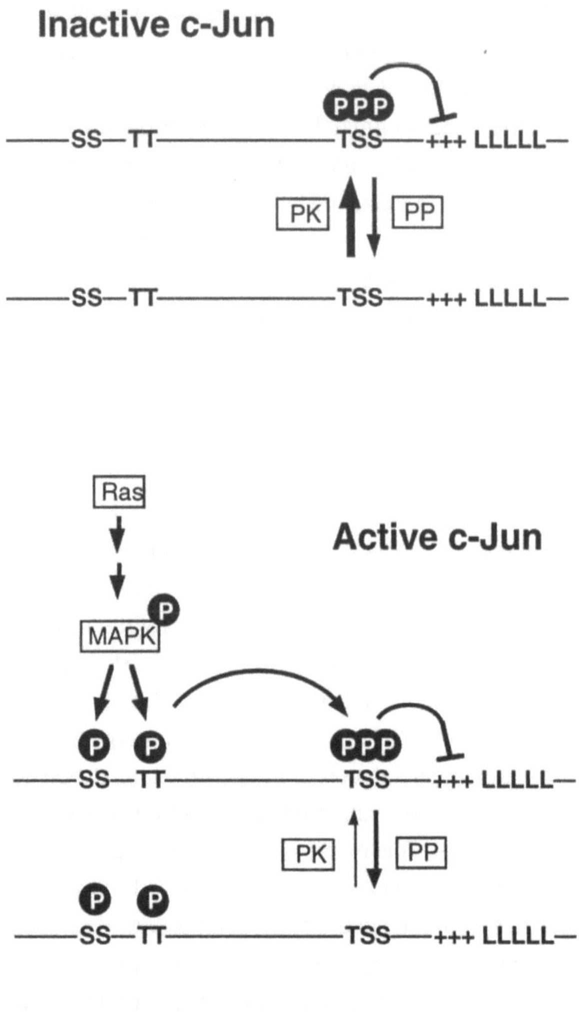

FIGURE 1. Intramolecular signal transduction in c-Jun.
Activation of Ras through growth factor receptor stimulation culminates in the activation of MAP-type kinases (MAPK) such as ERK (extracellular signal-regulated kinases), JNK (c-Jun NH_2-terminal kinases) or SAPK (stress-activated protein kinases) (PKC activation of MAPK cuts into the same pathway possibly by phosphorylation of Raf or a related enzyme). All these kinases potentially use c-Jun as a substrate (albeit with varying efficacy), causing transient phosphorylation of

serines 63 and 73 and of threonine 91 and/or 93. This event triggers a steric change in c-Jun and/or modulates its interaction with other proteins. As a consequence, c-Jun becomes a poor substrate for the protein kinase (PK) that targets the COOH-terminal cluster (and/or a better substrate for the cognate protein phosphatase (PP)) and therefore can no longer be phosphorylated efficiently at these sites, resulting in hypophosphorylation and enhancement of DNA-binding and transcriptional strength. Note that such a model poses that dephosphorylation at the COOH-terminus may be catalyzed by a constitutive PK/PP system whose activity is unregulated. Phosphorylatable serine (S) or threonine (T) residues, as well as the DNA-binding (+++) and dimerization (LLLLL = leucine zipper) domains of c-Jun are indicated.

REFERENCES

Binétruy B, Smeal T , Karin M (1991) Ha-Ras augments c-Jun activity and stimulates phosphorylation of its activation domain. Nature 351:122-127

Boyle WJ, Smeal T, Defize LHK, Angel P, Woodgett JR, Karin M, Hunter T (1991) Activation of protein kinase C decreases phosphorylation of c-Jun at sites that negatively regulate its DNA-binding activity. Cell 64:573-584

Carroll D, Santoro N, Marshak DR (1988) Regulating cell growth: Casein-kinase-II-dependent phosphorylation of nuclear oncoproteins. CSH Symp Quant Biol 53:91-95

Dérijard B, Hibi M, Wu I-H, Barrett T, Su B, Deng T, Karin M, Davis RJ (1994) JNK1: A protein kinase stimulated by UV light and Ha-Ras that binds and phosphorylates the c-Jun activation domain. Cell 76:1025-1037

Egan SE, Weinberg RA (1993) The pathway to signal achievement. Nature 365:781-783

Hibi M, Lin A, Smeal T, Minden A, Karin M (1993) Identification of an oncoprotein- and UV-responsive protein kinase that binds and potentiates the c-Jun activation domain. Genes Dev 7:2135-2148

Hurley JH, Dean AM, Sohl JL, Koshland Jr DE, Stroud RM (1990) Regulation of an enzyme by phosphorylation at the active site. Science 249:1012-1016

Karin M (1994) Signal transduction from the cell surface to the nucleus through the phosphorylation of transcription factors. Curr Opin Cell Biol 6:415-424

Kyriakis JM, Banerjee P, Nikolakaki E, Dai T, Rubie EA, Ahmad MF, Avruch J, Woodgett JR (1994) The stress-activated protein kinase subfamily of c-Jun kinases. Nature 369:156-160

Lin A, Frost J, Deng T, Smeal T, Al-Alawi N, Kikkawa U, Hunter T, Brenner D, Karin M (1992) Casein kinase II is a negative regulator of c-Jun DNA binding and AP-1 activity. Cell 70:777-789

Papavassiliou AG, Bohmann K, Bohmann D (1992) Determining the effect of inducible protein phosphorylation on the DNA-binding activity of transcription factors. Anal Biochem 203:302-309

Papavassiliou AG, Chavrier C, Bohmann D (1992) Phosphorylation state and DNA-binding activity of c-Jun depend on the intracellular concentration of binding sites. Proc Natl Acad Sci USA 89:11562-11565

Papavassiliou AG, Treier M, Bohmann D (1995) Intramolecular signal transduction in c-Jun. EMBO J 14:2014-2019

Pulverer BJ, Kyriakis JM, Avruch J, Nikolakaki E, Woodgett JR (1991) Phosphorylation of c-*jun* mediated by MAP kinases. Nature 353:670-674
Radler-Pohl A, Sachsenmaier C, Gebel S, Auer H-P, Bruder JT, Rapp U, Angel P, Rahmsdorf HJ, Herrlich P (1993) UV-induced activation of AP-1 involves obligatory extranuclear steps including Raf-1 kinase. EMBO J 12:1005-1012
Smeal T, Binetruy B, Mercola DA, Birrer M, Karin M (1991) Oncogenic and transcriptional cooperation with Ha-Ras requires phosphorylation of c-Jun on serines 63 and 73. Nature 354:494-496

The Generation of Cellular Diversity in the CNS

Jack Price

SmithKline Beecham Pharmaceuticals,

Harlow, Essex, CM19 5AW, UK.

Introduction

One of the tasks facing a developmental biologist interested in the nervous system is to explain the generation of cellular diversity; that is how the different neural cell types are generated in the correct time and place. In principle, this task is the same for the brain as for any other somatic tissue, but what makes the task so daunting for the neurobiologist is the enormous degree of cellular diversity in the brain. Whereas most somatic organs are composed of a small number of cell types, the number of types in the central nervous system (CNS) is so large we can only estimate it. Indeed, the diversity of cells is such that the problem becomes one of deciding first what constitutes a cell type: where one scientist sees ten types, another sees one hundred.

A more meaningful approach to diversity is to try to identify the broad features that have to be explained in as simple terms as possible, then to attempt to explain when and how the decision was made that led to the generation of that diversity. With this information in hand an explanation can be sought in

NATO ASI Series, Vol. H 99
Tumor Biology
Regulation of Cell Growth, Differentiation and Genetics in Cancer
Edited by Asterios S. Tsiftsoglou, Alan C. Sartorelli, David E. Housman and T. Michael Dexter

molecular terms, it being axiomatic that a major goal of modern biology be to explain the molecular mechanism that govern cell behaviour.

Features of Cellular Diversity

Moving towards this end, three features suggest themselves from what we know of CNS development and function. First, is diversity in the form that we have been considering it so far, in the sense of histological cell type. Avoiding the cul-de-sac described above, we can start with a minimum of five CNS cell types. Neurons we can split into two principle types: projection neurons, and interneurons. This distinction seems justified because it is based on whether a neuron projects to other brain regions or locally. This seems developmentally significant since the decision to project or not is acted upon early in a cell's history and is therefore likely to be a fundamental one. Moreover, the distinction correlates with a number of other fundamental neuronal features: projection neurons are usually excitatory, interneurons inhibitory; projection neurons often utilise glutamate as a transmitter, interneurons often use GABA; in many CNS tissues projection neurons are generated first, interneurons later. Not every CNS neuron falls clearly into one category or the other, but the difficult cases can be legitimately ignored, at least for a first approximation.

In addition to these two types, we can (again to a first approximation) identify three types of glial cells: grey matter (protoplasmic) astrocytes; white matter (fibrous astrocytes), and oligodendrocytes. In many ways, the developmental significance of this glial classification is less certain than that for neurons, but as we shall see later it fits with some of the most recent cell lineage data.

In addition to this histological diversity, there is a positional specification. In its entirety, the positional specification of the CNS is an enormous subject in which there has been considerable progress in recent years. In this article I only want to consider it in the limited context of cellular diversity. If we consider a particular area of the CNS, in our example the cerebral cortex, we can see that we can identify at least two positional features that contribute to cellular diversity. We can call them 'areal' fate and 'laminar' fate (Figure 1). The cerebral cortex is a continuous sheet of tissue, arranged with grey matter overlying white matter. But within that sheet we can identify the two positional features that I

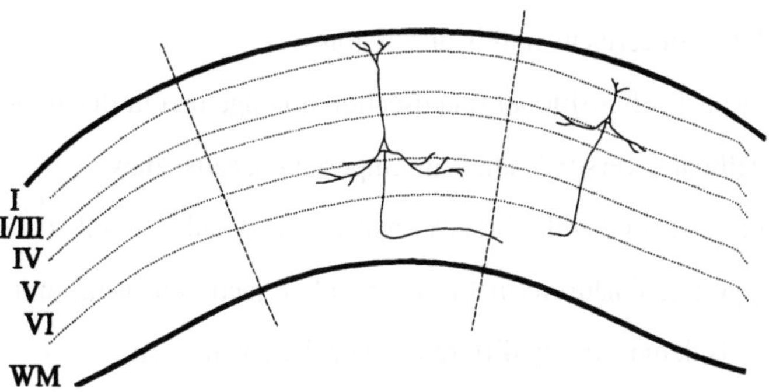

Figure 1: a schematic representation of the cerebral cortex. The cortex is a continuous sheet of tissue that is functionally organised into 'areas' as indicated by the dashed tangential lines, and 'layers' as indicated by the dotted circumferential lines. The layers are numbered I to VI as indicated, with the white matter (WM) underlying these. Two pyramidal neurons are drawn, one in lamina V and the other in lamina III of a neighbouring area.

referred to above. First, different areas--as indicated in Figure 1 by the dashed tangential lines--can be distinguished functionally and anatomically. We are familiar with terms such as 'visual cortex' or 'motor cortex'. These refer to the functional specification of different cortical areas. In addition to this 'areal' specification, we can see a 'laminar' specification. The cerebral cortex is arranged in layers as indicated by the dotted lines in Figure 1. The layers are numbered from lamina I, the outermost cell-sparse layer, through to lamina VI, lying just superficial to the white matter. These laminae are generated developmentally in an inside-out pattern whereby the deepest layers are generated first followed by the next more superficial in order. (The exceptions are lamina I and the deepest cells of lamina VI which are generated before all the others, but they need not concerns us in this discussion.)

How do these aspects of positional specification impinge on our discussion of the generation of cellular diversity? They are important because they constitute significant aspects of a cell's fate, as important as whether a cell is neuronal or glial. A pyramidal neuron (that is a cortical projection neuron) in, for example, motor cortex is demonstrably different--in projection pattern, connectivity, and physiology--from a similar cell in visual cortex, even though they share pyramidal cell properties. Similarly, pyramidal neurons are found in several cortical layers, most notably laminae III and V, but such cells are also demonstrably different in line with their different laminar specification.

The important point is this: if as developmental biologists we want to arrive at a meaningful explanation of the generation of cellular diversity, we must have a description and explanation of all of these three aspects of diversity. We need to understand how all three aspects of a cell's fate is determined. In this article for reasons of brevity, I shall concentrate on 'histological' cell fate, with some mention of 'lamina' fate. To start we need to consider the issues surrounding neural cell lineage.

Neural Cell Lineage

The primary fact concerning cell lineage is that all the different CNS cell types described above are derived from the population of precursor cells that compose the ventricular zone (Fig 2). These pseudostratified epithelial cells divide during early neural development to give more precursor cells, apparently like themselves. At some point in development, however, a transition occurs whereby some cells escape from the cell cycle, never to divide again. These are

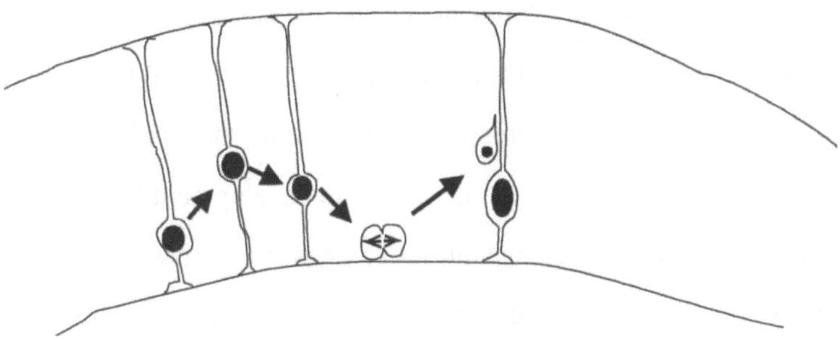

Figure 2. A schematic diagram of the neuroepithelium.

the first neurons, and in a number of neural areas, the cerebral cortex being an example, these cells migrate away from the neuroepithelial layer to form the neural plate, the nascent grey matter of the cortex. At some subsequent time point, overlapping with this period of neurogenesis, glial cells (or their immediate precursors) are generated from this same ventricular zone. First astrocytes appear, then oligodendrocytes.

This description tells us that all the neural cell types are derived from a single population of precursor cells, but it does not tell us what individual cells are doing. It could be--given the forgoing description--that each precursor cell in the VZ is developmentally equivalent; that is each is capable of giving rise to the full complement of cell types that composes any particular tissue (in this case the cerebral cortex). If that were true, each cell would switch the type of cell to which it gave rise as development proceeded, beginning with lamina VI neurons and finishing with oligodendrocytes. Alternatively, specific sub-populations of precursor cells could be established from an early stage, each with the limited potential to generate one or a small subset of cell types. According to this second model, no switching between the production of different cell types would be required during the course of development because the basis for the different cell types would have been established at an earlier stage. In general terms, what we are asking is: how many different precursor cell types are there; what is each one generating; and how is this potential changing with development?

In order to address this cell lineage question, the perfect experiment would be to label a cell *in situ* so that its fate could be followed. By repeating this experiment many times over at different stages of development, we would acquire a complete picture of all the possible outcomes for precursor cells, and how these outcomes change. This conceptually simple experiment is, however, technically difficult. The direct injection of a marker dye is feasible only in easily accessible embryos such as those of avian species (Wetts and Fraser, 1988). An alternative is provided by retroviral vectors. This approach to cell lineage has been reviewed elsewhere (see Price, 1993). Sufficient here to say that retroviruses are a naturally-evolved means of gene transfer. A normal wild-type virus transfers its own genome indelibly into the cell it infects by the integration of the provirus (the DNA form of the retroviral genome) into the chromosomal DNA of the host cell. Once integrated, the provirus is inherited unaltered by the progeny of the infected cell. Moreover, since the provirus constitutes an independent transcriptional unit, the viral genes are transcribed and translated using the cell's normal machinery. In a retroviral vector for cell lineage labelling, the viral genome is engineered to carry a marker gene--usually lacZ--and engineered also so that it can infect, integrate, and express, but cannot form new viral particles (as a wild type virus would do) and so cannot escape to infect neighbouring cells. If, therefore, neural precursor cells in vivo are infected with such a virus at such low titres that infective events are rare (say, less than ten in an entire brain), then the fate of the few labelled cells can be followed via their expression of the introduced marker gene.

The Retina

In a seminal paper, Turner and Cepko (1987) applied the retroviral technique to the rodent retina. They introduced virus into the vicinity of the precursor cells of the new-born rat retina, and analysed the tissue after histogenesis was complete to discover what had become of the labelled precursor cells. They discovered discrete *lacZ*-labelled clones, which were discovered on analysis frequently to be composed of multiple cell types. This result showed directly that precursor cells, even relatively late in development, were not restricted in their potential to the production of a single cell type, but it was not necessarily obvious how many types of precursor cells needed to be invoked to explain the data. The crucial test was to analyse whether the observed outcome was what could have been reasonably expected if there were just a single population of multipotential precursor cells generating differentiated cells by a stochastic mechanism.

The term 'stochastic mechanism' requires brief explanation. We know the final proportions of the different cell types, and from thymidine labelling studies we know when these different cell types arise in development. Therefore we know, for example, that roughly forty times as many rod photoreceptors are generated by the newborn precursor cells than any other cell type. A statistic model can be constructed predicting how frequently clones composed of each possible combination of cell types should appear. This model assumes that the probability of each cell in a clone turning into any of the possible cell types is the same as that cell type's relative frequency in the final tissue. It mimics the situation in which a single population of multipotential precursor cells are generating the entire range of progeny, and the fate of each individual cell is unaffected by the fate of other members of its clone.

The crucial finding of Turner and Cepko was that the outcome of their experiment was not significantly different from that expected from this model. This was taken by these authors to suggest that retinal development is dominated by a single population of multipotential precursor cells that generates the whole range of retinal cell types--with the exception of astrocytes, which migrate into the retina from the optic nerve (Watanabe and Raff, 1988).

This conclusion makes a prediction. If the determination of the fate of each cell is an independent event, then it must take place at or around the time the cell is first generated and becomes post-mitotic. If it were determined earlier, at the precursor cell stage, this would be equivalent to the precursor cell itself being determined, or at least having a bias in what it generates, and the Turner and Cepko result seems to preclude this. This implies that at the point at which a cell becomes post-mitotic, its fate should be determined.

This is a testable hypothesis, and indeed several groups have provided evidence that retinal cell fate can be altered by ambient conditions(Altschuler and Cepko, 1992; Harris and Messersmith, 1992; Reh, 1992; Watanabe and Raff, 1992). Perhaps the most elegant study is the recent one from Lillien, which showed that the fate of retinal cells can be altered by changing their sensitivity to EGF (epidermal growth factor) by changing the levels of receptor for that factor expressed by the cells(Lillien, 1995). One study in particular seems to indicate that cells retain this plasticity beyond the point at which they become post-mitotic. Adler and Hatlee (1989) showed that even 48 hours after chick retinal cells had left the cell cycle, the fate of presumptive neurons could be switched to that of photoreceptor. If the fate of cells were determined at the precursor cell level, this outcome would not have been expected. This is strong support for the conclusion drawn from the Turner and Cepko study. It also points quite clearly to a decision point in cell fate determination. Fate must be determined at or soon after the cell becomes post-mitotic. As we shall see, although cell lineage in the cerebral cortex follows a distinctly different pattern, it points to a similar decision point, but a different set of decisions.

The Cerebral Cortex

The experiments on the retina suggest a relatively simple model for cell lineage--a single population of precursor cells generates the entire range of cell types, along the lines of the first of the two models outlined above. Similar cell lineage studies of the cerebral cortex have suggested a much more complicated story.

An obvious place to start the analysis of the cerebral cortex is to ask whether that tissue contains a population of multipotential precursor cells analogous to those in the retina. Several lines of evidence suggest there is. Indirect evidence comes from various studies using cell lines(Snyder et al, 1992; Renfranz et al, 1991), but direct evidence comes from two groups, albeit that that evidence only comes from studies in culture (Davis and Temple, 1995; Williams and Price, 1995). Both describe a population of cells with the capacity to generate neurons, astrocytes, and oligodendrocytes. But these cells, which our group calls NE cells for 'neuroepthelial cells', have a clear characteristic which distinguishes them from their retinal counterparts: they occur at a high frequency only at early stages of development (Figure 3). Judging their frequency by the percentage of virally infected clones of this type that arise, we observed that whereas at E12 over 50% of clones are composed of NE cells, by E16 only 5% are of this type. To put this in context, neurogenesis in the cerebral cortex of the rat begins around E13/14 and finishes around the time of birth at E21/22. By E16 the cortical plate is only about five or six cells thick, and only a small fraction of the cortical neurons have been generated. The obvious question, therefore, is: if only 5% of the cortical precursor cells are multipotential NE cells at this relatively early stage of neurogenesis, what are the remaining 95%?

In answer to this question, we have much clearer data, both from in vivo and in vitro studies. Roughly 80% of the VZ precursor cells when labelled with retrovirus apparently generate a single cell type: neurons, grey matter astrocytes, white matter astrocytes, or oligodendrocytes (Price and Thurlow, 1988; Luskin et al, 1988; Grove et al, 1993). One other type of clone is composed of an astrocyte-like cell that is, however, negative for GFAP (glial fibrillary acidic protein, an astrocyte marker). The identity of this cell type is unknown. Some precursor cells do generate some other combination of cell types, the best characterised being the 6% of precursors that at E16 give both neurons and oligodendrocytes.

The in vivo results suggest that, unlike in the retina, cells in the cerebral cortex become specified before they become post-mitotic, but the results from similar studies in tissue culture played an important part in underlining this conclusion (Williams et al, 1991; Williams and Price, 1995). This was because of a problem with the in vivo study, namely that there was no independent means to ascertain whether the clusters of lacZ-labelled cells that were observed were true clones.

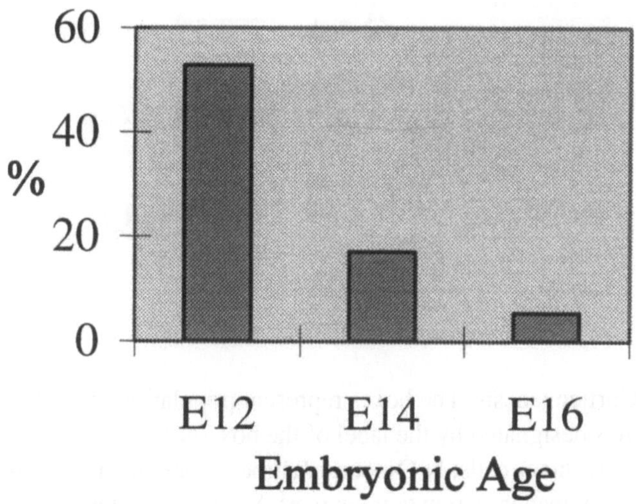

Figure 3. A graph showing the incidence of NE Clones (as a proportion of the total) in cultures derived from embryonic cerebral cortex from different embryonic ages.

We know that cortical cells disperse considerably during development, and it was possible that clones were dispersing to such an extent that they were being falsely identified as multiple clones. This could have considerably distorted the cell lineage picture that emerged.

Another limitation of the in vivo study was the extent to which determination of precursor cells can reasonably be concluded from cell lineage

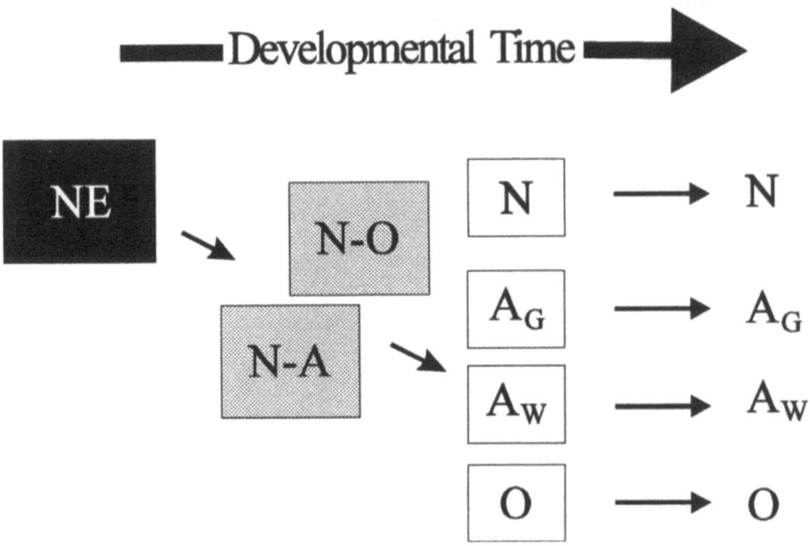

Figure 4. A Model for Corticogenesis. The boxes represent populations of VZ cells, restricted to the fates designated by the label of the box--N-O-neuron/oligodendrocyte precursor cells; N-O-neuron/astrocyte precursor cells; N-neuronal precursor; A_G-grey matter astrocyte precursor; A_W-white matter astrocyte precursor; O-oligodendrocyte precursor. The black box marked NE, represents the NE cell population. The unenclosed letters represent the final differentiated cell types.

data. Just the fact that a cell gives rise to restricted progeny does not provide reasonable grounds to conclude that it is restricted to that fate. It could be that the cell had a much broader developmental potential, but that it did not have the opportunity to express it full potential in that particular circumstance. The tissue culture studies were able to consider the fate of virally labelled cells undercircumstances where their degree of dispersion could be precisely monitored and where conditions could be varied considerably. Nonetheless, the cells in culture behaved exactly as they apparently had in vivo, generating in each case just a single cell type. The principal exception, as in vivo, was the cells that generated both neurons and oligodendrocytes(Williams et al, 1991). One difference between culture and in vivo experiments was that under certain culture conditions, these neuron-oligodendrocyte (N-O) clones failed to appear, but neuron-astrocyte (N-A) appeared instead (Williams and Price, 1995). Probably this constitutes a different outcome for the same population of precursor cells, although we have not proven that point.

Taken together, these data suggests a model for cell lineage in the cerebral cortex like that shown in Fig 4. At early time points the multipotential NE cells dominate. By mid corticogenesis, the VZ has become a mosaic of several different precursor cell types, each determined to generate a single cell type. We presume, but have not proven, that the N-O and N-A precursor cells lie somewhere between these two extremes of precursor cell types.

The Fate of the Cortical Neuronal Precursor Cells.

What we have observed so far is a stark contrast between the pattern of cell lineage in the retina compared to that in the cerebral cortex. In the cortex fate determination precedes the exit of cells from the cell cycle; in the retina the two events appear concomitant. Does this mean that the decision point we identified in the retina, at the point of exit from the cell cycle, does not exist in the cerebral cortex? The answer from a whole series of studies is clearly 'no'. The difference between the tissues seems to lie in the nature of the decisions that are made at that point.

What evidence is there that decisions are made in the cortex at about the time that cells become post-mitotic? Three types of evidence can be cited, all relating to the determined Neuronal precursor cell population--the cells represented by the N in a box in figure 4.

1. McConnell and her colleagues looked at the determination of lamina (O'Rourke et al, 1992). They performed heterotopic transplants of cells in the ferret retina. They labelled cells with tritiated ^{3}H-TdR in embryos at a stage when lamina VI neurons were being generated, and transplanted them into embryos at a later stage of development when lamina II cells were being generated. By looking for ^{3}H-TdR labelled cells in the host, they were able to analyse to which lamina the transplanted cells had migrated. They discovered that the answer depended on how long the cells had been left undisturbed in the donor animal before transplantation. If cells were transplanted immediately after ^{3}H-TdR labelling, they behaved like host cells and migrated to lamina II. If, however, the cells were left undisturbed in the donor for at least 4 hours prior to transplantation, they retained their donor fate and moved to lamina VI. This can be interpreted to mean that during the four hours after cells become labelled with ^{3}H-TdR (i.e. during the four hours after they enter 'S' phase for the final time), they become committed to a lamina fate. Four hours after the final 'S' phase is around the time of the final mitosis.

2. Retroviral lineage experiments conducted on cells in culture have indicated that neuronal clones can be composed of cells of both the two major cortical neurotransmitter phenotypes, glutamatergic and GABAergic (Goetz et al, 1995). These two phenotypes closely approximate to the two major neuronal types, pyramidal and non-pyramidal cells respectively. Cultured clones from E16 rat cortex can be composed purely of glutamatergic cells; purely of GABAergic cells; or a mixture of both phenotypes. If these results are analysed in a fashion similar to that described above for the retinal experiments cited above, to discover if the data fit a stochastic model involving a single population of bi-potential precursor cells, the result suggests that the data do fit such a model. In other words, the proportions of the different types of clones are as would be expected from a single population of such precursors. Using the same logic that was applied to the results on the retina, we take this result to mean that 'neurotransmitter' fate is probably determined around the time the cells become post-mitotic. This conclusion is supported by the finding of Götz and Bolz (1994) that the neurotransmitter fate is particularly sensitive to disturbances at the time cells are passing through their final cell cycle.
3. In a series of studies, Bolz and his colleagues have shown that the primary projection pattern of lamina V pyramidal cells must be determined at roughly the same point in time--as the cells are becoming post-mitotic (cited in Götz and Price, 1995). Most Lamina V pyramidal neurons project to one of two sites: when the cells' growth cones reach the sub-cortical white matter they either turn laterally and project sub-cortically, or they turn medially and project callosally. They basically have a binary decision: left or right. The cells' growth cones emerge soon after they leave the VZ, they always generate a single projection immediately--they do not exercise the option that neurons

sometimes take of projecting collaterals in multiple directions, and retracting the inappropriate ones at a later date. This means that this decision--whether their primary target is to be callosal or subcortical--must also have been made by the time the cell begins migration. Again, this suggests that the decision must have been made at, or soon after, the cell become post-mitotic.

A Unifying Perspective

To summarise this body of data, the cell lineage studies show that in the cortex primary cell fate--whether a cell will be a neuron, astrocyte, or oligodendrocyte--is determined at the precursor cell level before the cell passes through its final division. This contrasts with the situation in the retina when these events appear to be roughly simultaneous. Thus the cortical VZ during most of neurogenesis is a mosaic of different types of precursor cell types, unlike the single population of multipotential precursor cells that dominates the retina at similar stages. The decision point associated with the generate of individual post-mitotic cells does, however, exist in the cortex, but it seems to be associated with decision regarding 'lamina', 'neurotransmitter', and 'projection' fates. Whether these all constitute related aspects of a single overall decision, or whether they constitute a succession of decisions, remains to be worked out.

One could speculate on why two regions of the vertebrate CNS, which share many similarities, should have such fundamental differences in their patterns of cell lineage. A more useful application of these results is a gateway to meaningful questions about mechanism. A clear set of questions are now defined concerning the relatedness of the different 'decisions', and the factors--signalling molecules, receptors, transcriptional regulators--that influence the outcome of these decisions. There is good reason to believe that this phase of the research is well under way.

References

Adler, R. and Hatlee, M. (1989). Plasticity and differentiation of embryonic retinal cells after terminal mitosis. Science 243:391.

Altschuler, D. and Cepko C. (1992) A temporally regulated, diffusible activity is required for rod photoreceptor development in vitro. Development 114:947.

Davis, A.A. and Temple, S. (1994) A self-renewing multipotential stem cell in embryonic rat cerebral cortex. Nature 372:263.

Götz, M. and Bolz, J. (1994). Differentiation of transmitter phenotypes in rat cerebral cortex. Eur. J. Neurosci 6:18.

Götz, M. and Price, J. (1995) The determination of neuronal phenotype and axonal projection in the cerebral cortex. Seminars in Developmental Biology 5:359.

Götz, M., Williams, B.P., Bolz, J. and Price, J. (1995) The determination of Neuronal Fate: a common precursor for neurotransmitter subtypes in the cerebral cortex in vitro. Eur. J. Neurosci. 7:889.

Grove, E.A., Williams, B.P., Li, D-Q., Hajihosseini, M., Friedrich, A., and Price, J. (1993) Multiple restricted lineages in the embryonic rat cerebral cortex. Development 117:553.

Harris, W.A., Messersmith, S. L., (1992). Two cellular Inductions involved in photoreceptor determination in the xenopus retina. Neuron 9:357.

Lillien, L. (1995). Changes in retinal cell fate induced by overexpression of EGF receptor. Nature 377:158.

Luskin, M.B. Pearlman, A.L. and Sanes, J.R. (1988). Cell Lineage in the retrovirus. Neuron 1:635.

O'Rourke, N.A., Dailey, M.E., Smith, S.J., and McConnell, S.K. (1992). Diverse migratory pathways in the developing cerebral cortex. Science 258:299.

Price, J. (1993) Making sense of cell lineage. Perspectives in Developmental Neurobiology 1:139.

Price, J. & L. Thurlow (1988) Cell Lineage in the rat cerebral cortex: a study using retroviral-mediated gene transfer. Development 104:473.

Reh, T.A. (1992). Cellular Interactions determine neuronal phenotypes in rodent retinal cultures. Journal of Neurobiology 23:1067.

Renfranz, P.J., Cunningham, M.G., and McKay, R.D.G.(1991). Region-specific differentiation of the hippocampal stem cell line HIB5 upon implantation into the developing mammalian brain. Cell 66:713.

Snyder, E.Y., Deitcher, D.L., Walsh, C., Arnold-Aldea, S., Hartweig, E.A., and Cepko, C.L. (1992). Multopential neural cell lines can engraft and participate in the development of mouse cerebellum. Cell 68:33.

Turner, D., and Cepko, C. (1987). Cell lineage in the rat retina: a common progenitor for neurons and glia persists late in development. Nature 328:131.

Watanabe, T. and Raff, M.C. (1988). Retinal astrocytes are immigrants from the optic nerve. Nature 352:834.

Watanabe, T. and Raff, M.C. (1992). Diffusible rod-promoting signals in the developing rat retina. Development 114:899.

Wetts, R. and Fraser, S.E. (1988). Multipotent Precursors can give rise to all major cell types of the frog retina. Science 239:1142.

Williams, B.P., Read, J. and Price, J. (1991) The generation of neurones and oligodendrocytes in the cerebral cortex from a common precursor cell. Neuron 7:685.

Williams, B.P. and Price, J. (1995) Evidence for multiple precursor cell types in the embryonic rat cerebral cortex. Neuron 8:1181.

Stromal Cell Control of Haemopoiesis

Jane Qiu and Paul R. Harrison
Beatson Institute for Cancer Research
Switchback Rd., Bearsden
Glasgow G61 1BD, U.K.

Human peripheral blood constitutes 7% of body weight. About one-half of blood volume is occupied by erythrocytes. These cells have an average life span of 4 months. Some other blood cells have an even shorter life span, and the continuity of gas transport, immunity, and other vital functions is strictly dependent on the constant production of new cells. All the various types of blood cells are derived from a common haemopoietic stem cells and the first step in differentiation of stem cells involves the generation of progenitors committed to either the erythroid, granulocytic, megakaryocytic lineages or to lymphoid cells. Therefore haemopoiesis, the generation of blood cells, is a multistep process involving stem cell renewal, commitment, differentiation, maturation and consequent positioning of the cells within the tissue. Regulation of this process, particularly under conditions of perturbations, damage, and disease, is mediated by inducer colony-stimulating factors and interleukins counteracted by inhibitory cytokines. In addition, the haemopoietic microenvironment of bone marrow, spleen and foetal liver also play a pivotal role.

In this short article, the importance of haemopoietic microenvironment (HM) will be discussed and then several *in vitro* model systems which are utilised to study the stromal cell control of haemopoiesis will be introduced. The importance of HM on erythropoiesis will be discussed in particular because this is the research interest of our laboratory.

Haemopoietic microenvironment and haemopoiesis

The importance of haemopoietic growth factors in haemopoiesis has long been the centre of interest and only recently did people start to realise that haemopoietic microenvironment (HM), especially stromal cells, also play a vital role. In the site of haemopoiesis, that is, bone marrow and spleen in the adult and foetal liver, proliferation and differentiation of haemopoietic stem and progenitor cells occur in intimate contact with the marrow stromal cells and the associated extracellular matrix (ECM). Based on histological analysis, it was realised that haemopoietic cells are organised into discrete patterns. Electron microscopical histology revealed intimate interactions between haemopoietic cells and the stromal cells, which form secluded niches within

NATO ASI Series, Vol. H 99
Tumor Biology
Regulation of Cell Growth, Differentiation and
Genetics in Cancer
Edited by Asterios S. Tsiftsoglou, Alan C. Sartorelli,
David E. Housman and T. Michael Dexter

HM. This kind of defined micro-organisation of cells could be ascribed to inducer factors locally expressed by the stromal cells, which consist of a heterogeneous mixture of macrophages, endothelial cells, fibroblasts and adipocytes.

Although the exact role of the HM in determining haemopoietic stem and progenitor cell behaviour is not fully known, multiple possible functions have been postulated. As currently envisioned, the HM provides direct cell-cell contact between supporting and haemopoietic cells, provides anchorage for both growth factors and haemopoietic cells, supplies specific positive and negative growth regulatory factors, and probably provides for multiple cellular communications within what have been termed a "local area network". Within this network, stromal cells probably play a key role by both producing growth regulatory proteins and secreting complex extracellular matrix proteins for stabilisation of growth factors in high local concentration with adherent haemopoietic stem and progenitor cells. In particular, sequestration of cytokines on ECM components of the stromal tissue and expression of membrane forms of cytokines may provide the basis for the formation of microenvironment. The locally concentrated presentation of growth factors at the adherent stromal cell surface primes the stem or progenitors cells to respond to local or circulating combinations of other factors at physiologically relevant concentrations.

Erythropoiesis in bone marrow occurs in discrete, characteristic foci called erythroblastic islands (EBI) (Bessis, M., 1958). An EBI consists of a central macrophage (or more) (CM) surrounded by a group of erythroblasts that are at approximately the same stage of maturation. Careful three dimensional analysis have shown that the numbers of erythroblasts in EBI increase in multiple of two, to as many as thirty-two, suggesting that proliferation of the erythroid cells proceeds through at least five synchronous mitosis (Mohandas, N. and Prenant, M., 1978). When murine erythropoiesis was suppressed by hypertransfusion of the animals, EBI were observed to involute, with loss of the CM as well as the erythroid cells (Brookoff, D. and Weiss, L., 1982). Subsequent repeated bleeding of the animals was associated with a marked delay in the resumption of erythropoiesis, apparently because of the absence of supportive CM. Thus the control of erythropoiesis, to a significant degree, occurs communally at the level of the erythroblastic islands.

In vitro model systems for the study of stromal control of erythropoiesis

Although it is generally agreed that cellular interactions can regulate commitment, proliferation and differentiation in haemopoiesis, little is known about the cellular and molecular events that mediate or result from these interactions. Defining these molecular events is particularly difficult in complex tissue systems where the presence of heterogeneous cell populations interferes with the assessment of any specific cell-to-cell interactions. Studies on haemopoiesis have an additional complexity in that many of the interacting cells are not necessarily localised in the fixed architecture. Consequently, much of the work done on this system has examined the *in vitro* behaviour of explanted mixtures of cells, which may or may not approximate interactions that occur among individual cells *in vivo*. Most *in vitro* studies have been carried out either by long-term bone marrow cultures, or by using clonal stromal cell lines derived from bone marrow, spleen or foetal liver.

Long-term bone marrow culture (LTBMC)

When cultured in the appropriate conditions, bone marrow (BM) cells give rise to a complex adherent layer of stromal cells that closely mimic the BM microenvironment and support the self-renewal, proliferation and differentiation of totipotent haemopoietic stem cells (Dexter, T.M. *et al.*, 1990). In murine LTBMCs, the production of haemopoietic cells continues for up to a year, suggesting that multipotential stem cells with a high proliferative capacity are supported by the adherent layer. Spleen colony-forming units (CFU-S), granulocyte-macrophage colony-forming cells (CFU-GM) and erythroid burst-forming units (BFU-E), as well as a variety of intermediates of differentiating and mature granulocytes and macrophages, accumulate in these cultures (Zipori, D. and Tamir, M., 1989). The cells are periodically harvested and replaced by new cells that differentiate from a pool of stem cells residing within the adherent stromal layer. This *in vitro* phenomenon is apparently a close approximation of structures and functions in the bone marrow *in situ*. With respect to erythropoiesis, BFU-E cells are identifiable in these cultures, but no mature erythrocytes are produced unless erythropoietin is added (Eliason, J.F. *et al.*, 1979 and Tsai, S. *et al.*, 1986). Thus, all factors required to generate and support BFU-E cells are present in the adherent cell layer in LTBMCs, which probably quite accurately reflects the importance of stromal cells in erythropoiesis *in vivo*. Though this model system provides simplified conditions for studies of the control of stem cell renewal, commitment and differentiation, the complexity of this adherent layer hampers the understanding of the relationship and interactions between haemopoietic and stromal cells.

Study of erythropoiesis in vitro using clonal stromal cells

A series of studies carried out by Obinata *et al.* demonstrated that stromal endothelial (Obinata, M. *et al.*, 1989 and 1991) and epithelial-like cell lines (Obinata, M. *et al.*, 1990), derived from mouse spleen and foetal liver respectively, are able to support large erythroid colony formation and mimic erythropoiesis *in vivo*. The resulting mature erythrocytes and erythroblasts showing cytoplasmic budding were present in the large erythroid colonies on stromal cell lines. The morphological observation suggested that an erythrocyte expels its nucleus by the budding of the cytoplasm *in vitro*. Since erythroblasts are not enucleated in a usual semisolid culture, it is likely that the stromal cells may support their enucleation *in vitro* (Obinata, M. *et al.*, 1989). This kind of *in vitro* microenvironment seems to be created by direct contact and/or short range communication between the erythroid progenitor cells and the stromal cell layers since large erythroid colonies were not formed by separation of the progenitor cells from the stromal cells by a diffusion chamber or a nucleopore filter and the large erythroid colonies were formed only in the presence of semisolid support. Supplementing the conditioned medium of stromal cells did not show any effects on the colony formation. Neither do various combinations of soluble growth factors such as IL-3, GM-CSF and EPO. The study was extended by examining the role of adhesion molecules in erythropoiesis by blocking antibodies (Obinata, M. *et al.*, 1994). It was shown that the development of the erythroid cells on stroma cells was inhibited by anti-very late antigen-4 (VLA-4 integrin) antibody, but not by anti-VLA-5 antibody, although the erythroid cells express both VLA-4 and VLA-5. The extent of the inhibition of the large erythroid colony formation by anti-VLA-4 is higher than that by anti-*c-kit* antibodies, and no additive inhibition was observed by the simultaneous addition of the two antibodies. Furthermore, whereas high levels of expression of vascular cell adhesion molecule-1 (VCAM-1) and fibronectin (FN), ligands for VLA-4, were detected in the stroma cells, the adhesion and development of the erythroid progenitor cells were only partly inhibited by the blocking antibody against VCAM-1 but not FN. Therefore, though VLA-5 and FN could mediate adhesion of the erythroid progenitor cells to the stromal cells, the adhesion itself may not be sufficient for the stroma-supported erythropoiesis. The data also demonstrated that the stromal cells may support erythroid development by the adhesion through a new ligand molecule(s) for VLA-4 in addition of VCAM-1, and such collaborative interaction may provide adequate signalling for the erythroid progenitor cells in the erythropoietic microenvironment. This is one of the many reports demonstrating the importance of direct cell-cell contact and integrins in the modulation of proliferation, differentiation and apoptosis.

In contrast, there are also studies showing that bone marrow stromal cells can also preventing cell differentiation. Weber and Tykocinski (1994) reported the reversible blockade of leukaemic cell differentiation by bone marrow stromal cells, which may well be mediated by gap junction. Similarly, Verfaillie *et al.* (1993) found that an unknown factor(s) produced by human bone marrow stroma can increase proliferation and maturation or primitive haemopoietic progenitors while preventing their terminal differentiation. It is worthwhile noticing that the above data were from *in vitro* studies which is unlikely to fully recreate the *in vivo* physiological context. It will not be surprising that stromal cells have both positive and negative regulators of proliferation and differentiation. The important question is how they exert their effect in a temporal and spatial manner with great specificity and precision *in vivo*.

A Novel stroma-dependent in vitro murine erythroleukaemia systems

Though such studies using primary erythroid cells have the advantage of studying the normal process of erythropoiesis, they suffer from the difficulty in obtaining large number of cells for molecular biology studies. In order to dissect the molecular mechanisms of stromal cell interaction and the occurrence of stromal independent growth, we have utilised a novel *in vitro* system, the murine erythroleukaemia system (ELM). The ELM leukaemia was induced in a female C3H mouse by X-ray irradiation of 300 Rads. The ELM system is unusual in that the erythroleukaemia cells growing *in vivo* in the spleen cannot be maintained *in vitro* except in contact with bone marrow-derived stromal cells. They are also unusual in that they retain the ability to undergo erythroid differentiation in response to normal growth factors such as EPO or IL-3. Various stroma-dependent erythroleukaemia cell clones (ELM-Ds) have been isolated on stromal cells. By selecting for growth of ELM-D cells without stroma, we have obtained a series of 14 stroma-independent clones (ELM-Is). They are more tumourigenic when injected into mice compared with their dependent counterparts. All of the independent clones apart from one (i.e. ELM-I/1) are still able to differentiate into mature erythroid cells in response to EPO or IL-3. However, ELM-I/1 has lost the ability to differentiate under such conditions and furthermore displays the highest tumourigenicity *in vivo* (*Table 1*). The frequency of formation of truly stroma-independent variants is 10^{-5}, suggesting that further genetic changes are required. Similar to the studies by Obinata *et al.* (1989) using primary normal erythroid cells, various lines of evidences suggest that direct cell-cell contacts are required for the stromal cells to maintain long-term growth of ELM-D cells: (1) conditioned medium from either stromal cells or ELM-D/stromal cell co-cultures is unable to support long term growth of ELM-D cells; (2) long term survival is largely reduced if ELM-D cells are separated from stromal cells by an agar layer; (3) combinations of growth factors cannot replace stromal cell functions; and (4) co-cultures of ELM-D cells on stroma display morphological evidence of intimate contact according to morphological studies by electromicroscopy. Taken together the ELM cells retain certain phenotypes of normal erythroid cells (e.g. responsiveness to normal growth factors, stroma-dependence, etc.) in contrast to most of the various retrovirus-induced erythroleukaemia cell lines previously described. Therefore the ELM system provides us with a very useful cellular system that allows us to identify the molecular nature of cellular interactions and how this may contribute to the control of growth and differentiation of erythroid cells; it is also an *in vitro* model system of tumour progression and enables us to identify genetic changes involved in the occurrence of stroma-independence and differentiation arrest.

Table 1. ELM cell lines and their phenotypes.

	ELM-D	ELM-I/2,5,6, etc.	ELM-I/1
Stroma-dependence	+	-	-
Differentiation capacity in response to erythropoietin	+	+	-
Tumourigenicity	+	++	+++

In collaboration with Prof. Ostertag's group in Hamberg, it has been established that the stromal cell signal required for long-term growth of ELM-D cells seems to be normally generated by the SCF/SCF-R interaction, since antibodies against either SCF-R or SCF inhibit long-term growth (Itoh, K. *et al.*, 1995). The fact that soluble SCF (sSCF) supports only short term growth of ELM-D cells suggests that the membrane form of SCF (mSCF) is required and this is consistent with the fact that stromal cells from *Steel-Dickie* (*Sl/Sl^{d}*) mutant mice (which produce only sSCF) support only short-term ELM cell growth. This is most naturally interpreted to mean that the balance between long-term maintenance and short-term growth/clonal decline is controlled by the ratio of sSCF to mSCF presented to the SCF-R on ELM-D cells. This is confirmed by introducing the mSCF cDNA into *Sl/Sl^{d}* stroma and this restores its ability to maintain long term growth of ELM-D cells (Ostertag, W., personal communication). However, stromal cells from homozygous *Steel* (*Sl/Sl*) mice, lacking all forms of SCF, support long-term ELM-D growth which implies that the mSCF/SCF-R interaction is not absolutely required and that other stromal cell signals can substitute. It is possible that *Sl/Sl* stromal cells may produce alternative ligand of SCF-R. This line of work is being extended to the study of the roles of extracellular matrix (ECM) and cell adhesion molecules in the stromal maintenance of ELM-D cells. The ultimate aim is to reconstitute the stromal cell function by combinations of growth factors, ECM and cell adhesion molecules. This will enable us to identify the molecular events that are involved in maintaining the long term growth of erythroid cells.

The genetic changes responsible for the different phenotypes of the ELM cells have also been defined (Nibbs, R. *et al.*, 1993). It was found that at the *p53* locus in the primary leukaemic cells, one copy of the gene has been lost whereas the other contains an 18-bp deletion, implicating its mutation as an early step in the development of the leukaemia. Changes in the gene expression of *ets* family members have also been found. The *fli-1* gene region is rearranged in the primary tumour because of the insertion of a retrovirus inserted upstream of one *fli-1* allele, but this does not result in *fli-1* gene activation in any of the ELM-D or ELM-I cell lines except one (ELM-I/1).

It seems significant that this line is the only one to have lost the ability to differentiate in response to erythropoietin. In addition, up-regulation of *erg* is associated with stromal cell-independent growth, since all ELM-I mutants have moderate levels of *erg* mRNA, whereas only low or undetectable levels are found in primary leukaemic cells *in vivo* or in ELM-D cells *in vitro*. This up-regulation of *erg* mRNA seems to be important for stromal cell-independent growth, since ELM-D cells show elevated expression of the *erg* gene after separation from stromal cells, This seems to be made permanent in ELM-I mutants, since they do not down-regulate *erg* mRNA when grown in contact with stromal cells. We therefore proposed that *ets* family members regulate both the survival and differentiation of erythroid cells. The ongoing functional analysis will allow us to provide more direct evidence.

References

Bessis, M. (1958), *Rev. Haemotol. 13*, 8-11.

Brookoff, D. and Weiss, L. (1982), *Blood 60*, 1337-1344.

Dexter, T.M., Coutinho, L.H., Spooncer, E., Heyworth, C.M., Daniel, C.P., Schiro, R., Chang, J., and Allen, T.D. (1990), *Stromal cells in haemopoiesis*. Molecular control of haemopoiesis. p76.

Eliason, J.F., Testa, N.G. and Dexter, T.M. (1979), *Nature 281*, 382-384.

Itoh, K., Klinger, K., Laker, G., Just, U., Mannweiler, K., Mori, J., Nishikawa, S.-I., Fang, L., Wagener, C., Nibbs, R., Harrison, P.R. (1995), submitted.

Mohandas, N. and Prenant, M. (1978), *Blood 51*, 633-643.

Nibbs, R., Itoh, K., Ostertag, W. and Harrison, P.R. (1993), *Mol. Cell. Biol. 13*, 5582-5592.

Ohneda, O., Yanai, N. and Obinata, M. (1990), *Development 110*, 379-384.

Tsai, S., Sieff, C.A. and Nathan, D.G.(1986), *Blood 67*, 1418-1426.

Weber, M.C. and Tykocinski, M.L. (1994), *Blood 83*, 2221-2229.

Yanai, N., Matsuya, Y. and Obinata, M. (1989), *Blood 74*, 2391-2397.

Yanai, N., Satoh, T. and Obinata, M. (1991), *Cell Structure and Function 16*, 87-93.

Yanai, N., Sekine, C., Yagita, H. and Obinata, M. (1994), *Blood 83*, 2844-2850.

Zipori, D. and Tamir, M. (1989), *Int. J. Cell Cloning 7*, 281-291.

The Insulin-Like Growth Factor I Receptor and Its Role in Tumorigenesis and Cell Survival

Mariana Resnicoff and Renato Baserga
Department of Microbiology and Immunology
Thomas Jefferson University
233 S. 10 th. Street
Philadelphia, PA 19107
USA

Key words: IGF-1R/ growth/ tumorigenesis/ apoptosis/ antisense strategies/ antitumor response

Abstract:
The IGF-1R is required for optimal growth in vivo and in vitro; it plays a crucial role in the establishment and maintenance of the transformed phenotype and it protects tumor cells from apoptosis both in vivo and in vitro.
Using antisense strategies to the IGF-1R RNA, we establish a correlation between levels of the IGF-1R, apoptosis and tumorigenesis of C6 rat glioblastoma cells. In addition to the apoptotic effect, an antitumor response is triggered, protecting the rats from subsequent tumor challenge and causing regression of established tumors.

The IGF-1R in growth:

The insulin-like growth factor I receptor (IGF-1R) is a tyrosine kinase receptor with 70% homology to the insulin receptor. It is synthesized as a single precursor polypeptide of 1367 aminoacids, including a 30 aminoacid signal peptide. After removal of the signal peptide, the proreceptor is cleaved to form the alpha and the beta subunits, linked by disulfide bonds. The alpha subunit is entirely extracellular and contains the ligand binding domain. The beta subunit has a transmembrane domain and contains the ATP binding site and the tyrosine kinase catalytic domain, which are entirely intracellular.
An IGF-1R activated by its ligands (IGF-I, IGF-II and insulin at supraphysiological concentrations) is required for the growth of many cell types, including fibroblasts, epithelial cells, hemopoietic cells (stem cells, myeloid cells and T lymphocytes),

NATO ASI Series, Vol. H 99
Tumor Biology
Regulation of Cell Growth, Differentiation and Genetics in Cancer
Edited by Asterios S. Tsiftsoglou, Alan C. Sartorelli, David E. Housman and T. Michael Dexter

chondrocytes, bone cells and smooth muscle cells. The mitogenic signal is transduced via IRS-1 and Shc (Myers et al., 1993; White and Kahn, 1994), through ras and raf, all the way to the nucleus (Crews and Erikson, 1993).

The importance of the IGF-1R in cell growth has been confirmed in vivo by the findings of Efstratiadis and his group: mouse embryos with a targeted disruption of the IGF-1R and the IGF-II genes have a size at birth that is only 30% the size of wild-type littermates (Liu et al., 1993; Baker et al, 1993). This is a formal demonstration that the IGF-1R is responsable for 70% of developmental murine growth but it also demonstrates that there is 30% of this growth that is unaffected by the IGF-1R.

3T3-like cells derived from these mouse embryos devoid of IGF-1Rs can grow in 10% serum but at a rate that is 40% lower than the cells obtained from the wild-type littermates (Sell et al., 1994). A cell cycle analysis shows that all the phases of the cell cycle are elongated, suggesting that the IGF1R is required in all phases (Sell et al., 1994; Valentinis et al., 1994)

The IGF-1R in transformation:

Cells derived from these mouse embryos are refractory to transformation by SV40 T antigen, by an activated and overexpressed Ha-ras or by a combination of both (Sell et al., 1993 and 1994), or by overexpressed growth factor receptors, such as the epidermal growth factor receptor (Coppola et al., 1994) and the platelet-derived growth factor receptor (De Angelis et al., 1995), all of which transform very efficiently the corresponding wild-type cells or other 3T3 cells. This resistance to transformation is abolished if a plasmid expressing the wild-type but not a mutant IGF-1R cDNA is stably transfected into these cells, showing that the phenotype is specifically due to the lack of IGF-1R (Sell et al., 1994; Coppola et al., 1994).

If the absence of IGF-1R precludes the establishment of transformation, then, interference with its expression could possibly revert the transformed phenotype. In fact, the reversal of the transformed phenotype has been achieved by several methods in different laboratories. They have aimed at interfering with the IGF-1R or its ligands by the use of antisense plasmids or oligodeoxynucleotides targeting the IGF-1R (Baserga et al., 1994, Resnicoff et al., 1993, 1994 a and b; Shapiro et al., 1994) or IGF-I (Trojan et al., 1992 and 1993) or IGF- II (Christofori et al., 1994), by the use of antibodies (Arteaga, 1992; Kalebic et al., 1994) and by dominant negative mutants of the IGF-1R (Prager et al., 1994).

We have achieved a reduction in the number of IGF-1Rs by two basic strategies: 1) stable transfection of a plasmid expressing an antisense RNA to the IGF-1R RNA; 2) treatment of the cells with antisense oligodeoxynucleotides against the IGF-1R

RNA. All of these experiments have unequivocally shown that a decrease in the number of IGF-1Rs causes a reversal of the transformed phenotype, as measured by colony formation in soft agar (Baserga et al., 1994; Resnicoff et al., 1993, 1994 a and b).

In in vivo tumorigenesis, the most complete and dramatic effects were observed with C6 rat glioblastoma cells expressing an antisense IGF-1R RNA.

The generation of these cells has already been described in detail elsewhere (Resnicoff et al., 1994 a). The number of animals has been increased since publication of our results. Table 1A constitutes an update of the observations already published.

a) C6 cells expressing antisense IGF-1R RNA are no longer tumorigenic: C6 wild-type and sense cells injected s.c. at 107 cells/rat developed palpable tumors after 5 days of injection and the rats needed to be sacrificed after 3 weeks following the development of bulky tumors. None of the 52 rats injected with antisense IGF-1R RNA transfected cells developed tumors. Some of these animals remained tumor-free for as long as 1 year.

b) C6 antisense cells are able to confer protection against wild-type tumor challenge: When antisense cells were injected 1-3 weeks prior to injection of C6 wild-type cells in opposite flanks, no tumors appeared in any of the animals. Sense cells did not provide any protection against challenge with C6 wild-type cells and in fact, bilateral tumors developed in all animals 5 days after each injection, as shown in table 1A, line 4.

Simultaneous injection of C6 wild-type and antisense cells in opposite flanks resulted in no tumor formation for more than 2 months. Only 1 rat out of 10 developed a small wild-type tumor after 9 days that regressed completely within 2 weeks with no recurrence (table 1A, line 7)

c) C6 antisense cells are able to affect the growth of established tumors: When antisense cells were injected 1-3 weeks after injection of C6 wild-type cells in opposite limbs, a significant reduction in the size of wild-type tumors was observed a week later and within 2 weeks, all tumors completely regressed with no recurrence during the 3 month-follow up (table 1A, line 8). Injection of sense cells had no effect on established tumors.

In order to verify if these effects were specifically due to antisense IGF-1R RNA or to non-viable or non-proliferating cells, sense and antisense cells were irradiated with 5000 rads of ^{60} Co. As shown in table 1A, line 9, when irradiated sense cells were injected 6 days before the challenge with C6 wild-type cells, no protection was observed and in fact, all animals developed wild-type tumors. Irradiated antisense cells were effective in promoting protection against wild-type tumor

challenge. Moreover, irradiated antisense cells could regress established tumors but they were less effective than antisense viable cells: complete regression took 3 weeks vs. 2 weeks when non-irradiated antisense cells were injected. Irradiated sense cells did not affect the growth of established wild-type tumors.
Both protection against wild-type tumor challenge and complete regression of established wild-type tumors (with no recurrence) could also be achieved by treating C6 wild-type cells with antisense oligodeoxynucleotides to the IGF-1R RNA , while treatment with random oligodeoxynucleotides had no effect (table 1B).

Table 1: Effect of antisense strategies to the IGF-1R RNA on tumorigenesis of C6 cells in syngeneic BD IX rats.

A) Expression plasmids.

Cell type		Tumor development
Injection #1(right)	Injection #2 (left)	No. of animals
C6 wild-type cells		45/45
C6 sense cells		34/34
C6 antisense cells		0/52
C6 sense cells	C6 w.t. (day 6)	13/13 bilaterally
C6 antisense cells	C6 w.t. (day 6)	0/13
C6 wild-type cells	C6 sense (simultaneous)	10/10 bilaterally
C6 wild-type cells	C6 antisense (simult.)	0/10
C6 wild-type cells	C6 antisense (1-3 weeks)	18/18 total regression
Irrad. sense cells	C6 w.t. (day 6)	6/6 left flank
Irrad. antisense cells	C6 w.t. (day 6)	0/6
C6 wild-type cells	Irrad. sense (1-3 weeks)	6/6 right flank
C6 wild-type cells	Irrad. antisense (1-3 weeks)	6/6 total regression

10^{7} C6 cells were injected s.c. above the hind leg of 7-week-old male BD IX rats. Palpable tumors appeared after 5 days of injection. In the case of tumor regression, it was complete after 2 weeks of antisense cell injection. Cells were irradiated by exposure to 5000 rads of ^{60}Co.

Table 1 (cont.)
B) Oligodeoxynucleotides

Injection #1(right)	Injection #2(left)	Tumor development No. of animals
C6 + none		9/9
C6 + random sequence		9/9
C6 + antisense oligo		0/9
C6 + random	C6 w.t. (day 5)	9/9 bilaterally
C6 + antisense	C6 w.t. (day 5)	0/9
C6	C6 + random (day 7)	9/9 bilaterally
C6	C6 + antisense (day 7)	9/9 total regression

C6 cells were treated with oligodeoxynucleotides (120 µg/ml) for 24 hours prior to injection.

Altogether these experiments clearly show that a decrease in the number of IGF-1Rs has 3 major effects: a) abrogation of tumorigenicity;
b) protection against tumor challenge;
c) regression of established tumors

The IGF-1R in cell survival:
The abrogation of tumorigenicity observed in the antisense cells suggested that these cells could be dying in vivo but this was difficult to demonstrate after injection of the cells into the subcutaneous tissue. Therefore, in order to overcome this problem, we decided to encapsulate the cells in a bio-diffusion chamber (Abraham et al., 1993; Lanza et al., 1994). Due to the 0.1 µm pore-sized membranes, intact cells cannot leave or enter the chamber; only soluble factors can go through. The cells are placed in the chambers which are then inserted into the subcutaneous tissue of BD IX rats. Once the bio-diffusion chambers are removed from the animals, the cells can be quantitatively recovered and analyzed. Using this device, we were able to demonstrate that the antisense cells die very rapidly in vivo, being most of the cells dead after 3 hours. A time-course experiment for survival of C6 antisense cells in vivo is shown in table 2. Wild-type and sense cells, on the contrary, were not only able to survive in the chambers but they doubled in number after 24 hours in vivo (table 2, lines 1 and 2).

We also analyzed the cells by flow cytometry and showed that C6 cells expressing an antisense RNA to the IGF-1R RNA are dying through apoptosis when placed in vivo whereas in vitro they are arrested in G1 phase of the cell cycle (Resnicoff et al., 1995 a).

We then wanted to determine if the antitumor response could also be triggered when the antisense cells were placed in bio-diffusion chambers. As shown in table 2, antisense cells placed in the chambers for only 3 hours in vivo (at which time, the survival was 3.5%) were able to confer protection against C6 wild-type tumor challenge. On the contrary, when wild-type or sense cells had been placed in the chambers, even for 24 hours, no protection against tumor challenge was observed and all animals developed tumors after 5 days of injection.

We also determined whether antisense cells placed in the chamber were able to induce regression of established tumors. For this purpose, 107 C6 cells were injected s.c. in BD IX rats. After a week, when the tumors were 1 cm. in diameter, antisense cells were implanted in the chamber into the subcutaneous tissue of the rats. After 24 hours, the chambers were removed from the rats. Nine days later, complete regression of established tumors was achieved with no recurrence observed during the 3 month follow-up.

Table 2: Antitumor response triggered by cells expressing an antisense IGF-1R RNA encapsulated in bio-diffusion chambers.

Cells in chambers	Recovery (%)	Inject.(day7)	Tumor development
C6 w.t. (24 hs)	200.0	C6 w.t.	9/9 (day 5)
C6 sense (24 hs)	200.0	C6 w.t.	3/3 (day 5)
C6 antisense (24 hs)	0.0	C6 w.t.	0/6 (day 95)
C6 antisense (3 hs)	3.5	C6 w.t.	0/3 (day 95)
C6 antisense (45 min.)	65.0	C6 w.t.	6/6 (day 5)
C6 antisense (15 min.)	95.0	C6 w.t.	6/6 (day 5)
C6 antisense (5 min.)	100.0	C6 w.t.	6/6 (day 5)

Recovery is defined as the percentage of viable cells recovered from the chambers at the indicated time-points. Viability was assessed by trypan blue exclusion and was over 90% in all cases.

5×10^5 cells were resuspended in 0. 2 ml of phosphate buffered saline (calcium/ magnesium -free) before being placed in the chambers. A week after removal of the chambers, 10^7 C6 wild-type cells were injected s.c. in the rats. The animals were monitored for tumor development. The days indicated in parenthesis refer to the lag in tumor appearance.

Correlation between levels of IGF-1Rs, survival and tumorigenesis:

In order to verify if there was a correlation between the levels of IGF-1Rs, cell survival and tumorigenesis of C6 cells, antisense oligos were used to decrease the number of receptors. As shown in table 3, a decrease of only 50% in the number of IGF-1Rs was enough to induce massive apoptosis in vivo. This effect was achieved with 20 µg/ml of antisense oligo whereas a random sequence did not affect the number of IGF-1Rs and cell survival when tested at 200 µg/ml.

Table 3: Correlation between levels of IGF-1R and cell survival in vivo.

Dose of AS oligo (µg/ml)	Levels of IGF-1R (% of wild-type)	Recovery (%)
0 (or R at 200 µg/ml)	100	200
1	77	54
10	80	35
15	82	35
20	56	1.3
40	56	0.4
60	56	0.4
80	42	0.010
100	43	0.008

If the percentage recovery reflects the real extent of apoptosis in vivo, then, increasing doses of antisense oligo should affect the tumorigenicity of C6 cells, proportionally delaying the appearance of palpable tumors after subcutaneous injection of the treated cells. We decided to test first tumorigenicity in athymic nude mice where the antitumor response does not take place. The results presented in table 4 clearly show a delay in tumor appearance which can be correlated to the percentage of surviving cells. Actually, the tumors appeared even later than we had expected; we have several possibilities to explain this: it could be due to underestimation of the extent of apoptosis in vivo, since we only determined it after 24 hours or it could be due to a slower cell growth when only few cells survive. The expected tumor appearance was estimated based on a doubling time of 24 hours in vivo. Cells treated with random oligo even at 200 µg/ml behaved as controls and tumors were palpable after 4 days of injection.

Table 4: Tumorigenesis of antisense-treated C6 cells in nude mice

Dose of AS oligo (ug/ml)	Recovery (%)	Expected tumor appearance (days)	Tumorigenesis (days)
0	200	4	4
10	35	6	6
20	1.3	12	17
40	0.5	13	17
60	0.4	13	17
100	0.010	19	24

We then determined the tumorigenicity of the antisense-treated C6 cells in immunocompetent animals (syngeneic BD IX rats). The results are presented in table 5. Due to the strong antitumor response elicited by the antisense-treated cells, no tumors appeared in the rats when the dose of oligo was 20 μg/ml or higher. For the lower concentrations, tumors became initially palpable at day 5 but completely regressed within 3 weeks, with no recurrence observed during the 3 month follow-up.

Table 5: Tumorigenesis of antisense-treated C6 cells in immunocompetent rats.

Dose of oligo (μg/ml)	Levels of IGF-1R (% of control)	Tumor development
0	100	palpable (day 4) ; kept growing
1-15	80	palpable (day 5); regressed
20-60	56	no tumors appeared (day 95)
80-120	43	no tumors appeared (day 95)

Analysis of the antitumor response:

The antitumor response elicited by the antisense cells can hardly been neglected since it leads to tumor resistance and can even induce regression of established tumors. Therefore, we decided to investigate the nature of this response and for that purpose, two questions were addressed:

a) Is the antitumor response an immune process?

b) In such case, what are the effectors involved?

We considered the possibility of a humoral response induced by the C6 antisense cells. For this purpose, the sera of the immunized rats were incubated with C6 wild-type cells followed by FACS analysis with a fluoresceinated antibody. There were no significant differences between values obtained from the different groups, indicating that no antibodies directed against C6 wild-type cells could be detected in the sera of antisense cell injected rats.

We also evaluated the involvement of a cellular response and for that purpose, two types of assays were performed: a lymphocyte proliferation assay and a cytotoxicity assay. In a lymphocyte proliferation assay, the lymphocytes from C6 antisense injected rats stimulated with C6 wild-type cells showed a doubling in the number of $CD8^+$ cells, compared to lymphocytes from the sense injected rats. The proportion of $CD4^+$ cells remained unchanged in either group. In a cytotoxicity assay, there was a 2-3 fold increase in the cytotoxic index observed with lymphocytes from antisense injected rats over the lymphocytes from sense injected rats.

Based on these results, $CD8^+$ cells appear to be involved in the antitumor response triggered by the antisense cells. These observations are in agreement with other reports in the literature in which expression of antisense IGF-1 RNA (Trojan et al., 1993) or s-myc (Asai et al., 1994) in C6 rat glioblastoma cells elicits an immune response by $CD8^+$ cells, which results in eradication of established tumors.

Concluding remarks:

We have shown that interference with the IGF-1R by antisense strategies causes massive apoptosis of C6 rat glioblastoma cells. The antisense cells die very rapidly in vivo, eliciting an antitumor response that takes care of the few surviving cells and protects the rats from subsequent tumor challenge. Moreover, when implanted in tumor-bearing rats, the antisense cells are able to induce regression of established tumors.

The use of antisense strategies is frequently and severely critized due to non-specific effects. Because we are fully aware of the pitfalls in the use of antisense oligodeoxynucleotides, we have included appropiate controls: 1) we demonstrate a decrease in the target protein by Scatchard plot analysis, being this decrease proportional to the concentration of antisense oligo used; 2) the differences between the antisense and the random sequences are dramatic: the antisense sequence is able to induce apoptosis in vivo at a dose of 1 µg/ml whereas the random sequence has no effect on survival even at 200 µg/ml; 3) the antisense sequences were carefully designed to avoid G-quartets and palindromic sequences involving more than 6 nucleotides that could account for artifactual antiproliferative effects; 4) different antisense sequences have different effects. We have tested sequences starting at different positions of the IGF-1R cDNA (at the methionine codon, at 3 or 9 nucleotides downstream or at the middle) and have found that the sequences targeted at the 3' end are less active in inducing apoptosis in vivo.

Recent reports in the literature (Krieg et al., 1995; Holt, 1995) discuss the induction of a B cell-mediated immune response due to activation by CpG motifs in synthetic oligos. However, we can also induce the antitumor response by C6 cells expressing an antisense IGF-1R RNA, therefore ruling out an artifactual effect. Preliminary observations seem to indicate that this strategy can even discriminate between normal and tumor cells, making it a promising agent for antitumor therapy.

REFERENCES

Abraham, D., Lange, A.M., Yutanaviboonchai, W., Trpis, M., Dickerson, J.W., Swenson, B. and Eberhard, M.L. (1993) Survival and development of larval Onchocerca volvulus in primate and rodent hosts. J. Parasitol. 79: 571-582

Arteaga, C.L. and Osborne, C.K. (1989) Growth inhibition of human breast cancer cells in vitro with an antibody against the type 1 somatomedin receptor. Cancer Res. 49: 6237-6241

Arteaga, C.L. (1992) Interference of the IGF system as a strategy to inhibit breast cancer growth. Breast Cancer Res. and Treat. 22: 101-106

Asai, A., Miyagi, Y., Hashimoto, H., Lee, S.H., Mishima, K., Sugiyama, A., Tanaka, H., Mochizuki, T., Yasuda, T. and Kuchino, Y. (1994) Modulation of tumor immunogenicity of rat glioma cells by s-myc expression: eradication of rat gliomas in vivo. Cell Growth & Diff. 5: 1153-1158

Baker, J., Liu, J.P., Robertson, E.J. and Efstratiadis, A. (1993) Role of insulin-like growth factors in embryonic and postnatal growth. Cell 75: 73-82

Baserga, R. and Rubin, R. (1993) Cell cycle and growth control. Crit. Rev. Eukaryot. Gene Expr. 3: 47-61

Baserga, R., Sell, C., Porcu,P.L. and Rubini, M. (1994) The role of the IGF-1 receptor in the growth and transformation of mammalian cells. Cellular Prolif. 27: 63-71

Christofori, G., Naik, P. and Hanahan, D. (1994) A second signal supplied by insulin-like growth factor II in oncogene-induced tumorigenesis. Nature 369: 414-418

Coppola, D., Ferber, A., Miura, M., Sell, C., D'Ambrosio, C., Rubin, R. and Baserga, R. (1994) A functional insulin-like growth factor I receptor is required for the mitogenic and transforming activities of the epidermal growth factor receptor. Mol. Cell. Biol. 14: 4588-4595

Crews, C.M. and Erikson, R.L. (1993) Extracellular signals and reversible protein phosphorylation: what to MEK of it all. Cell 74: 215-217

Evan, G., Wyllie, A., Gilbert, C., Littlewood, T., Land, H., Brooks, M., Waters, C., Penn, L. and Hancock, D. (1992) Induction of apoptosis in fibroblasts by the c-myc protein. Cell 69: 119-125

Fisher, D.E. (1994) Apoptosis in cancer therapy: crossing the threshold. Cell 78: 539-542

Harrington, EA., Bennett, M.R., Fanidi, A. and Evan, G. (1994) c-myc-induced apoptosis in fibroblasts is inhibited by specific cytokines. EMBO J. 13: 3286-3295

Kalebic, T., Tsokos, M. and Helman, L.J. (1994) In vivo treatment with antibody against IGF-1 receptor suppresses growth of human rhabdomyosarcoma and down-regulates p34 cdc2. Cancer Res. 54: 5531-5534.

Lanza, R.P., Beyer, A.M. and Chick, W.L. (1994) Xenogeneic humoral responses to islets transplanted in biohybrid diffusion chambers. Transplantation 57: 1371-1375

Liu, J.P., Baker, J., Perkins, A.S., Robertson, E.J. and Efstratiadis, A. (1993) Mice carrying null mutations of the genes encoding insulin-like growth factor I (igf-1) and type I IGF receptor (igf 1 r). Cell 75: 59-72

Myers, M.G., Sun, X.J., Cheatham, B., Jachna, B.R., Glasheen, E.M., Backer, J.M. and White, M.F. (1993) IRS-1 is a common element in insulin and insulin-like growth factor I signaling to the phosphatidilinositol 3' kinase. Endocrinology 132: 1421-1430

Pietrzkowski, Z., Lammers, R., Carpenter, G., Soderquist, A.M., Limardo, M., Phillips, P.D., Ullrich, A. and Baserga, R. (1992) Constitutive expression of IGF-1 and IGF-1 receptor abrogates all requirements for exogenous growth factors. Cell Growth & Diff. 3: 199-205

Prager, D., Li, H.L., Asa, S. and Melmed, S. (1994) Dominant negative inhibition of tumorigenesis in vivo by human insulin- like growth factor I receptor mutant. Proc. Natl. Acad. Sci. USA 91: 2181-2185

Resnicoff, M., Ambrose, D., Coppola,D. and Rubin, R. (1993) Insulin-like growth factor 1 and its receptor mediate the autocrine proliferation of human ovarian carcinoma cell lines. Lab. Invest. 69: 756-760

Resnicoff, M., Sell, C., Rubini, M., Coppola, D., Ambrose, D., Baserga, R. and Rubin, R. (1994 a) Rat glioblastoma cells expressing an antisense RNA to the insulin-like growth factor I (IGF-I) receptor are non-tumorigenic and induce regression of wild-type tumors. Cancer Res. 54: 2218-2222

Resnicoff, M., Coppola, D., Sell, C., Rubin, R., Ferrone, S. and Baserga, R. (1994 b) Growth inhibition of human melanoma cells in nude mice by antisense strategies to the type 1 insulin-like growth factor receptor. Cancer Res. 54: 4848-4850

Sell, C., Rubini, M., Rubin, R., Liu,J.P., Efstratiadis, A. and Baserga, R. (1993) Simian virus 40 large tumor antigen is unable to transform mouse embryonic fibroblasts lacking type I IGF receptor. Proc. Natl. Acad. Sci. USA 90: 11217-11221

Sell, C., Dumenil, G., Deveaud, C., Miura, M., Coppola,D., De Angelis, T., Rubin, R., Efstratiadis, A. and Baserga, R. (1994) Effect of a null mutation of the type I IGF receptor gene on growth and transformation of mouse embryo fibroblasts. Mol. Cell. Biol. 14: 3604-3612

Sell, C., Baserga, R. and Rubin, R. (1995) Insulin-like growth factor I (IGF-I) and the IGF-I receptor prevent etoposide-induced apoptosis. Cancer Res. (in press)

Trojan, J., Blossey, B.K., Johnson, T.R., Rudin, S.D., Tykocinski, M., Ilan, J. and Ilan, J. (1992) Loss of tumorigenicity of rat glioblastoma directed by episome-based antisense cDNA transcription of insulin-like growth factor I. Proc. Natl. Acad. Sci. USA 89: 4874-4878

Trojan, J., Johnson, T.R., Rudin,S.D., Ilan, J., Tykocinski, M. and Ilan, J. (1993) Treatment and prevention of rat glioblastoma by immunogenic C6 cells expressing antisense insulin-like growth factor I RNA. Science 259: 94-97

Ullrich,A., Gray, A., Tam, A.W., Yang-Feng, T., Tsubokawa, M., Collins, C., Henzel, W., Le Bon, T., Kahuria, S., Chen, E., Jakobs, S., Francke,U., Ramachandran, J., Fujita-Yamaguchi,Y. (1986) Insulin-like growth factor I receptor primary structure: comparison with insulin receptor suggests structural determinants that define functional specificity. EMBO J. 5: 2503-2512

Valentinis, B., Porcu, P.L., Quinn, K. and Baserga, R. (1994) The role of the insulin-like growth factor I receptor in the transformation by simian virus 40 T antigen. Oncogene 9: 825-831

White, M.F. and Kahn, C.R. (1994) The insulin signaling system. J. Biol. Chem. 269: 1-4

Inhibition of Megakaryocyte Differentiation *in vivo* by E2F-1

Murray O. Robinson
Amgen, Inc.
Thousand Oaks, CA 91320

Abstract

The transcription factor E2F-1 plays a central role in the cell cycle through its ability to activate genes involved in cell division (La Thangue, 1994; Nevins, 1992). E2F-1 activity is regulated directly by retinoblastoma (Flemington et al., 1993; Helin et al., 1993) and indirectly by the cyclin dependent kinases (Skapek et al., 1995) and their inhibitors (Halevy et al., 1995; Parker et al., 1995), regulators that are known to affect differentiation. Here we investigated a potential role of E2F-1 in differentiation by assaying the ability of megakaryocytes to form platelets in an *in vivo* transgenic model. E2F-1 expression in megakaryocytes blocked differentiation during maturation resulting in severe thrombocytopenia and apoptosis. Furthermore, E2F-1 caused massive megakaryocyte accumulation in both normal and ectopic sites, first evident in E15 embryonic liver. These data indicate that E2F-1 can prevent terminal differentiation directly through its cell cycle stimulatory activity.

Introduction

Decisions of proliferation and differentiation require the activity of genes involved in cell cycle regulation. One such gene, the retinoblastoma suseptibility gene (Rb) functions both as a negative regulator of proliferation and as a mediator of differentiation in certain

NATO ASI Series, Vol. H 99
Tumor Biology
Regulation of Cell Growth, Differentiation and Genetics in Cancer
Edited by Asterios S. Tsiftsoglou, Alan C. Sartorelli, David E. Housman and T. Michael Dexter

cell types. Rb is thought to regulate growth at a key point in G1 to S phase transitions where it is known to interact with a number of cellular proteins. Among these associated proteins, the E2F transcription factor appears to be a biologically relevant target for the action of Rb. Rb as well as Rb-related family members complex with E2F, resulting in an inhibition of E2F transcriptional activity. This specific inhibition by Rb might suggest a potential role for E2F in the control of differentiation.

The E2F transcriptional activity has been implicated as a positive regulator of cell proliferation. For example, E2F binding sites are found in the promoters of several genes that are transcriptionally responsive to proliferative signals. In addition, the overexpression of E2F-1 cDNA can activate DNA synthesis in quiescent cells (Johnson et al., 1993) and transform primary fibroblasts in culture (Singh et al., 1994). Although these effects of E2F on proliferation are known, E2F's effects on differentiation are not elucidated.

To address the role of E2F-1 in differentiation, we examined the consequences of deregulation of E2F-1 expression *in vivo* using the well-defined differentiative process of platelet formation as a model (Gewirtz and Poncz, 1991). Megakaryocyte precursors proliferate and then exit the cell cycle as diploid megakaryoblasts. These cells, readily identifiable due to their large size, mature over several days, finally fragmenting into blood platelets. Experimental manipulation of this process can be easily assayed by monitoring circulating platelet levels in peripheral blood.

Results and Discussion

Transgenic mice were generated expressing the human E2F-1 gene under the control of the megakaryocyte-specific platelet factor 4 (PF4) promoter (Ravid et al., 1991). Expression of this promoter is thought to initiate during the megakaryoblast stage. Following microinjection, four lines of transgenic mice were established and assayed for expression of the human E2F-1 mRNA in bone marrow by semi-quantitative RTpcr (Fig. 1). All four transgenic lines expressed the transgene, albeit at different levels. Bone marrow RNA from line PE82 appeared to have the highest levels of expression, followed by PE51 > PE50 $\geq$ PE42. Immunohistochemical analysis of bone marrow from PE82 using antibodies directed against E2F-1 protein revealed punctate nuclear staining only in transgenic megakaryocytes and not in megakaryocytes from non transgenic sections (data not shown).

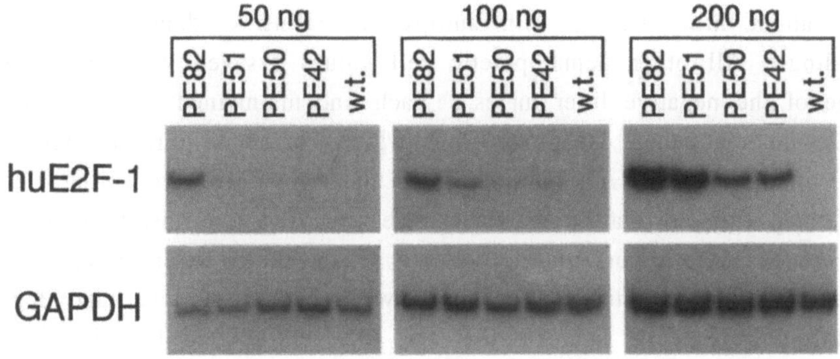

Figure 1.
Expression of Human E2F-1 transgene.
RTpcr analysis of human E2F-1 expression in transgenic bone marrow RNA. The amplified products for huE2F-1 and GAPDH were 321 bp and 352 bp respectively. Methods. For the RTpcr, 50, 100 or 200 nanograms of total RNA from bone marrow of the indicated transgenic (PE#) or wild type (w.t.) mice was reverse transcribed, amplified for 25 cycles using human-specific E2F-1 primers and analyzed as described previously (Robinson and Simon, 1991), modified by substituting rTth polymerase (Perkin Elmer, Norwalk, CT) for MoMuLV Reverse transcriptase and Taq polymerase. Reactions using glyceraldehyde phosphate dehydrogenase (GAPDH) primers were included to demonstrate uniformity of RNA amount and efficiency of amplification among samples. Oligonucleotide primers used for RTpcr were: huE2F-1, forward: 5'-ACCTTCGTAGCATTGCAGACC-3', reverse 5'-TTCTTGCT-CCAGGCTGAGTAG-3'; GAPDH, forward: 5'-GGGTGGAGCCAAACGGGTCATC-3', reverse GCCAGTGAGCTTCCCGTTCAGC-3'. Exposure times were: E2F-1, 72 hr; GAPDH, 1 hr. Quantitation of the amount of radioactivity in each band was performed using a Molecular Dynamics Phosphorimager (Sunnyvale, CA).

To examine the effects of E2F-1 expression in megakaryocytes, blood samples from several mice of each line were analyzed by blood smears and by a Technicon H-1E hematology analyzer (Tarrytown, NY) for alterations in peripheral blood parameters. All expressing lines presented various degrees of thrombocytopenia or platelet deficiency, the most affected line (PE82) exhibited levels only 7% of wild type (Fig. 2A). Cytologic examination of peripheral blood from this line revealed a complete lack of blood platelets (data not shown). The thrombocytopenia in PE82 was so severe that the mice lived on average only about nine weeks, with autopsy revealing evidence of internal bleeding. All other hematopoietic cell values assayed were within the range of the negative litter mates.. Each line maintained a specific level of thrombocytopenia, the severity of which correlated directly with the level of E2F-1 mRNA. Homozygous mice generated from line PE50, confirmed this correlation as such mice, carrying two insertions of the transgene, had lower platelet counts than their heterozygote litter mates (Fig. 2A). This relationship between expression level and degree of thrombocytopenia suggests that the transgene is exerting its effects in a dose-dependent manner.

The severe thrombocytopenia pointed to a defect in terminal differentiation of megakaryocytes. The extent and degree of differentiation in megakaryocytes can be assayed and classified by morphological criteria which, in the mouse, mature through three defined stages (Odell and Jackson, 1968). To pinpoint the block in differentiation, the maturation stage of the transgenic megakaryocytes were examined. The majority (54%) of the megakaryocytes evaluated in the transgenic bone marrow were morphologically judged to be stage two, characterized by a lobated nucleus, a lightly basophilic cytoplasm and little to no granulation (Fig. 2B). Normal control mice had equal numbers of the three stages. We conclude that E2F-1 expression in megakaryocytes acts to block differentiation midway through the post-mitotic maturation phase. This block in differentiation precludes the formation of platelets, causing extreme thrombocytopenia.

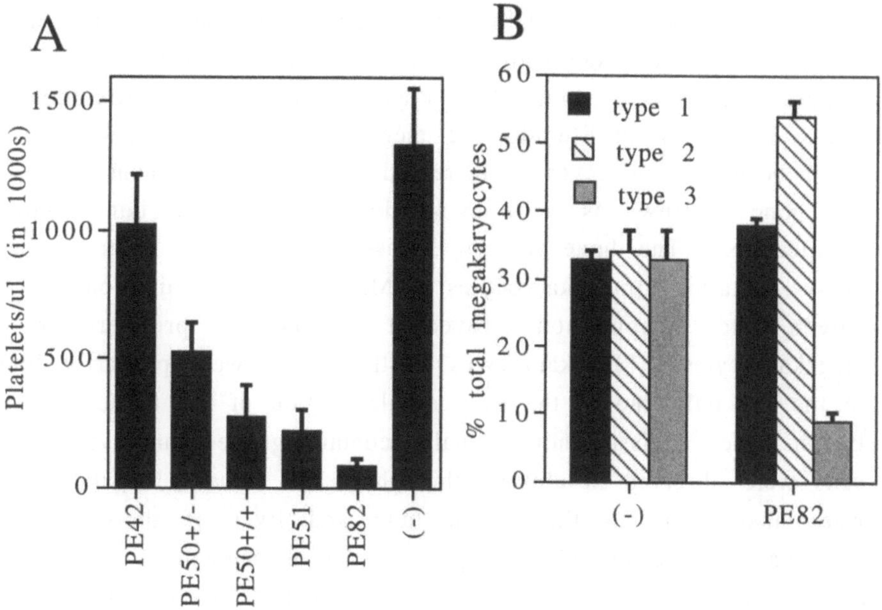

Figure 2.
Differentiation block in E2F-1 transgenic mice.
(A) Thrombocytopenia in E2F-1 expressing mice. Platelet counts from several mice (N $\geq$ 11) of each indicated transgenic and normal lines were determined. The data are presented as the mean (+/- S.D.) of individual determinations. The p value < 0.001 for the difference between the platelet counts of PE50+/+ and PE50+/-. (B) Maturation block in megakaryocytes expressing E2F-1. The data are presented as the mean percent +/- S.D. Methods. Thirty microliters of blood from the lateral tail vein was mixed with 1/10 volume of 3% EDTA (w:v) and examined on an Tecknicon H1-E/hematology analyzer (Myles, Tarrytown, NY). Homozygous mice were identified as previously described (Robinson et al., 1994). To type megakaryocytes, bone marrow from each of three animals of each group was extruded by gentle pneumatic pressure from femurs with the ends removed. The marrow was then smeared onto glass slides, stained using an Aerospray Hematology Slide Stainer (model 7120, Wescor, Inc. Logan, UT). The megakaryocytes were identified and staged according to criteria established by Odell and Jackson (Odell and Jackson, 1968). A minimum of 135 identifiable cells was counted for each mouse.

E2F-1 expression in cultured fibroblasts is known to stimulate cell division (Johnson et al., 1993) and cellular transformation (Singh et al., 1994). Cytologic examination of mice from line PE82 (the highest expressing line) demonstrate that in addition to blocking differentiation, E2F-1 is driving proliferation of megakaryocytes. Histopathology of PE82 bone marrow revealed increased numbers of megakaryocytes in the bone marrow, as well as a disruption of the bone marrow architecture. The bone marrow sinuses were greatly enlarged, with many containing megakaryocytes. Microscopic examination of the spleen and lymph node revealed a massive proliferation of megakaryocytes. Megakaryocytes in the spleen were primarily found in the red pulp and at the sub capsular region of the tissue. In the lymph node, a tissue not normally containing megakaryocytes, these cells were observed in clusters throughout the tissue. Megakaryocytes were also found in the blood, lung and liver of these animals. Examination of embryonic liver, the site of hematopoiesis during fetal development, also revealed increased numbers of megakaryocytes, similar to the phenotype seen in adult tissues. This demonstrates that E2F-1 expression causes proliferation of the megakaryocyte lineage at least as early as fetal hematopoiesis, suggesting that the transgene alone is sufficient to cause the proliferative stimulus.

The extent of megakaryocytosis in the transgenic mice argues that the early megakaryocytes are proliferating. However, it is possible that the megakaryocyte accumulation observed is due to increased survival of cells and not from increased proliferation. To address the fate of the transgenic megakaryocytes further studies were performed. Microscopically, approximately 10% of the megakaryocytes in the spleen and the lymph nodes appeared to have nuclei with altered staining patterns, suggestive of the appearance of apoptosis. Using a modification of the TUNEL method (Surh and Sprent, 1994) to detect DNA fragmentation, the presence of apoptosis was confirmed in about 1/10 of the megakaryocytes in the transgenic spleen (data not shown). Examination of 60 identifiable megakaryocytes in normal spleen sections revealed no evidence of apoptosis, suggesting that apoptosis, if it occurs at all in normal splenic megakaryocytes, is a rare event. Preliminary ultra structure observations using electron microscopy are consistent with this conclusion.

The decision between proliferation and differentiation requires the concerted activity of genes that regulate the cell cycle. Although, E2F-1 is known to have a positive effect on cell division *in vitro*, the data presented here demonstrate that E2F-1 can both stimulate cell division and modulate the differentiative potential of a cell *in vivo*. E2F-1 expression in the megakaryocyte lineage causes proliferation of early PF4 positive cells, leading to greatly increased numbers of mature megakaryocytes. However, it appears that the differentiative capacity of these cells is reduced, causing developmental arrest during the maturation phase. Because these megakaryocytes cannot terminally differentiate into platelets they appear to die through an apoptotic pathway, a fate that under certain conditions E2F-1 has been shown to stimulate (Wu and Levine, 1994). The complete penetrence and the early onset of the phenotype suggest that the proliferation is likely due solely to the effects of E2F-1 expression and not due to secondary events or mutations.

In addition to the effects of E2F-1 shown here, previous experiments suggest that Rb may also be important for differentiation of this lineage. Megakaryocytes from mice expressing a PF4 promoter/temperature-sensitive SV40 large T antigen transgene were unable to terminally differentiate (Robinson et al., 1994). A possible mechanism to explain the phenotypes observed in both of these mice stipulates an interaction between Rb and E2F-1 to control differentiation. We would propose that Rb, known to be expressed abundantly in megakaryocytes (Szekely et al., 1992), allows the cell to differentiate through inhibition of E2F activity. An increase of active E2F-1 (either through overexpression or by blocking Rb) prevents the cell from entering a differentiative state. This model, in which the suppression of E2F activity is required for differentiation, may apply to other terminally differentiating cell types. Rb deficiency in the developing mouse lens causes increased proliferation and apoptosis of lens cells (Morgenbesser et al., 1994), and myoblast differentiation in culture requires functional Rb protein (Gu et al., 1993). Furthermore, evidence from Qin et al.(Qin et al., 1995) that suggests that E2F and Rb form an inhibitory complex to repress transcription through E2F sites is consistent with such a model.

References

Flemington, E. K., Speck, S. H., and Kaelin, W. G., Jr. (1993). E2F-1-mediated transactivation is inhibited by complex formation with the retinoblastoma susceptibility gene product. Proc Natl Acad Sci U S A *90*, 6914-8.

Gewirtz, A. M., and Poncz, M. (1991). Megakaryocytopoiesis and platelet production. In Hematology: Basic principles and practice, R. Hoffman, E. J. Benz, S. J. Shattil, B. Furie and H. J. Cohen, eds. (New York: Churchill Livingstone), pp. 1148-1157.

Gu, W., Schneider, J. W., Condorelli, G., Kaushal, S., Mahdavi, V., and Nadal Ginard, B. (1993). Interaction of myogenic factors and the retinoblastoma protein mediates muscle cell commitment and differentiation. Cell *72*, 309-24.

Halevy, O., Novitch, B. G., Spicer, D. B., Skapek, S. X., Rhee, J., Hannon, G. J., Beach, D., and Lassar, A. B. (1995). Correlation of terminal cell cycle arrest of skeletal muscle with induction of p21 by MyoD. Science *267*, 1018-1021.

Helin, K., Harlow, E., and Fattaey, A. (1993). Inhibition of E2F-1 transactivation by direct binding of the retinoblastoma protein. Mol Cell Biol *13*, 6501-8.

Johnson, D. G., Schwarz, J. K., Cress, W. D., and Nevins, J. R. (1993). Expression of transcription factor E2F1 induces quiescent cells to enter S phase. Nature *365*, 349-52.

La Thangue, N. B. (1994). DRTF1/E2F: an expanding family of heterodimeric transcription factors implicated in cell-cycle control. TIBS, 108-114.

Morgenbesser, S. D., Williams, B. O., Jacks, T., and DePinho, R. A. (1994). p53-dependent apoptosis produced by Rb-deficiency in the developing mouse lens. Nature *371*, 72-74.

Nevins, J. R. (1992). E2F: a link between the Rb tumor suppressor protein and viral oncoproteins. Science *258*, 424-9.

Odell, T. T., and Jackson, C. W. (1968). Polyploidy and maturation of megakaryocytes. Blood *32*, 102-110.

Parker, S. B., Eichele, G., Zhang, P., Rawls, A., Sands, A. T., Bradley, A., Olson, E. N., Harper, J. W., and Elledge, S. J. (1995). p53-independent expression of p21/cip1 in muscle and other terminally differentiating cells. Science *267*, 1024-1027.

Qin, X., Livingston, D. M., Ewen, M., Sellers, W. R., Arany, Z., and Jr., W. G. K. (1995). The transcription factor E2F-1 is a downstream target of Rb Action. Mol. Cell. Biol. *15*, 742-755.

Ravid, K., Beeler, D. L., Rabin, M. S., Ruley, H. E., and Rosenberg, R. D. (1991). Selective targeting of gene products with the megakaryocyte platelet factor 4 promoter. Proc Natl Acad Sci U S A *88*, 1521-5.

Robinson, M. O., and Simon, M. I. (1991). Determining transcript number using the polymerase chain reaction: Pgk-2, mP2, and PGK-2 transgene mRNA levels during spermatogenesis. Nucleic Acids Res *19*, 1557-62.

Robinson, M. O., Zhou, W., Hokum, M., Danilenko, D. M., Hsu, R.-Y., Atherton, R. E., Xu, W., Mu, S., Saris, C., Swift, S., Elliot, G., Del Castillo, J., Hunt, P., and Bosselman, R. A. (1994). The tsA58 simian virus 40 large tumor antigen disrupts megakaryocyte differentiation in transgenic mice. Proc. Natl. Acad. Sci. USA *91*, 12798-12802.

Singh, P., Wong, S. H., and Hong, W. (1994). Overexpression of E2F-1 in rat embryo fibroblasts leads to neoplastic transformation. Embo J *13*, 3329-38.

Skapek, S. X., Rhee, J., Spicer, D. B., and Lassar, A. B. (1995). Inhibition of Myogenic differentiation in proliferating myoblasts by cyclin D1-dependent kinase. Science *267*, 1022-1024.

Surh, C. D., and Sprent, J. (1994). T-cell apoptosis detected *in situ* during positive and negative selection in the thymus. Nature *372*, 100-103.

Szekely, L., Jiang, W. Q., Bulic Jakus, F., Rosen, A., Ringertz, N., Klein, G., and Wiman, K. G. (1992). Cell type and differentiation dependent heterogeneity in retinoblastoma protein expression in SCID mouse fetuses. Cell Growth Differ *3*, 149-56.

Wu, X., and Levine, A. J. (1994). E2F-1 and p53 cooperate to mediate apoptosis. Proc Natl Acad Sci U S A *91*, 3602-6.

Regulation of c-*myc* mRNA Half-Life by an RNA-Binding Protein

Jeffrey Ross, Philip Bernstein[1], Rebecca D. Prokipcak[2], and David J. Herrick[3]

McArdle Laboratory for Cancer Research and Department of Pathology
University of Wisconsin - Madison
1400 University Ave.
Madison, WI 53706

----

[1]The Rockefeller University, Box 168, 1230 York Ave., New York, NY 10021-6399

[2]Department of Pharmacology, Rm. 4308 Medical Sciences Building, 8 Taddle Creek Rd., University of Toronto, Toronto, Ontario M5S 1A8 CANADA

[3]Children's Hospital of Philadelphia, Division of Pulmonary Medicine, 34th St. and Civic Center Blvd., Philadelphia, PA 19104

KEYWORDS/ABSTRACT: mRNA stability/ c-*myc*/ coding region determinant/ RNA-binding protein/ cell-free extracts/ regulation endoribonuclease
c-*myc* mRNA contains at least two discrete sequence elements that account for its short half-life, one in the 3'-untranslated region, the other in the carboxy terminal coding region (coding region determinant). A combination of experiments with cell-free extracts and intact cells suggests that the coding region determinant interacts with an abundant RNA-binding protein that, when bound to the mRNA,

NATO ASI Series, Vol. H 99
Tumor Biology
Regulation of Cell Growth, Differentiation and Genetics in Cancer
Edited by Asterios S. Tsiftsoglou, Alan C. Sartorelli, David E. Housman and T. Michael Dexter

protects it from endonucleolytic attack. The mRNA-protein complex might serve as a fail-safe mechanism to permit rapid c-*myc* mRNA destruction at certain times.

INTRODUCTION

The c-*myc* proto-oncogene encodes phosphoproteins that influence cell proliferation, differentiation, and neoplastic transformation (reviewed in Cole, 1990; Luscher and Eisenman, 1990; Spencer and Groudine, 1991). Perhaps because of its central role in essential cell processes, the gene is regulated both transcriptionally and post-transcriptionally. Depending on cell growth conditions, the mRNA half-life can vary from approximately 30 to 120 minutes. Thus, the steady-state level of c-*myc* mRNA, like that of many other mRNAs, is determined to a large extent by its half-life, which is not fixed (reviewed in (Beelman and Parker, 1995; Ross, 1995; Sachs, 1993) One reason for this variability is that the mRNA contains at least two half-life determinants, one in the 3'-untranslated region (3'-UTR), the other in the carboxy terminal portion of the coding region (the coding region determinant or CRD). Deleting either segment prolongs the mRNA half-life, but only by a factor of five or less. The CRD was first detected indirectly through the analysis of 3'-UTR deletions. c-*myc* mRNA lacking most of its 3'-UTR was more stable than wild-type c-*myc* mRNA but was still significantly less stable than most other mRNA's. Therefore, the 3'-truncated mRNA retained an instability determinant (Aghib et al., 1990; Bonnieu et al., 1990; Jones and Cole, 1987; Laird-Offringa et al., 1991; Wisdom and Lee, 1990). Subsequent experiments demonstrated directly that the carboxy terminal coding region of the mRNA contains an instability determinant (Wisdom and Lee, 1991).

These observations raised three questions. *i.*) Why does c-*myc* mRNA contain two instability-determining segments? *ii.*) Does each instability determinant specify a unique degradation pathway? *iii.*) What *trans*-acting regulatory factors interact with each segment, and how and under what circumstances do they control mRNA stability? The experiments described here were designed primarily to address the latter question and exploit an *in vitro* mRNA decay system, which includes polysomes from the human erythroleukemia cell line, K562. Previous studies indicated that this system provides a reasonable picture of mRNA decay in intact cells, since relative mRNA half-lives are the same *in vitro* as in cells, and since mRNA regulatory factors can be identified and purified using this system (Brewer and Ross, 1988 and 1989; Brewer, 1991; Ross and Kobs, 1986). We reasoned that *trans*-acting regulatory proteins that function by binding to specific regions of c-*myc* mRNA can be identified *in vitro* by adding a competitor RNA corresponding to that region, or its complement, to the polysome-containing reactions. The competitor would titrate off any binding proteins and might thereby affect the mRNA half-life.

Using this approach, we find that a 70 kDa polysomal protein binds with high specificity to the c-*myc* mRNA coding region determinant. Competitor RNA corresponding to this region induces c-*myc* mRNA destabilization *in vitro*, apparently by exposing an endonuclease cleavage site in the mRNA. We describe the purification of this protein and confirm the role of the coding region determinant in intact cells and cell-free extracts.

RESULTS

Destabilization and Endonucleolytic Cleavage of Polysome-Bound c-*myc* mRNA in Cell-Free mRNA Decay Reactions Supplemented with Coding Region Determinant RNA Competitor

c-*myc* mRNA is unstable in cell-free mRNA decay reactions containing polysomes (Ross and Kobs, 1986; Brewer and Ross, 1988 and 1989). To determine whether the polysomes contain RNA-binding proteins that influence the mRNA half-life, either as stabilizers or destabilizers, various competitor RNAs were added to the reactions. If a competitor sequestered a regulatory factor, by virtue of binding to it and effectively titrating it, and if that factor functioned to affect the mRNA half-life, then the stability of the mRNA should be altered by the competitor. If the factor were specific, then only c-*myc* mRNA stability should be affected by a c-*myc* competitor.

A variety of sense and anti-sense competitors were added to decay reactions, and c-*myc* mRNA half-life was determined by Northern blotting. The only competitor with any effect was a 182 nucleotide (nt) sense RNA corresponding the last 182 nt's of the mRNA coding region (Bernstein et al., 1992). To test the specificity of this result further, we prepared chimeric genes encoding mRNAs with one or both putative c-*myc* mRNA instability determinants (CRD and 3'-UTR) located within human ß-globin mRNA (Fig 1). Globin mRNA was considered appropriate for these studies for two reasons. First, the half-life of human ß-globin mRNA is 20 to 29 hr in erythroid cells (Ross and Sullivan, 1985). Therefore, instability determinants from c-*myc* should induce measurable destabilization of a globin-*myc* hybrid mRNA, assuming that the globin sequences do not repress c-*myc* destabilization signals. Second, since HeLa cells, into which the genes were transfected, lack endogenous globin mRNA, cross hybridization of globin probes to the chimeric mRNAs is avoided. Each chimeric mRNA maintained the normal ß-globin or c-*myc* open reading frames and used the normal ß-globin initiation codon as well as the normal ß-globin or c-*myc* termination codon (Fig. 1).

Polysomes from HeLa cells expressing the Glob-Myc-Glob mRNA

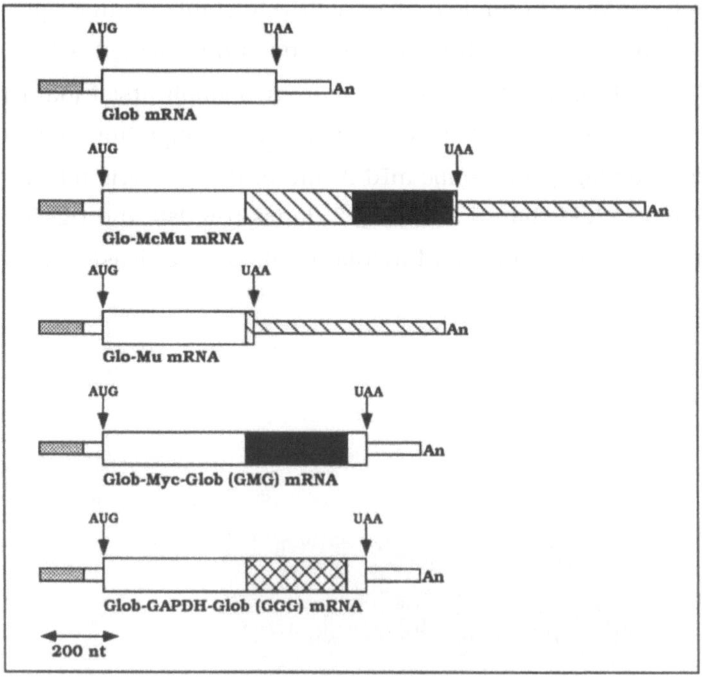

Figure 1. mRNAs transcribed by transfected genes driven by the CMV or c-*fos* promoter. mRNAs from the CMV promoter contained 122 nt derived from CMV 5'-untranslated sequences (shaded rectangles). Each chimeric mRNA contained ß-globin sequences (unfilled rectangle) extending from the ß-globin 5' terminus to the ß-globin EcoR I site. Sequences downstream from the EcoR I site were from globin, c-*myc* (hatched and black rectangles), glyceraldehyde phosphate dehydrogenase (GAPDH; cross-hatched rectangles), or a combination of sequences. The black rectangle designates the c-*myc* mRNA coding region determinant. The DNA sequence of each junction was confirmed using double-stranded sequencing. Each chimeric mRNA was translated in-frame from the indicated AUG to UAA sites. Reprinted with permission from The American Society of Microbiology.

(Fig. 1) were incubated without competitor or with the 182 nt CRD competetor RNA (denoted as "RNA d" in Fig. 2). RNA prepared from the reactions was electrophoresed, transferred to a filter, and hybridized with a probe that would recognize both globin and c-*myc* sequences (Fig. 2, left) or only globin sequences 5' of the c-*myc* insert (Fig. 2, right). The chimeric Glob-Myc-Glob mRNA was realtively stable in unsupplemented reactions (Fig. 2, lanes 1-4) or in reactions supplemented with "non-specific" RNA from plasmid vectors or from ß-globin itself (data not shown). In contrast, "RNA d", the c-*myc* CRD competitor, reduced the half-lives of endogenous c-*myc* mRNA and of the chimeric mRNA by at least 20-fold to less than 5 min (Fig. 2, lanes 5-8, left and right). Therefore, the c-*myc* CRD can function on its own, at least *in vitro*, as a

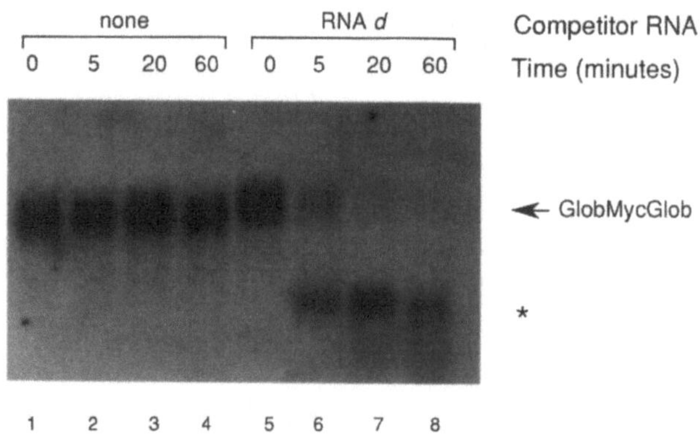

Figure 2. Competitor RNA-induced destabilization of chimeric GlobMycGlob mRNA in cell-free mRNA decay reactions. Polysomes from HeLa cells transfected with the CMV-driven GlobMycGlob gene (Fig. 1) were incubated with or without excess competitor RNA *d*, which was made *in vitro* and corresponds to the coding region determinant. RNA was extracted from reactions incubated for the idicated times, blotted, and hybridized to a probe for the 5' one-half of ß-globin mRNA. Reprinted with permission from The Cold Spring Harbor Laboratory Press.

determinant of mRNA half-life, even in the context of a stable mRNA. Moreover, in reactions supplemented with CRD RNA, an apparent endonucleolytic decay product was readily observed (Fig. 2, asterisks). The size of this product is consistent with it being an endonucleolytic cleavage product, with the cleavage occurring within (approximately in the middle of) the CRD.

These and other observations are consistent with a model (summarized in Fig. 3) in which a shielding protein, the CRD-binding protein or CRD-BP, is associated with polysomes and perhaps with c-*myc* mRNA itslef. Additon of CRD competitor RNA ("RNA d") titrates the protein away from the polysomes (or mRNA). As a result, the shield is removed, exposing the mRNA to endonucleolytic attack. To investigate this model, gel mobility shift assays were performed. "^{32}P-RNA d" was uniformly radiolabeled and incubated with polysomes in a mock mRNA decay reaction, to allow RNA-protein (RNP) complexes to form. The reaction mixture was then analyzed by standard methods involving

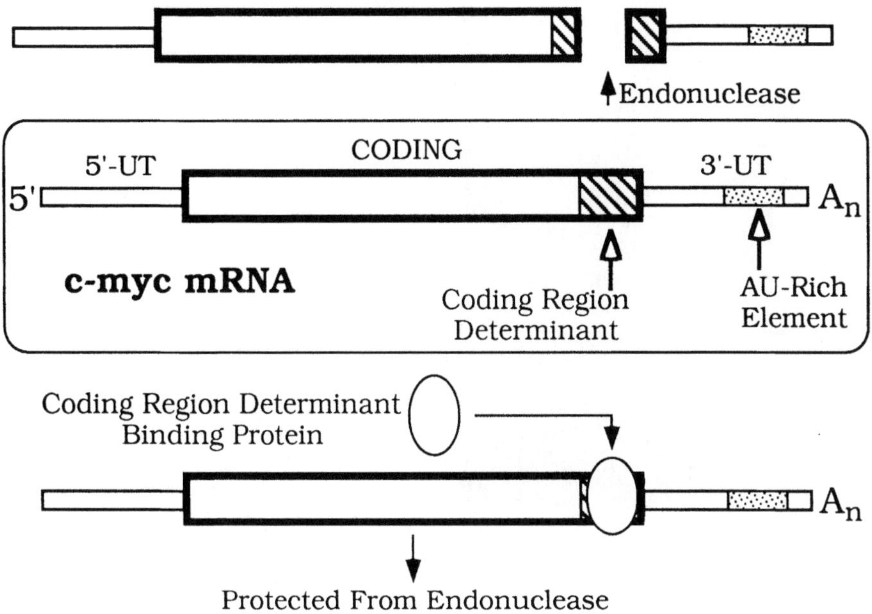

Figure 3. Model for protection of the c-*myc* mRNA coding region determinant (CRD) by the coding region determinant-binding protein (CRD-BP). In the absence of the CRD-BP, the CRD is susceptible to rapid cleavage by a ribosome- or mRNA-associated endonuclease, accounting for the appearance of a prominent endonucleolytic deay product in cell-free reactions supplemented with competitor RNA, the function of which is to titrate the CRD-BP from the mRNA (Fig. 2). In the presence of the CRD-BP (unfilled oval), the mRNA is shielded from the endonuclease. Note the presence of two mRNA half-life determinants in c-Imyc mRNA (coding region and AU-rich element).

electrophoresis in a non-denaturing gel (Bernstein et al., 1992). As a control, parallel reactions were performed with radiolabeled globin RNA of the same size.

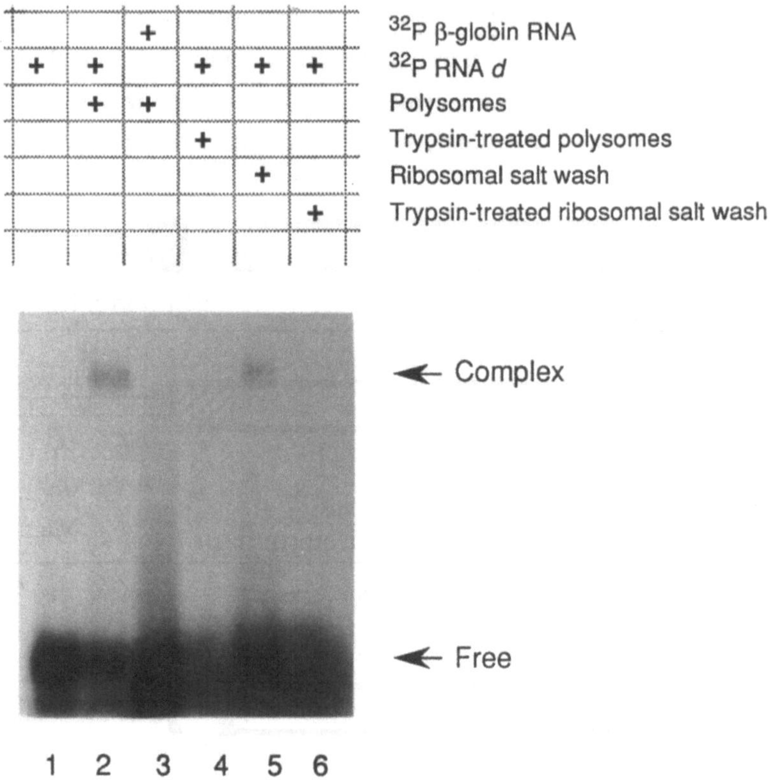

Figure 4. Binding of polysomal protein to coding region determinant RNA. ^{32}P-RNA corresponding to the c-*myc* mRNA coding region determinant or to ß-globin mRNA was incubated with polysomes, polysomes pre-treated with trypsin, or a high salt, ribosome-free extract of polysomes (ribosomal salt wash). RNA-protein complexes were detected by electrophoresing the reactions in a 6% nondenaturing gel and autoradiography. Free and Complex refer to free RNA and RNA in a RNA-protein complex. Reprinted with permission from The Cold Spring Harbor Laboratory Press. Reprinted with permission from The Cold Spring Harbor Laboratory Press.

DISCUSSION

The abundance and phosphorylation state of the c-*myc* proteins can have significant effects on cell growth and differentiation. c-*myc* protein levels, in turn, are determined by a network of transcriptional and post-transcriptional control processes. Perhaps the two c-*myc* mRNA stability determinants also respond to different control factors as part of this multi-level regulatory strategy. Our results suggest that the coding region of the mRNA contains a segment that has the capacity to interact with a protein, which we have purified, and to be protected by that protein.

It is important to note that we have not proved that the protein is actually bound to the mRNA in polysomes. All we know is that the protein is relatively abundant in polysomes, is scarce in the post-polysomal supernatant and in nuclei (data not shown), and is readily release by high salt. We also do not know the function of the protein with respect to mRNA decay or whether each stability determinant (the one in the coding region and the one in the 3'-UTR) has a unique role in cells at different stages of differentiation or in different environments. For example, is the CRD a fail-safe mechanism to allow the cell rapidly and efficiently to eliminate c-*myc* mRNA following some insult, such as exposure to a toxin or removal of growth factors? If so, does it affect mRNA stability only during those times? Constitutive expression of c-*myc* mRNA in fibroblasts following serum withdrawal causes the cells to undergo apoptosis, instead of entering a quiescent state (Bissonnette et al., 1992; Evan et al.,1992; Fanidi et al.,1992). Perhaps this or other insults to cells triggers rapid c-*myc* mRNA destabilization via the CRD and endonucleolytic pathway. Future experiments exploiting immunologic and molecular probes for the CRD-BP and the endoribonuclease should help to clarify the role of the CRD and its binding protein.

A RNP band was observed in reactions containing polysomes plus "RNA d" (Fig. 4A, lanes 1 and 2) but not globin RNA (lane 3). The RNP band was not observed when the polysomes were pretreated with trypsin, indicating that formation of the gel-shifted RNP band required polysomal protein. The protein(s) responsible for the gel shift could be solubilized by extracting the polysomes with high salt. Thus, when "^{32}P-RNA d" was incubated with the high salt extract (ribosomal salt wash or RSW), a RNP complex of the same mobility was observed (Fig. 4, lane 5). The RSW RNP complex was also trypsin-sensitive (lane 6). These data are consistent with the model and indicate that a polysomal protein is capable of binding with apparent high specificity to the c-*myc* mRNA CRD. Subsequent experiments involving purification of the CRD-BP confirmed this conclusion (see below).

Effect of the c-*myc* mRNA Coding Region Determinant of the mRNA Half-Life in Cells

Several experiments were performed to assess further how the CRD influences mRNA half-life in intact cells. As determined by comparing the steady-state expression of various chimeric mRNAs (Fig. 1) in HeLa cells, it appeared as though mRNAs containing the 3'-UTR were the least stable, while the mRNA containing the c-*myc* CRD was more stable than that containing the c-*myc* 3'-UTR but less stable than a control chimeric mRNA containing glyceraldehyde phosphate dehydrogenase (GAPDH) sequences in place of c-*myc* sequences (Herrick and Ross, 1994). Similar conclusions were derived from experiments using actinomycin D.

To confirm these result and to obtain an independent measur of mRNA half-lives in a way that avoids possible drug effects on mRNA stability, mRNA half-lives were measured following serum induction of c-*fos* promoter-driven genes. Unstable mRNAs under c-*fos* promoter control are scarce in serum-starved 3T3 cells, because the c-*fos* promoter is silent and the mRNA is short-lived (Greenberg and Ziff, 1984; Kabnick and Housman, 1988; Shyu et al., 1989 and 1991; Wellington et al., 1993).

Adding serum transiently activates the promoter, causing a burst of transcription lasting only 30 to 60 minutes. The level of the mRNA then increases, and its subsequent rate of disappearance reflects its half-life.

The mRNA generated from the *fos*-Glob-Myc-Glob (GMG) gene was clearly less stable than that of the control *fos*-Glob-GAPDH-Glob (GGG) mRNA (Fig. 5). Therefore, although the c-*myc* CRD region does not have as significant effect on mRNA half-life as does the c-*myc* 3'-UTR (Fig. 5, compare Glo-Mu with GMG), the c-*myc* sequence clearly has a reproducible and measurable effect in destabilizing the otherwise very stable globin sequences. Thus, each c-*myc* instability determinant (3'-UTR and CRD) functioned independently, even when embedded within a very stable mRNA. The c-*myc* CRD does influence mRNA stability in intact cells, because GMG mRNA was 2- to 3-fold less stable than GGG mRNA. These data confirm previous results obtained with transfected myoblasts (Wisdom and Lee, 1990 and 1991) and with cell-free extracts (Bernstein et al., 1992).

Purification of the c-*myc* mRNA Coding Region Determinant-Binding Protein (CRD-BP)

The CRD-BP co-sediments with polysomes but can be solubilized with high salt and assayed by RNA-protein gel shift, using ^{32}P-RNA corresponding to the c-*myc* CRD (Fig. 4). Optimal solubilization of the CRD-BP was achieved by incubating polysomes in 1.0 M NaCl, pelleting out the polysomes, and harvesting the supernatant ribosomal salt wash (RSW). The binding activity in the RSW showed specificity for the c-*myc* CRD sequence, since neither ß-globin RNA nor pGEM4 vector RNA competed for binding as effectively as CRD RNA (data not shown). Binding was consistently more efficient with RSW than with polysomes, and RSW was used as the starting material for a 4-step purification of the CRD-BP (Prokipcak et al., 1994). High performance liquid chromatography (HPLC) using a C4 reverse phase column was chosen as the final step because of its high resolution and high capacity, and, fortunately, because binding activity could be renatured from acetonitrile-trifluoroacetic acid (TFA). Preparative fractionation was achieved with a shallow gradient, and approximately 90% of the recovered RNA-binding activity was detected in fractions corresponding to 37% acetonitrile. After the last step, approximately 70-80% of the protein migrated in a single 70 kDa band (Fig. 6, lane 6). When this band was cut out of a gel and renatured and

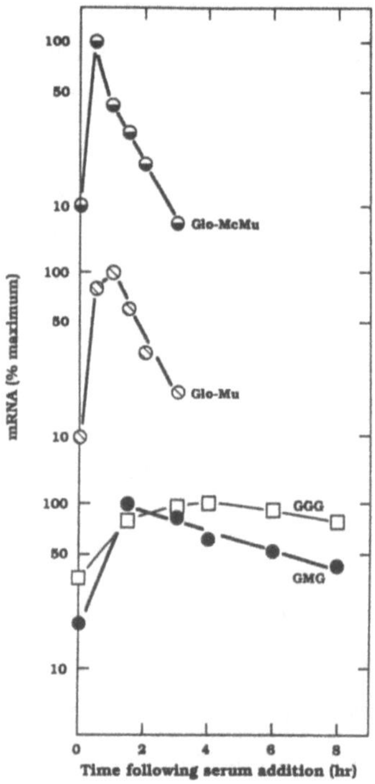

Figure 5. Kinetics of expression of globin-*myc* and globin-GAPDH chimeric mRNAs in serum-induced 3T3 cells. Stable pools of 3T3 cells transfected with the indicated *fos*-driven genes (see Fig. 1) were cultured in separate dishes for 25 hr. in medium containing 0.5% serum. The medium was then aspirated and replaced with medium containing 15% serum. At the indicated times thereafter, cells from individual dishes were harvested, cytoplasmic RNA was extracted, and 10 µg of RNA was analyzed by northern blotting and hybridization with a ß-globin cDNA probe. The intensities of gel bands from two or three experiments were quantitated by densitometry and averaged, and data were plotted as percentage of maximum mRNA level following serum induction. Reprinted with permission from The American Society for Microbiology.

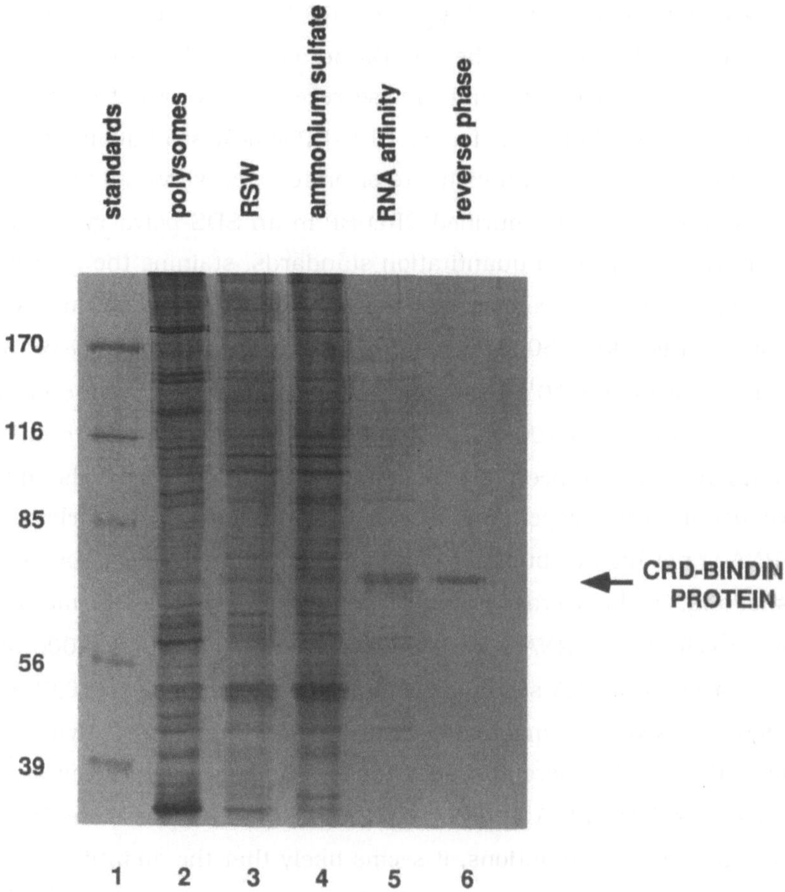

Figure 6. Analysis of coding region determinant-binding protein (CRD-BP) purification by SDS-PAGE. CRD-BP was purified from 132 mg of ribosomal salt wash protein derived from 30 liters of K562 cells. Proteins from each step were electrophoresed in an 8% SDS-polyacrylamide gel, which was stained with silver. The protein amounts indicated below represent the percentage of total protein recovered at each step that was loaded onto each lane. Lane 1: protein molecular weight standards; sizes in kDa are indicated on the left. Lane 2: 2.5 μg polysomal protein (0.002% of total). Lane 3: 2.5 μg ribosomal salt wash (RSW) protein (0.005% of total). Lane 4: 2.5 μg 40-60% ammounium sulphate precipitated protein (0.02% of total). Lane 5: 0.5 μg CRD RNA affinity-purified material (0.05 % of total). Lane 6: 3.5 μl of material purified by reverse phase HPLC (0.4% of total). Reprinted with permission from The American Society for Biochemistry and Molecular Biology.

analyzed by gel shift with ^{32}P-labeled CRD RNA, astrong gel shift was observed, confirming that the 70 kDa protein was the CRD-BP.

Specific activities and precise recoveries were not obtained, because the gel shift assay for the CRD-BP is only semi-quantitative (data not shown). However, an estimate of protein recovery was made by electrophoresing HPLC-purified CRD-BP in an SDS-polyacrylamide gel along with two protein quantitation standards, staining the gel with Coomassie blue, and comparing band intensities. Based on this assay, approximately 30 to 60 μg (2.5 to 5 x 10^{14} molecules) of CRD-BP was recovered from 1 x 10^{10} cells. Assuming 50% recovery of the protein, there are 50,000 to 100,000 molecules per cell of CRD-BP. c-*myc* mRNA abundance has not been quantitated carefully in cultured cells, in part because it varies, depending on growth conditions. Nevertheless, c-*myc* mRNA is not highly abundant, and its steady-state level can be estimated, as follows: If the average K562 cell contains 5 x 10^5 mRNA molecules [20 pg per cell of total RNA x 2.5% mRNA, average 6 x10^5 Da (2000 nt)], and if the average mRNA sequence complexity corresponds to 10,000 different mRNAs per mammalian cell (Bishop et al., 1974; Ryffel and McCarthy, 1975), there are, on average, 50 molecules per cell of each mRNA (5 x 10^5 mRNA molecules per 10^4 mRNAs). Therefore, based on a few reasonable assumptions, it seems likely that the quantity per cell of CRD-BP exceeds that of c-*myc* mRNA by as much as 1500-fold (75,000 CRD-BP molecules per 50 c-*myc* mRNA molecules). For comparison, each cell would contain ≈ 2 to 5 x 10^6 ribosomes, giving a ratio of approximately one molecule of CRD-BP per 100 ribosomes.

We do not understand why the CRD-BP seems to be in such an excess over c-*myc* mRNA. Nevertheless, two observations illustrate the considerable specificity with which the CRD-BP might interact with c-*myc* mRNA. *i.*) The CRD-BP does not bind to prokaryotic vector RNAs or to globin RNA (Prokipcak et al., 1994). *ii.*) Exogenous c-*myc* CRD competitor RNA destabilizes c-*myc* mRNA *in vitro* but not polysome-associated c-*myb* or AP4 mRNAs, even though AP4 mRNA encodes a basic, helix-loop-helix, leucine zipper motif similar to that encoded by the c-*myc* CRD (Bernstein et al., 1992). These data imply a limited function for the CRD-BP, perhaps directed specifically at c-*myc* mRNA. On the other hand, the quantity per cell of CRD-BP might be in excess of the amount of c-*myc* mRNA by as much as 1500-fold (see above). If this is so, three important questions will need to be answered. *i.*) Is the CRD-BP actually bound to polysome-associated c-*myc* mRNA? *ii.*) If so, does more than one CRD-BP molecule bind to each c-*myc* mRNA? *iii.*) Are any excess CRD-BP molecules bound to other mRNAs or to ribosomes?

ACKNOWLEDGEMENTS

This work was supported by grants CA23076, CA07175, and CA63676 from the National Cancer Institute.

REFERENCES

Aghib, D F, JM Bishop, S Ottolenghi, A Guerrasio, A Serra, and G Saglio (1990) A 3' truncation of MYC caused by chromosomal translocation in a human T-cell leukemia increases mRNA stability. Oncogene 5:707-711.

Beelman, CA and R Parker (1995) Degradation of mRNA in eukaryotes. Cell 81:179-183.

Bernstein, PL, DJ Herrick, RD Prokipcak, and J Ross (1992) Control of c-*myc* mRNA Half-Life *in Vitro* by a Protein Capable of Binding to a Coding Region Stability Determinant. Genes Dev. 6:642-654.

Bishop, JO, JG Morton, M Rosbash, and M Richardson (1974) Nature 250:199-204.

Bissonnette, RP, F Echeverri, A Mahboubi, and DR Green (1992) Apoptotic cell death induced by c-*myc* is inhibited by *bcl-2*. Nature 359:552-554.

Bonnieu, A, P Roux, L Marty, P Jeanteur, and M Piechaczyk (1990) AUUUA motifs are dispensable for rapid degradation of the mouse c-*myc* mRNA. Oncogene 5:1585-1588.

Brewer, G (1991) An A+U-rich element RNA-binding factor regulates c-*myc* mRNA stability in vitro. Mol. Cell. Biol. 11:2460-2466.

Brewer G and J Ross (1988) Poly(A) shortening and degradation of the 3' A+U-rich sequences of human c-*myc* mRNA in a cell-free system. Mol. Cell. Biol. 8:1697-1708.

Brewer, G and J Ross (1989) Regulation of c-*myc* mRNA stability *in vitro* by a labile destabilizer with an essential nucleic acid component. Mol. Cell. Biol. 9:1996-2006.

Cole, MD (1990) The *myb* and *myc* nuclear oncogenes as transcriptional activators. Curr. Opin. Cell Biol. 2:502-508.

Evan, GI, AH Wyllie, CS Gilbert, TD Littlewood, H Land, M Brooks, CM Waters, LZ Penn, and DC Hancock (1992) Induction of apoptosis in fibroblasts by c-*myc* protein. Cell 69:119-128.

Fanidi, A, EA Harrington, and GI Evan (1992) Cooperative interaction between c-*myc* and *bcl-2* proto-oncogenes. Nature 359:554-556.

Greenberg, ME and EB Ziff (1984) Stimulation of 3T3 cells induces transcription of the c-fos proto-oncogene. Nature 311:433-437.

Herrick, DJ and J Ross (1994) The half-life of c-*myc* mRNA in growing and serum-stimulated cells: influence of the coding and 3'-untranslated regions and the role of ribosome translocation. Mol. Cell. Biol. 14: 2119-2128.

Jones, TR and MD Cole (1987) Rapid cytoplasmic turnover of c-*myc* mRNA: requirement of the 3' untranslated sequences. Mol. Cell. Biol. 7:4513-4521.

Kabnick, KS and DE Housman (1988) Determinants that contribute to cytoplasmic stability of human c-*fos* and ß-globin mRNAs are located at several sites in each mRNA. Mol. Cell. Biol. 8:3244-3250.

Laird-Offringa, IA, P Elfferich, and AJ van der Eb (1991) Rapid c-*myc* mRNA degradation does not require (A+U)-rich sequences or complete translation of the mRNA. Nucl. Acids Res. 19:2387-2394.

Luscher B and RN Eisenman (1990) New light on Myc and Myb. Part I. Myc. Genes Dev. 4:2025-2035.

Prokipcak, RD, DJ Herrick, and J Ross (1994) Purification and properties of a protein that binds to the C-terminal coding region of human c-*myc* mRNA. J. Biol. Chem. 269:9261-9269.

Ross, J (1995) mRNA Stability in Mammalian Cells, in "Microbiological Reviews," W. K. Joklik, ed. (American Society for Microbiology, Washington, DC), 59:423-450.

Ross, J and G Kobs (1986) H4 histone messenger RNA decay in cell-free extracts initiates at or near the 3' terminus and proceeds 3' to 5'. J. Mol. Biol. 188:579-593.

Ross, J and T Sullivan (1985) Half-lives of beta and gamma globin messenger RNAs and of protein synthetic capacity in cultured human reticulocytes. Blood 66:1149-1154.

Ryffel, GU and BJ McCarthy (1975) Biochemistry 14:1379-1385.

Sachs, AB (1993) Messenger RNA degradation in eukaryotes. Cell 74:413-421.

Shyu, A-B, ME Greenberg, and JG Belasco (1989) The c-*fos* transcript is targeted for rapid decay by two distinct mRNA degradation pathways. Genes Dev. 3:60-72.

Shyu, A-B, JG Belasco, and ME Greenberg (1991) Two distinct destabilizing elements in the c-*fos* message trigger deadenylation as a first step in rapid mRNA decay. Genes Dev. 5:221-231.

Spencer, CA and M Groudine (1991) Control of c-*myc* regulation in normal and neoplastic cells. Adv. Cancer Res. 56:1-48.

Wellington, CL, ME Greenberg, and JG Belasco (1993) The destabilizing elements in the coding region of c-*fos* mRNA are recognized as RNA. Mol. Cell. Biol. 13:5034-5042.

Wisdom, R and W Lee (1990) Translation of c-*myc* mRNA is required for its post-transcriptional regulation during myogenesis. J. Biol. Chem. 265:19015-19021.

Wisdom, R and W Lee (1991) The protein-coding region of c-*myc* mRNA contains a sequence that specifies rapid mRNA turnover and induction by protein synthesis inhibitors. Genes Dev. 5:232-243.

Role of Intracellular Interleukin-6 in Growth Factor-Induced Cell Proliferation

Michael Roth
Laboratory of Pneumology
Department of Research
University Hospital Basel
Hebelstrasse 20
CH-4031 Basel
Switzerland

KEYWORDS/INTRODUCTION:

cell proliferation/ interleukin-6/ growth factor/ human fibroblasts/ human vascular smooth muscle cells/ interleukin-6 antisense oligonucleotides/ cell cycle control.

The tight control of cell proliferation is essential during embryogenesis, development, wound repair and maintenance of tissue structure. These control mechanisms involve the communication of each single cell with its neighbouring cells via interleukins and cytokines. These molecules exert their functions through specific membrane receptors which induce activation of intracellular signalling proteins upon activation by their specific agonists. Disturbance or disruption of this control mechanisms are believed to result in abnormal cell proliferation as observed in various diseases that correlate with proliferative disorders such as hyperdermia, fibrosis, atherosclerosis, glomerulonephritis, and tumorigenesis. Regardless whether the dividing cell is of transformed or non-transformed origin, the basic cellular events triggering cell proliferation are the same.

Usually, cell division is initiated by a mitogen binding to its corresponding membrane receptor, thereby inducing activation of a cascade of intracellular signalling proteins, resulting in transition of the cell from the $G_{0/1}$- into the S-phase. Increasing evidence suggests that interleukin-6 (IL-6) plays a key role in the control of proliferation. Earlier, we reported that intracellular IL-6 functions as a central modulator of platelet-derived growth factor (PDGF)-induced cell

NATO ASI Series, Vol. H 99
Tumor Biology
Regulation of Cell Growth, Differentiation and Genetics in Cancer
Edited by Asterios S. Tsiftsoglou, Alan C. Sartorelli, David E. Housman and T. Michael Dexter

proliferation in various non-transformed human cell types. Since IL-6 is induced by other growth factors such as interleukin-1 (IL-1), epithelial growth factor (EGF), tumor necrosis factor (TNF) and platelet-activating factor (PAF) in different cell types, we hypothesized that an intracellular form of IL-6 has eventually to be considered as a major regulator of cell division.

BIOLOGICAL FUNCTIONS OF INTERLEUKIN-6

IL-6 has been characterized as an important modulator of the host immune response (reviewed by Wong and Clark 1988, Akira et al. 1993). Based on its broad spectrum of actions, the following synonyms for IL-6 have been created: interferon-beta2 (Weissenbach et al. 1980), 26-kDa protein (Haegeman et al. 1986), B cell hybridoma/plasmacytoma growth factor (Van Damme et al. 1987), T-cell activating factor (Wong and Clark 1988), B-cell stimulatory factor (Hirano et al. 1986), hepatocyte stimulatory factor (Gauldie et al. 1987), B-cell differentiation factor (Sehgal et al. 1987), and macrophage-granulocyte inducing factor 2A (Akira et al. 1993). IL-6 has also been considered to be an acute phase cytokine during inflammatory reactions.

The human gene coding for IL-6 is localized on chromosome 7p21 and consists of five exons and four introns (Yasukawa et al. 1987). The IL-6 gene encodes a 212 amino acid residue precursor that undergoes several post-translational modifications. Natural human IL-6 is a glycoprotein containing two N-glycosylation sites, several O-glycosylation sites, and IL-6 can also undergo sialysation (Santhanam et al. 1989). IL-6 has a number of serines which can be differentially phosphorylated in a cell type specific manner (May al. 1988); it possesses a N-terminal hydrophobic secretory signal of 28-amino acid residues. Due to the post-translational modifications, at least six different IL-6 species have been identified with molecular weights ranging from 23- to 30-kDa (Santhanam et al. 1989). Increasing evidence suggests that the post-translational modifications of IL-6 are cell type and stimulus specific (May et al. 1988, May et al. 1989, Sehgal et al. 1988, Akira et al. 1993). Santhanam et al. (1989) demonstrated different immuno-chemical properties of the various IL-6 species. However, no data is available on specific actions of the various forms of IL-6. Remarkably, IL-6 itself is capable of modu-lating N-glycosylation of several plasma proteins during pregnancy and inflammation (MacKiewicz et al. 1987, MacKiewicz and Kushner 1989).

IL-6 exerts its biological functions via binding to a specific receptor which exists both in a membrane bound and soluble form. Still, binding of IL-6 to its receptor with the subsequent intracellular signalling is not yet completely understood (Kishimoto et al. 1994). Several models have been established to explain the observation that binding of IL-6 to its specific 80 kDA receptor is not sufficient to induce cellular signalling, the IL-6/Il-6 receptor complex additionally needs further binding to the membrane bound signal transducer gp130 (Hibi et al. 1990, Saito et al. 1992, Kishimoto et al. 1994, Ward et al. 1994). Interestingly, gp130 serves not only as a signal transducing membrane bound protein for the IL-6/IL-6 receptor-complex, it also mediates signalling of other cytokines which share structural homologies to IL-6, such as IL-11, IL-12, oncostatin and LIF (Taga and Kishimoto 1992, Yien et al. 1992, Kishimoto et al. 1994). The tertiary structure IL-6 forms a four helix-bundle, alpha-helix cytokine, which is related to Epo, granulocyte-colony stimulating factor (G-CSF), oncostatin M, IL-11, leukocyte inhibitory factor (LIF), melanoma growth factor (Bazan 1990; Akira et al. 1993), and IL-12 (Merberg et al. 1992). The C-terminal region of all mentioned proteins is responsible for binding to their corresponding receptors (Taga and Kishimoto 1992, Akira et al. 1993).

REGULATION OF IL-6 GENE EXPRESSION.

The de novo synthesis of IL-6 can be induced by a large number of mitogens or antigenic stimuli such as: lymphotoxin (Mantovani et al. 1990), interferons (IFN) (Akira et al. 1993), IL-1 (Walther et al 1988), TNF (Van Damme et al. 1987, Kishimoto et al. 1994), PDGF (Akira et al. 1993, Roth et al. 1995), PAF (Braquet et al. 1991, Roth unpublished data) and granulocyte inhibitory protein (Ziesche et al. 1994). Therefore, it is not surprizing that nearly all cell types found in humans are capable of producing IL-6 (reviewed by: Akira et al. 1993, Kishimoto et al. 1994).

To understand the fine tuned regulation of IL-6 expression, a number of studies characterizing the 5'-promotor region of the IL-6 gene are undertaken. So fare, the known regulatory elements include supressor elements such as: glucocorticoid-responsive element (GRE), retinoblastoma control element (RCE) (Hirano 1994), p53 responsive element; and a variety of enhancer elements such as: cAMP-responsive element (CRE) (Hirano 1994), c-fos/serum responsive element (c-fos/SRE) (Keul et al. 1995), multiresponse element, activating protein-1- (AP-1), nuclear factor IL-6- (NFIL-6) (Ray et al. 1989), and NF_kB-binding sites (Shimizu et al. 1990).

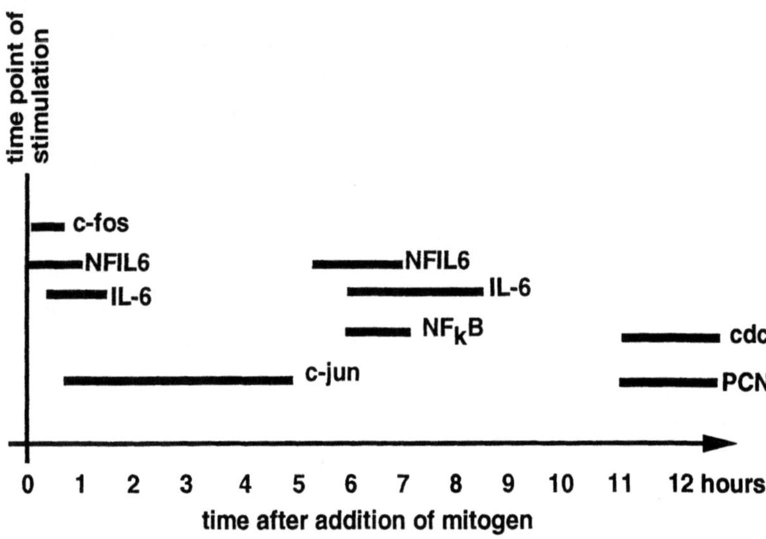

Figure 1: Sheme of mitogen-induced kinetics of gene activation in cultivated human pulmonary fibroblasts or VMSCs. Duration of maximal transcription of the genes is indicated by length of the respective barrs.

In cultivated human fibroblasts, vascular smooth muscle cell (VSMC) and mesangial cells the mitogen-induced transcription of NFIL6 and IL-6 display a biphasic time course; for all other investigated factors only a single signal of the respective mRNA was observed (Fig. 1). The first peak of NFIL6 and IL-6 mRNAs was detectable between 30 to 60 minutes after stimulation, declining thereafter to the basal level; a second peak for both NFIL6 and IL-6 mRNA was observed 6-8 hours after stimulation (Fig. 1). Using c-fos, NFIL-6, NF_kB, and IL-6 antisense phosphothioate modified oligonucleotides (PTOs), we studied the possible involvement of these factors in mitogen-induced gene activation in cultivated human pulmonary fibroblasts, VSMCs, MCs (Roth et al. 1995), T-lymphocytes, B-lymphocytes and granulocytes (manuscript in preparation). We demonstrated that c-fos, NF_kB, and IL-6 antisense PTOs suppressed mitogen-induced transcription of c-fos, c-fos/SRE, NF_kB, and IL-6 in all cell types studied (table 1). In contrast, c-fos, NF_kB, and IL-6 antisense PTOs had no effect on the transcription of c-jun and that of NFIL6 (manuscript in preparation). The specificity of various PTOs has already been reported partly (Levy et al. 1991, Roth et al. 1995). From our data we conclude that the transcription of the IL-6 gene in cultivated human fibroblasts, VSMCs and MC is mainly regulated by the action of c-fos, NF_kB and AP-1 (Ziesche et al. 1994, Keul et al. 1995). However, there may be additional regulatory elements in the IL-6 promotor which have not yet been identified.

antisense PTOs	mRNAs	c-fos	NFIL-6	NF_kB	IL-6	cdc2	PCNA
c-fos		---	(+)	---	---	---	---
NFIL6		+	---	+	---	(+)	(+)
NF_kB		+	+	---	---	---	---
IL-6		+	---	+	---	---	---

Table 1: Effects of antisense PTOs on mitogen-induced transcription of genes associated with IL-6 synthesis and proliferation in cultivated human pulmonary fibroblasts or VSMCs. "+" indicates normal transcription in the presence of mitogen and antisense PTOs; "---" indicates abolished transcription in the presence of mitogen and antisense PTOs; (+) indicates partly inhibition of transcription in the presence of antisense PTOs..

INTERLEUKIN-6 AND ITS ROLE IN CELL PROLIFERATION

During recent years it became obvious, that cell proliferation is paralleled by an accumulation of IL-6. Overexpression of IL-6 has been reported in various tumors, IL-6 transgenic animals, and transformed cell lines. In these cases, IL-6 overexpression is always correlated with enhanced proliferation. We have shown that cultivated human fibroblasts, VSMCs and MCs produce large amounts of IL-6 upon stimulation with various mitogens (Ziesche et al. 1994, Roth et al. 1995). These data are confirmed by findings of Loppnow and Libby (1990) who described that proliferating VSMCs secrete copious IL-6 following stimulation with IL-1. In murine pulmonary fibroblasts, IL-6 has also been defined as an autocrine growth factor (Fries et al. 1994).

IL-6 has also been reported as an enhancer of cell proliferation in various diseases correlating with proliferation disorders and in tumors. In IL-6 transgenic mice, frequent development of plasmacytomas, myelomas, and glomerulonephritis has been reported by Suematsu et al. (1992). DiCosmo et al. (1994) showed major structural changes in the lung of mice overexpressing IL-6 in Clara cells. Furthermore, overexpression of Il-6 has also been observed following experimentally induced angiogenesis in mice (Motro et al. 1990), in atherosclerosis in hyperlipidemic rabbits (Ikeda et al. 1992), in human sarcoidosis (Bost et al. 1994), and in chronic renal diseases (Ziesche et al.1994). IL-6 also seems to play a role in the pathogenesis of Paget's disease (Roodman et al. 1992). Several investigators suggested IL-6 being an autologous growth factor in melanomas (Lu and Kerbel 1993), hairy cell leukemia (Barut et al. 1993), myeloma (Levy et al. 1991), and glioblastoma (Hirano 1994).

In stimulated non-transformed cells, we observed a marked increase of secreted IL-6 protein during the first twelve hours. C-fos, NF_kB, and IL-6 antisense PTOs, but not NFIL-6 or c-jun antisense PTOs inhibited the mitogen-induced secretion of IL-6 in a dose-dependent manner. Histochemical immunofluorescence staining for IL-6 protein revealed a diminished amount of immunoreactive IL-6 protein in the cytoplasma of cells treated with antisense c-fos, NF_kB, or IL-6 PTOs (data not shown). By determining incorporation of [^{3}H]-thymidine into newly synthesized DNA and by counting cell numbers, we demonstrated that c-fos, c-jun, NF_kB, or IL-6 antisense PTOs, but not NFIL6 or neutralizing anti-IL-6 antibodies, completely inhibited mitogen-inducible de novo synthesis of DNA (Roth et al. 1995, Keul et al. 1995). Our data on proliferation inhibiting effects of IL-6 antisense PTOs are similar to those of Levy et al. (1991) who reported inhibition of myeloma cell proliferation in the presence of IL-6 antisense PTOs. Barut et al. (1993) demonstrated that intracellular IL-6 is involved in the growth control of hairy cell leukemia cells. Interestingly, Lu et al. (6) suggested that IL-6 may undergo transition from a paracrine growth inhibitor to an intracellular autocrine stimulator during progression of human melanomas. All these groups agree to assign a crucial role in the control of cell proliferation to an intracellular form of IL-6 .

In contrary, Brach et al. (1990) reported that proliferation of leukemic megakaryoblasts was dependent on exogenous IL-6 because proliferation of these cells was abolished in the presence of anti IL-6 antibodies. To test a possible autocrine feedback mechanism of secreted IL-6 on cell proliferation via its corresponding receptor, we used monoclonal neutralizing anti IL-6 antibodies to block secreted IL-6. The effectiveness of the neutralizing antibodies was assessed by enzyme linked immunosorbent assay. However, the missing potency of secreted IL-6 to alter mitogen-dependent cell proliferation *in vitro* does not exclude a possible intracellular action of IL-6.

INTERLEUKIN-6 AND CELL CYCLE CONTROL

In earlier studies we investigated the effect of IL-6 antisense PTOs on the expression of two genes which are tightly associated with the proliferative state of a cell, cdc2 and the proliferating cell nuclear antigen (PCNA) (Dalton 1992, Jaskulski et al. 1988a). The cdc2 is a cell cycle regulated protein kinase essential for initiation of DNA replication and entry of the cell into mitosis (Dalton 1992). PCNA is a nuclear protein which is required for the synthesis of the leading strand DNA by DNA-polymerase d (Jaskulski et al. 1988b). Both genes are activated during transition from the $G_{0/1}$ to S-phase (Dalton 1992, Loppnow and Libby 1990). Consistent with the effects of antisense PTOs on cell proliferation and incorporation of [^{3}H]-thymidine, the mitogen-induced expression of both genes, cdc2 and PCNA, was diminished in the presence of c-fos, NF_kB, or IL-6 antisense PTOs, but remained unaffeced in the presence of sense PTOs as well as neutralizing anti IL-6 antibodies. Interestingly, endogenous IL-6 has been reported to shorten the G_0-phase in murine IL-3-stimulated blast cells (Ikebuchi et al. 1987).

Based on the fact that IL-6 antisense PTOs were effective to block mitogen-induced proliferation in three different non-transformed cell types, and on similar reports of IL-6 antisense PTOs on the proliferation of transformed cell lines or tumors, we postulate that the growth-mediating action of an intracellular product of the IL-6 gene apparently represents a general principle in cell cycle control (Fig. 2). Therefore, we suggest a new action of a yet unidefined form of intracellular IL-6 controlling transition from the $G_{0/1}$- to the S-phase of the cell cyle.

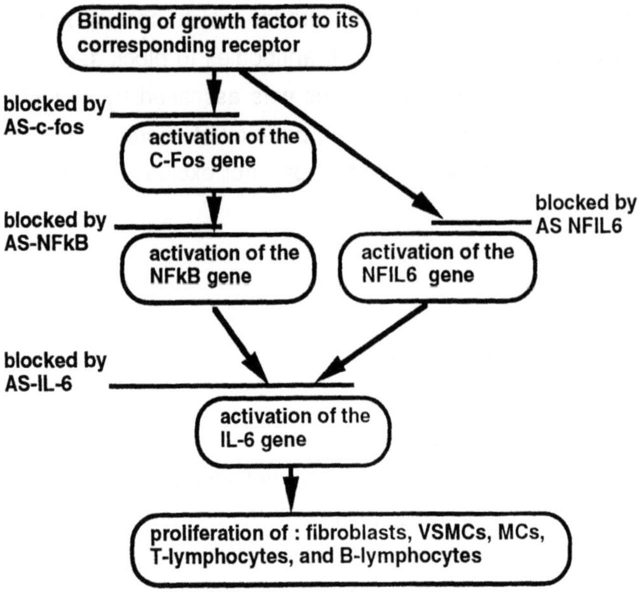

Figure 2: Hypothesized regulatory pathway of intracellular IL-6 and transcription factors in mitogen-induced cell proliferation

FUTURE ASPECTS OF INTEREST

Summarizing the knowledge about IL-6 and its variable forms of appearence poses new questions on the post-translational modifications of the molecule inside of the cell. More detailed analyses of the cellular distribution of the various N- and O-glycosylated, the sialysated, and the phosphorylated forms of intracellular IL-6 is essential to specify their biological activities on cellular metabolism and on the cell cycle. An apparent discrepancy of the mode of action of IL-6 is evident when comparing cells derived from the hematopoetic system to those forming solid tissues. While proliferation of cells of the hematopoetic system seems to involve extracellular IL-6, tissue forming cells obviously need an intracellular form of IL-6 for cell division. This effect might be linked to the observation that different cell types produce different forms of post-translational modified IL-6. Since new data on IL-6 suggest a crucial role of the cytokine in development and proliferation, its initially addressed functions during inflammation and acute phase response may be explained by the involvement of IL-6 in tissue repair and maintenance of tissue structure.

REFERENCES:

Akira S, Taga T, Kishimoto T (1993) Interleukin-6 in biology and medicine. Advances Immunol 54:1-78

Barut B, Chauhan D, Utchiyama H, Anderson KC (1993) Interleukin-6 functions as an intracellular growth factor in hairy cell leukemia in vitro. J Clin Invest 92: 2346-2352

Bazan JF (1990) Haemopoietic receptors and helical cytokines. Immunol Today 11: 350-354

Brach MA, Loewenberg B, Mantovani L, Schwulera U, Mertelsmann R, Herrmann F (1990) Interleukin-6 (IL-6) is an intermediate in IL-1-induced proliferation of leukemic megakaryoblasts. Blood 76:1972-1979.

Braquet P, Pignol B, Maisonnet T, Menica-Huerta JM (1991) Platelet-activating factor modulates interleukin-6 production by mouse fibroblasts. Int. Arch. Allergy Appl. Immunol 94:165-166

Bost TW, Riche DWH, Schumacher B, Carreé PC, Khan TZ, Martinez JAB, Newman LS (1994) Alveolar macrophages from patients with berylium disease and sarcoidosis express increased levels of mRNA for tumor necrosis factor-alpha and interleukin-6 but not interleukin-1beta. Am J Respir Cell Mol Biol 10:506-513

Chomczynski P, Sacchi N (1987) Single step method of RNA isolation by acid guanidinium isothiocynanate-phenol-chloroform extraction. Anal Biochem 162:156-159

Dalton S (1992)Cell cycle regulation of the human cdc2 gene. EMBO J 11:1797-1802

DiCosmo BF, Geba GP, Picarella D, Elias JA, Rankin JA, Stripp BR, Whitsett JA, Flavelli RA (1994) Airway epithelial cell expression of interleukin-6 in transgenic mice. J Clin Invest 94:2028-2035

Fries KM, Felch ME, Phipps RP (1994) Interleukin-6 is an autocrine growth factor for murine lung fibroblast subsets. Am J Respir Cell Mol Biol 11:552-560

Gauldie J, Richards C, Harnish D, Lansdorp P, Baumann H (1987) Interferon beta 2/B-cell stimulating factor type 2 shares identity with monocyte-derived hepatocyte stimulatory factor and regulates the major acute phase protein response in liver cells. Proc Natl Acad Sci USA 84:7251-7255

Haegeman G, Content J, Volkaert G, Dreynck R, Tavenir J, Fiers W (1986) Structural analysis of the sequence coding for an inducible 26-kDa protein in human fibroblasts. Eur J Biochem 159:625-638

Hibi M, Murakami M, Saito T, Hirano T, Taga T, Kishimoto T (1990) Molecular cloning and expression of an IL-6 transducer, gp130. Cell 63:1149-1157

Hirano T, Yasukawa K, Harada H, Taga T, Watanbae Y, Matsuda T, Kashiwamura S, Nakajima K, Koyama K, Iwamatsu A, Tsunasawa S, Sakiyama F, Matsui H, Takahara Y, Taniguchi T, Kishimoto T (1986) Complementary DNA for a novel human interleukin (BSF-2) that induces B lymphocytes to produce immunoglobulin Nature 324:73-76

Hirano T (eds) (1994) Interleukin 6. The cytokine Handbook, 2nd ed. Academic Press, New York

Ikeda U, Ikeda M, Seino Y, Takahashi M, Kano S, Shimada K (1992) Interleukin-6 gene transcripts are expressed in atherosclerotic lesions of genetically hyperlipdemic rabbits. Atherosclerosis 92:213-218

Jaskulski D, Gatti C, Travali S, Calabretta B, Baserga R (1988a) Regulation of the proliferating cell nuclear antigen cyclin and thymidine kinase mRNA levels by growth factors. J Biol Chem 263:10175-10179

Jaskulski D, DeRiel JK, Mercer WE, Calabretta B, Baserga R (1988b) Inhibition of cellular proliferation by antisense oligodeoxynucleotides to PCNA cyclin. Science 240:1544-1546

Keul R, Roth M, Nauck M, Papakonstantinou E, Tamm M, Perruchoud AP, Block LH (1995) Broncho-Vaxom (OM85-BV) induces the expression of interleukin-6 and interleukin-8 via C-Fos/serum responsive element. Thorax (in press)

Kishimoto T, Taga T, Akira S (1994) Cytokine signal transduction. Cell 76:253-262

Levy Y, Tsapis A, Brouet JC (1991) Interleukin-6 antisense oligonucleotides inhibit the growth of human myeloma cell lines. J Clin Invest 88:696-699

Loppnow H, Libby P. (1990) Proliferating or interleukin-1 activated human vascular smooth muscle cells secrete copious interleukin-6. J Clin Invest 85:731-738

Lu C, Kerbel RS (1993) Interleukin-6 undergoes transition from paracrine growth factor inhibitor to autocrine stimulator during human melanoma progression. J Cell Biol 120:1281-1288

MacKiewicz A, Ganapathi MK, Schultz D, Kushner I (1987) Monokines regulate glycosylation of acute phase proteins. J Exp Med 166:253-258

MacKiewicz A, Kushner I (1989) Interferon-beta2/B cell stimulating factor 2/interleukin-6 affects glycosylation of acute phase proteins in human hepatoma cell lines. Scand J Immunol 29:265-271

Mantovani L, Henschler R, Brach MA, Wieser R, Lübber M, Lindemann A, Mertelsmann RH, Herrmann F (1990) Differential regulation of interleukin-6 expression in human fibroblasts by tumor necrosis factor -alpha and lymphotoxin. FEBS Letters 270:152-156

May LT, Gharyeb J, Santhanam U, Tatter SB, Sthoegger Z, Helfgott DC, Chiorazzi N, Grieninger G, Sehgal PB (1988) Synthesis and secretion of multiple forms of beta 2-interferon/B-cell differentiation factor 2/hepatocyte stimulating factor by human fibroblasts and monocytes. J Biol Chem 263:7760-7766

May LT, Torica G, Cozzolino F, Ray A, Tatter SB, Santhanam U, Sehgal PB, Stern D (1989) Interleukin-6 gene expression in human endothelial cells: RNA start sites, multiple IL-6 proteins and inhibition of proliferation. Biochem Biophys Res Commun 159:991-998

Merberg DM, Wolff SF, Clark SC (1992) Sequence similarities between NKSF and the IL-6/G-CSF family. Immunol Today 13:77-78

Motro B, Itin A, Sachs L, Keshet E (1990) Pattern of interleukin-6 gene expression in vivo suggests a role for this cytokine in angiogenesis. Proc Natl Acad Sci USA 87: 3092-3096

Ray A, Sassone-Corsi P, Sehgal PB (1989) A multiple cytokine- and second messenger-responsive element in the enhancer of the human interleukin-6 gene: Similarities with c-fos gene regulation. Mol Cell Biol 9:5537-5547

Roodman GD, Kurihara N, Ohsaki Y, Kukita A, Hosking D, Demulder A, Smith JF, Singer FR. (1992) Interleukin-6: A potential autocrine/paracrine factor in Paget's disease of bone. J Clin Invest 89:46-52

Roth M, Nauck M, Tamm M, Perruchoud AP, Ziesche R, Block LH (1995) Intracellular interleukin-6 mediates platelet-derived growth factor-dependent cell proliferation. Proc Natl Acad Sci USA 92:1312-1316

Saito M, Yoshida K, Hibi M, Taga T, Kishimoto T (1992) Molecular cloning of a murine IL-6 receptor-associated signal transducer gp130, and its regulated expression in vivo. J Immunol 148:4066-4071
Sambrook J, Fritsch EF, Maniatis T (eds) (1989) Molecular cloning. A laboratory manual. Cold Spring Harbor Press, New York.
Santhanam U, Ghraveb J, Sehgal PB, May L (1989) Post-translational modifications of human interleukin-6. Arch Biochem Biophys 274:161-170
Sehgal PB, May LT, Tamm I, Vilcek J (1987) Human beta2-interferon and B-cell differentiation factor BSF-2 are identical. Science 235:731-732
Sehgal PB, Helfgott DC, Santhanam U, Tatter SB, Clarick RH, Ghayeb J, May LT. (1988) Regulation of the acute phase and immune response in viral disease. J Exp Med 167:1951-1956
Shimizu H, Mitomo K, Watanabe T, Okamoto S, Yamamoto KI (1990) Involvement of a NF-kB-like transcription factor in the activation of the interleukin-6 gene by inflammatory lymphokines. Mol Cell Biol 10:561-568
Suematsu S, Matsuksaka T, Matsuda T, Ohno S, Miyazaki J, Yamamura K, Hirano T, Kishimoto T (1992) Generation of plasmacytomas with the chromosomal translocation t(12;15) in interleukin-6 transgenic mice. Proc Natl Acad Sci USA 89:232-235
Taga T, Kishimoto T (1992) Cytokine receptors and signal transduction. FASEB J 6: 3387-3396
Van Damme J, Opdenakker G, Simpson RJ, Rubira MR, Cayphas S, Vink A, Billiau A, Van Snick J (1987) Identification of the human 26-kD protein, interferon-beta2, as a B cell hybridoma/plasmacytoma growth factor induced by interleukin-1 and tumor necrosis factor. J Exp Med 165:914-922
Walther Z, May LT, Sehgal PB (1988) Transcriptional regulation of the interferon-beta2/B cell differentiation factor BSF-2/Hepatocyte-stimulatory factor gene in human fibroblasts by other cytokines. J Immunol 140:974-977
Ward LD, Howlett GJ, Discolo G, Yasukawa K, Hammacher A, Moritz RL, Simpson RJ (1994) High affinity interleukin-6 receptor is a hexameric complex consisting of two molecules of each of interleukin-6, interleukin-6 receptor and gp 130. J Biol Chem 269:23286-23289
Weissenbach J, Chernajovsky Y, Zeevi M, Shulmann L, Soreq H, Nir U, Wallach D, Perrichaud M, Tiollais P, Revel M (1980) Two interferon mRNAs in human fibroblasts: in vitro translation and Escherichia coli cloning studies. Proc Natl Acad Sci USA 77:7152-7156
Wong GG, Clark SC (1988) Multiple actions of interleukin-6 within a cytokine network. Immunology Today 9:137-140
Yasukawa K, Hirano T, Watanabe Y, Muratahi K, Matsuda T, Nakai S, Kishimoto T 1987) Structure and expression of human B cell stimulatory factor-2 (BSF-2/ IL-6) gene. EMBO J 6:2939-2945
Yin T, Taga T, Tsang MLS, Yasukawa T, Kishimoto T, Yang YC (1993) Involvement of interleukin-6 signal tranducer gp130 in interleukin-11-mediated signal transduction. J Immunol 151:2555-2561
Ziesche R, Roth M, Papakonstantinou E, Nauck M, Hörl W, Kashgarian M, Block LH. (1994) A granulocyte inhibitory protein overexpressed in chronic renal disease regulates expression of interleukin-6 and interleukin-8. Proc Natl Acad Sci USA 91:301-305

Differentiation of Malignant Cells as a Therapeutic Approach

Alan C. Sartorelli, Kimiko Ishiguro, Jianming Li, Debbie C. Koay and John A. Sokoloski
Department of Pharmacology and Developmental Therapeutics Program, Cancer Center
Yale University School of Medicine
New Haven, CT 06520 USA

Abstract

The hematological malignancies may be considered to be diseases of altered maturation in which the rate of proliferation is increased relative to that of terminal differentiation. Many neoplastic cells, however, retain the capacity to mature to end-stage cells with a finite life-span. Differentiation to a nonmalignant state can be produced in many different malignant cell types by exposure to a variety of structurally diverse biological and chemical agents, including cytokines, solvents, hormones, vitamins, tumor promoters and cancer chemotherapeutic agents. The induced transformation to a terminally differentiated phenotype demonstrates the reversibility of the malignant state.

A number of structurally diverse cancer chemotherapeutic agents, such as the anthracyclines, epipodophyllotoxins, alkylating agents, vinca alkaloids, antifolates, purine antimetabolites, and pyrimidine nucleoside analogs, are capable of inducing differentiation, making it conceivable that the antineoplastic effects of these agents are the result of a combination of both cytodestruction and terminal differentiation. The alterations responsible for the termination of proliferation by these two mechanisms in some cases are the result of different metabolic events. We have shown that for the antileukemic agent 6-thioguanine, the free base is the form that initiates the maturation process by interacting with benzodiazepine receptors present on the surface of the leukemic cells, while conversion to the nucleotide forms of the 6-thiopurine is responsible for cytotoxicity. In an analogous manner the trisaccharide-containing anthracyclines, aclacinomycin and marcellomycin, are capable of inducing differentiation, whereas the monosaccharide-containing anthracyclines, adriamycin and pyromycin, cause cytodestruction. The granulocyte colony-stimulating factor (G-CSF) and its receptor constitute an example of the importance of a physiological cytokine to the differentiation process.

Termination of proliferation through the induction of differentiation is a programmed event and the production of a differentiated phenotype results in a partial normalization of cellular events. This has been shown by the demonstration that leukemia cells primarily express a facilitated diffusion transport system for the pyrimidine nucleoside uridine, which tends to equilibrate internal and external concentrations of this ribonucleoside in a manner characteristic of most neoplastic tissues. The induction of differentiation by dimethylsulfoxide or by 12-O-tetradecanoylphorbol 13-acetate causes a down-regulation of the facilitated transport mechanism and the predominance of a Na^+-dependent active transporter characteristic of normal tissue. This results in an accumulation of free uridine in the mature cells. Studies conducted with the HL-60 human promyelocytic leukemia and the WEHI-3B monomyelocytic leukemia have been used to provide the examples of these phenomena.

NATO ASI Series, Vol. H 99
Tumor Biology
Regulation of Cell Growth, Differentiation and Genetics in Cancer
Edited by Asterios S. Tsiftsoglou, Alan C. Sartorelli, David E. Housman and T. Michael Dexter

Introduction

Significant advances have been made towards the cure and palliation of cancer with existing chemotherapeutic agents; however, since the fundamental mechanism of action of these drugs is dependent upon the cytodestruction of neoplastic cells, their beneficial effects are most often accompanied by significant morbidity. This suggests that approaches to cancer therapy should be sought that do not involve cell kill; one such approach envisions the conversion of malignant cells through induced differentiation to benign forms with no proliferative potential. Such an approach assumes that malignancy is not an irreversible state as has been demonstrated with a variety of tumor types, including teratocarcinomas, neuroblastomas, squamous cell carcinomas, leukemias and adenocarcinomas of the breast (see for a review, Reiss et al., 1986).

Cancer has been characterized as a disease of altered differentiation in which the carcinogenic process may be visualized as a series of genetic changes that alter the normal regulatory processes that govern the progression of a developing cell through the maturation pathway, such that it interferes with movement through the differentiation pathway. When the transition to a malignant form occurs at the stem cell level or at the level of a transitional (i.e., developing) cell that has not yet lost its proliferative potential, an amplification of the altered clone can occur, as well as progression to more anaplastic forms. These phenomena result in an increase in tissue mass.

There is considerable evidence that many neoplastic cells retain the capacity to mature to adult, end-stage cells and lose the immortality characteristic of most malignant cells. Thus, for example, there is direct evidence that leukemic blast cells can leave the neoplastic cell population by spontaneously differentiating to form mature cells (Fearon et al., 1986), as well as evidence that a small number of tumors of early life that have undergone spontaneous complete regression do so in a process that appears to be the result of terminal differentiation (Bolande, 1985).

The conversion of malignant cells to end-stage cells with a finite life-span by the induction of differentiation can be accomplished by a wide variety of structurally diverse biological and chemical agents, including in the latter category solvents, hormones, vitamins, tumor promoters and cancer chemotherapeutic agents (Koeffler, 1983; Reiss et al., 1986). The clinical validation of the use of differentiation inducing agents as a therapeutic approach was provided by the finding that all-trans retinoic acid produced complete remissions in a majority of patients with acute promyelocytic leukemia (Huang et al., 1988; Castaigne et al., 1990; Warrell et al., 1991). The success of differentiation therapy as an approach to the treatment of various forms of cancer requires the discovery of compounds that act by the induction of terminal differentiation rather than by the cytodestructive mechanisms that characterize currently used cancer therapeutics.

Mechanism(s) of Differentiation Produced by Antineoplastic Agents

The fact that a large number of antineoplastic agents are capable of initiating the differentiation of neoplastic cells (Schwartz et al., 1983) makes it reasonable to assume that the beneficial effects produced by anticancer agents in patients may be due to a mixture of cytodestruction and terminal differentiation. Since it is conceivable that the fundamental mechanism by which at least some of these drugs induce maturation is different from that responsible for cell kill, it also seems reasonable to assume that if such a mechanism is operative, that it might be possible to factor out the cytotoxic properties of the molecule by synthetic modification.

We have demonstrated with at least two classes of antineoplastic agents that the metabolic events which lead to cytotoxicity and to terminal differentiation are different. The first agent studied is the antileukemic agent, 6-thioguanine (TG); this purine antimetabolite requires conversion to 6-thioguanosine 5'-monophosphate (TGMP) to express its cytotoxicity (Ishiguro et al., 1984). Gusella and Housman (1976) reported that TG had the capacity to induce the erythroid differentiation of Friend murine leukemia cells that lacked hypoxanthine-guanine phosphoribosyltransferase (HGPRT) activity and postulated that the formation of TGMP and the subsequent incorporation of TG residues into DNA, which are required for cytodestruction, were not requisite for the maturation process. We independently confirmed these findings in the Friend murine erythroleukemia (Schwartz et al., 1983) and expanded the studies to a mutant of the HL-60 human promyelocytic leukemia deficient in HGPRT and a double mutant deficient in both HGPRT and deoxycytidine kinase (Ishiguro et al., 1984). Examination of the metabolism of the purine antimetabolite and its capacity to induce differentiation in wild-type parental HL-60 cells and in both mutant cell clones provided evidence that the free base TG was the form that was required for the induction of maturation, while the formation of TG nucleotides not only was necessary for the expression of cytotoxicity, but also appeared to antagonize the maturation process.

Further studies with HL-60 cells demonstrated the presence of both high (K_D = 7.3 nM, T_B = 14.5 pmol/mg protein with Ro5-4864) and low (K_D = 28.6 μM, T_B = 199 pmol/mg protein with diazepam) affinity benzodiazepine binding sites (Ishiguro et al., 1987). Benzodiazepines were effective initiators of maturation in the concentration range of 50 to 150 μM, suggesting that low affinity benzodiazepine receptors may be involved in the induction process. TG caused inhibition of the binding of benzodiazepines to both high and low affinity sites; the binding of TG appeared to be through the formation of a mixed disulfide between the 6-thiopurine and membrane protein thiols. Thus, the occupancy of benzodiazepine receptors by TG is assumed to be the initial event in transmitting the signal by the thiopurine to enter a differentiation pathway.

The finding that the free base itself (i.e., TG) is the form that induces differentiation allowed the use of the physiological nucleoside inosine in combination with TG to minimize the quantity of TGMP formed by parental wild-type HL-60 cells (Ishiguro and Sartorelli, 1985). This resulted in a decrease in the formation of the nucleotide form of the 6-thiopurine that is responsible for cytotoxicity while maintaining the purine antimetabolite as the free base form responsible for the induction of maturation. Under these conditions the terminal differentiation of HL-60 cells occurred, suggesting that such modulation may have utility in the treatment of some cancers by the induction of differentiation.

The anthracyclines form a second class of antineoplastic agents in which the metabolic lesion(s) responsible for cytotoxicity is distinct from that producing cellular differentiation. In this series of agents, the trisaccharide-containing anthracyclines, aclacinomycin (ACM) and marcellomycin (MCM), were shown to be potent inducers of the differentiation of HL-60 cells, whereas the monosaccharide-containing anthracyclines, adriamycin (ADM) and pyrromycin (PYM), were inactive and significantly less potent, respectively, as initiators of the maturation process (Schwartz and Sartorelli, 1982; Schwartz et al., 1983; Morin and Sartorelli; 1984).

Several differences exist between the trisaccharide- and monosaccharide-containing anthracyclines with respect to the biochemical alterations that they produce in intact cells including: (a) the binding of these agents to DNA being determined in part by the length of the oligosaccharide side-chain (DuVernay et al., 1979); (b) the inhibition of nucleolar RNA synthesis compared to DNA synthesis being much more sensitive to ACM and MCM (the 50% inhibitory concentrations for nucleolar RNA synthesis being less than 40 nM) than to ADM and PYM (50% inhibitory concentrations being on the order of 6 μM) (Crooke et al., 1978); and (c) the potent inhibition of the synthesis of asparagine-linked glycoproteins by the trisaccharide anthracyclines, ACM and MCM, in the absence of interference with the biosynthesis of these oligonucleotides by the monosaccharide-containing anthracyclines (Morin and Sartorelli, 1984).

In keeping with these findings, Sato et al. (1992) demonstrated that patients with acute myelomonocytic leukemia refractory to various kinds of intensive chemotherapy responded to low doses of ACM in a process that was suggestive of the induction of differentiation.

These findings demonstrate that structural modification of certain classes of antineoplastic agents may permit the introduction of biochemical events that would lead to the initiation of terminal differentiation, with the loss or minimization of cytodestructive properties.

Role of Granulocyte Colony-Stimulating Factor (G-CSF) snd its Receptor in the Induction of Terminal Differentiation

G-CSF is a member of a family of growth factors that control hematopoiesis. It stimulates the formation of granulocytic colonies from murine and human bone marrow, enhances the survival of mature peripheral blood neutrophils, increases antibody-dependent cell-mediated cytotoxicity of neutrophils, and primes neutrophils to undergo enhanced oxidative metabolism in response to the chemotactic peptide *N*-formyl-methionyl-leucyl-phenylalanine (fMLP). G-CSF has been used clinically in patients who are myelosuppressed as a result of cancer chemotherapy, irradiation therapy, or bone marrow transplantation. Recent studies have also shown that G-CSF has anticancer activity. Thus, introduction of the G-CSF gene into colon adenocarcinoma C-26 cells suppresses the tumorigenicity of these cells through the recruitment and targeting of neutrophilic granulocytes to the G-CSF-releasing cells (Colombo et al., 1991). In addition, G-CSF has been found to inhibit the metastatic spread of hematogenous and non-hematogenous tumors in mice (Matsumato et al., 1991). Of particular significance to the use of G-CSF in differentiation therapy are the reports that patients with acute promyelocytic leukemia and acute myelocytic leukemia have achieved remissions through the administration of recombinant G-CSF and that the remissions are probably the result of leukemia cell differentiation induced by the cytokine (Toki et al., 1989; Yamasaki et al., 1991).

The G-CSF receptor (G-CSFR) is expressed predominantly in normal progenitor and mature neutrophils and in various myeloid leukemia cells that can be induced by G-CSF either to differentiate along the granulocytic pathway or to undergo growth (Nicola and Metcalf, 1984, 1985; Nicola et al., 1985; Begley et al., 1987; Park et al., 1989). Some non-hematopoietic cells, such as human endothelial cells, placental cells and carcinoma cells, have also been shown to express the G-CSFR (Berdel et al., 1989; Bussolino et al., 1989; Uzumaki et al., 1989).

WEHI-3B D+ murine myelomonocytic leukemia cells express a low level of G-CSF receptors (Burgess and Metcalf, 1980; Fukunaga et al., 1990) and form granulocytic colonies in semi-solid culture in the presence of G-CSF (Burgess and Metcalf, 1980; Metcalf, 1980). Accompanying a morphological change in colony structure is a suppression of the clonogenicity of the cells. Similar effects are also observed with M1 and HL-60 leukemia cells (Tsuda et al., 1986; Begley et al., 1987). These results suggest that G-CSF is functioning as an inducer of the differentiation of these leukemic cells. To define the effects of G-CSF on WEHI-3B D+ cells and to establish a better understanding of the role of the G-CSFR in the control of cellular proliferation and differentiation, we have established a series of WEHI-3B D+ clones that express the G-CSFR at high levels by transfection of an expression vector containing the G-CSFR gene (Li et al., 1993). We have found that a high level of expression of the G-CSFR by itself does not promote or suppress cellular proliferation or initiate differentiation; however, exposure of G-CSFR transfected cells to G-CSF causes a large percentage of the population to enter a differentiation pathway, while parental cells respond poorly under the same experimental conditions. The growth rate of the transfected cells is also decreased by exposure to G-CSF, presumably due to the terminal differentiation of the cells. These findings collectively demonstrate that the predominant function of G-CSF and its receptor in WEHI-3B D+ cells is to mediate differentiation, and that the level of the G-CSFR portion of the signal transduction mechanism in this malignant cell line is important in the elicitation of a response to the maturation inducing function of the cytokine.

WEHI-3B D- cells, derived from WEHI-3B D+ cells which have a near diploid content of DNA, have assumed a near tetraploid karyotype and are no longer responsive to inducers of differentiation such as G-CSF and retinoic acid (Metcalf and Nicola, 1982). The inability of WEHI-3B D- cells to respond to G-CSF appears to be due to the lack of G-CSFRs on the cell surface (Nicola and Metcalf, 1984). To provide information of the role of the G-CSFR in the control of proliferation and differentiation, we have transfected WEHI-3B D- with a G-CSFR expression plasmid and have evaluated the effects of the overexpression of the G-CSFR on cellular growth and maturation (Li et al., 1995). Three major changes have been observed: (a) G-CSFR expressing clones became sensitive to the induction of differentiation by RA, as well as by G-CSF; (b) the morphology of G-CSFR transfected colonies changed from that of a tightly aggregated type to that of a dispersed type; and (c) the DNA ploidy of G-CSFR expressing WEHI-3B D- cells changed from near tetraploid to near diploid. The results suggest that a diploid phenotype is required for WEHI-3B leukemia cells to respond to certain inducers of differentiation.

The finding of a relationship between G-CSF and its receptor and the induction of differentiation both in experimental systems and in patients with non-lymphocytic leukemias

(Toki et al., 1989; Yamaski et al., 1991) makes it reasonable to consider the use of this cytokine in combination with other initiators of the maturation process. To this end, we have demonstrated that combinations of G-CSF and retinoic acid produced supra-additive differentiation of the WEHI-3B D^+ leukemia (Li and Sartorelli, 1992) suggesting that consideration be given to the addition of G-CSF to all-trans retinoic acid in the therapy of patients with acute promyelocytic leukemia.

Differentiation of Leukemia Cells Results In a Normalization of Cellular Events

A variety of changes that are characteristic of the normal phenotype occur when malignant cells are induced to enter a differentiation pathway. We have chosen to study the transport of uridine in HL-60 cells induced to undergo both granulocytic and monocytic differentiation. Previous studies by Darnowski and Handschumacher (1986) had shown that normal tissues such as spleen, liver, kidney and intestine contain pools of free uridine up to 13-fold greater than the concentration of uridine in the plasma. In suspensions of murine splenocytes, Darnowski et al. (1987) demonstrated that intracellular uridine pools were created by a Na^+-dependent uridine transport mechanism. A similar Na^+-dependent uridine transport system(s) was also detected in rat, rabbit and bovine renal (LeHir and Dubach, 1985; Lee et al., 1988; Jarvis et al., 1989; Lee et al., 1990) and rabbit intestinal (Jarvis, 1989) brush border membrane vesicles and in guinea pig enterocytes (Schwenk et al., 1984), suggesting a role for this transporter in the accumulation of uridine in a variety of tissues and species.

In contrast, the accumulation of free uridine from medium was not observed in a variety of neoplastic cell lines, presumably because of an overriding facilitated diffusion mechanism which results in equilibrium with the medium (Darnowski and Handschumacher, 1986). Thus, it was postulated by these investigators that the loss or decrease in the activity of the Na^+-dependent uridine transporter might well coincide with the development of the malignant state. Consistent with this postulate, uridine transport in undifferentiated HL-60 cells was found to occur primarily by facilitated diffusion (Lee et al. 1991, 1994). Treatment with dimethylsulfoxide (DMSO) or 12-O-tetradecanoylphorbol 13-acetate (TPA) caused a concentration-dependent decrease in facilitated uridine transport attributable to a decrease in the number of transporter molecules. The exposure to the inducers of differentiation produced a corresponding increase in the Na^+-dependent transporter, the mechanism being a change in the affinity of the uridine for the transporter, with no change in the maximum velocity. The consequence of these changes was the generation of an increase in the intracellular concentration of uridine relative to the medium. The changes are consistent with a transition from a malignant phenotype to a more normal one.

Conclusions

Terminal differentiation of leukemia cells can be obtained by the biological agent G-CSF in a process dependent upon the concentration of the G-CSFR and the ploidy of the cells, as well as by a variety of antineoplastic agents. Separation of the molecular events required to initiate the maturation process by the antileukemic agent TG and by the trisaccharide-containing anthracyclines from the biochemical alterations responsible for the cytotoxicity produced by these agents have been shown. The findings suggest that combinations of biological and chemical agents have different mechanisms of action and, therefore, might be combined to achieve more effective conversion of malignant cells to mature benign forms.

References

Begley, CG, Metcalf, D, Nicola, NA (1987) Primary human myeloid leukemia cells: comparative responsiveness to proliferative stimulation by GM-CSF or G-CSF and membrane expression of CSF receptors. Leukemia 1:1-8.

Begley, CGD, Metcalf, D, Nicola, NA (1987) Purified colony-stimulating factors (G-CSF and GM-CSF) induce differentiation in human HL-60 leukemic cells with suppression of clonogenicity. Int J Cancer 39:99-105

Berdel, WE, Danhauser-Riedl, S, Steinhauser, G, Winton, EF (1989) Various human hematopoietic growth factors (interleukin-3, GM-CSF, G-CSF) stimulate clonal growth of nonhematopoietic tumor cells. Blood 73: 80-83

Bolande, RP (1985) Spontaneous regression and cytodifferentiation of cancer in early life: the oncogenic grace period. Survey and Synthesis of Pathology Research 4:296-311

Burgess, AW, Metcalf, D (1980) Characterization of a serum factor stimulating the differentiation of myelomonocytic leukemic cells. Int J Cancer 39: 647-654

Bussolino, F, Wang, JM, Defilippi, P, Turrini, F, Sanavio, F, Edgell, C-JS, Aglietta, M, Arese, P, Mantovani, A (1989) Granulocyte and granulocyte-macrophage-colony stimulating factors induce human endothelial cells to migrate and proliferate. Nature 337:471-473

Castaigns, S, Chomienne, C, Daniel, MT, Ballerini, P, Berger, R, Fenaux, P, Degos, L (1990) All-trans retinoic acid as a differentiation therapy for acute promyelocytic leukemia. I. Clinical Results. Blood 76:1704-1709

Colombo, MP, Ferrari, G, Stoppacciaro, A, Parenza, M, Rodolfo, M, Mavilio, F, Parmiani, G (1991) Granulocyte colony-stimulating factor gene transfer suppresses tumorigenicity of a murine adenocarcinoma in vivo. J Exptl Med 173:889-897.

Crooke, ST, DuVernay, VH, Golvan, L, Prestayko, AW (1978) Structure-activity relationships of anthracyclines relative to effects on macromolecular synthesis. Mol Pharmacol 14:290-298

Darnowski, JW, Handschumacher, RE (1986) Tissue uridine pools: evidence in vivo of a concentrative mechanism for uridine uptake. Cancer Res 46:3490-3494

Darnowski, JW, Holdridge, C, Handschumacher, RE (1987) Concentrative uridine transport by murine splenocytes: kinetics, substrate specificity and sodium dependency. Cancer Res 47:2614-2619

DuVernay, VH, Pachter, JA, Crooke, ST (1979) Deoxyribonucleic acid binding studies on several new anthracycline antitumor antibiotics. Sequence preference and structure activity relationships of marcellomycin and its analogues as compared to adriamycin. Biochemistry 18:4024-4030

Fearon, ER, Burke, PJ, Schiffer, CA, Zehnbauer, BA, Vogelstein, B (1986) Differentiation of leukemia cells to polymorphonuclear leukocytes in patients with acute nonlymphocytic leukemia. New Eng J Med 315:15-24

Fukunaga, R, Ishizaka-Ikeda, E, Seto, Y, Nagata, S (1990) Expression cloning of a receptor for murine granulocyte colony-stimulating factor. Cell 61:341-350

Gusella, JF, Houseman, D (1976) Induction of erythroid differentiation in vitro by purines and purine analogues. Cell 8:263-269

Huang, M-E, Ye, Y-C, Chen, S-R, Chai, J-R, Lu, J-X, Zhoa, L, Gu, L-J, Wang, Z-Y (1988) Use of all-trans retinoic acid in the treatment of acute promyelocytic leukemia. Blood 72:567-572

Ishiguro, K, Sartorelli, AC (1985) Enhancement of the differentiation-inducing properties of 6-thioguanine by hypoxanthine and its nucleosides in HL-60 promyelocytic leukemia cells. Cancer Res 45:91-95

Ishiguro, K, Schwartz, EL, Sartorelli, AC (1984) Characterization of the metabolic forms of 6-thioguanine responsible for cytotoxicity and induction of differentiation of HL-60 acute promyelocytic leukemia cells. J Cell Physiol 121:383-390

Ishiguro, K, Taft, WC, DeLorenzo, RJ, Sartorelli, AC (1987) The role of benzodiazepine receptors in the induction of differentiation of HL-60 leukemia cells by benzodiazepines and purines. J Cell Physiol 131:226-234

Jarvis, SM (1989) Characterization of sodium-dependent transport in rabbit intestinal brush-border membrane vesicles. Biochim Biophys Acta 979:132-138

Jarvis, SM, Williams, TC, Lee, CW, Cheeseman, CI (1989) Active transport of nucleosides and nucleoside drugs. Biochem Soc Trans 17:448-449

Koeffler, HP (1983) Induction of differentiation of human acute myelogenous leukemia cells: therapeutic implications. Blood 62:709-721

Lee, C-W, Sokoloski, JA, Sartorelli, AC, Handschumacher, RE (1991) Induction of the differentiation of HL-60 cells by phorbol 12-myristate 13-acetate activates a Na^+-dependent uridine-transport system. Biochem J 274:85-90

Lee, C-W, Sokoloski, JA, Sartorelli, AC, Handschumacher, RE (1994) Differentiation of HL-60 cells by dimethylsulfoxide activates a Na^+-dependent nucleoside transport system. In vivo 8:795-802

Lee, CW, Cheeseman, CI, Jarvis, SM (1988) Na^+- and K^+-dependent uridine transport in rat renal brush-border membrane vesicles. Biochim Biophys Acta 942:139-149

Lee, CW, Cheeseman, CI, Jarvis, SM (1990) Transport characteristics of renal brush-border Na^+- and K^+-dependent carriers Am J Physiol 25:F1203-F1210

LeHir, M, Dubach, VC (1985) Uphill transport of pyrimidine nucleosides in renal brush border vesicles. Pflugers Arch 404:238-243

Li, J, Koay, DC, Sartorelli, AC (1995) The role of DNA ploidy in the differentiation of WEHI-3B D^- leukemia cells transfected with the granulocyte colony-stimulating factor receptor gene. Exptl Cell Res 219:579-588

Li, J, Koay, DC, Xiao, H, Sartorelli, AC (1993) Regulation of the differentiation of WEHI-3B D^+ leukemia cells by granulocyte colony-stimulating factor receptor. J Cell Biol 120:1481-1489

Li, J, Sartorelli, AC (1992) Synergistic induction of the differentiation of WEHI-3B D^- myelomonocytic leukemia cells by retinoic acid and granulocyte colony-stimulating factor. Leukemia Res 16:571-576

Matsumoto, Y, Saiki, I, Murata, J, Okuyama, H, Tamura, M, Azuma, I (1991) Recombinant human granulocyte colony-stimulating factor inhibits the metastasis of hematogenous and non-hematogenous tumors in mice. Int J Cancer 49:444-449

Metcalf, D (1980) Clonal extinction of myelomonocytic leukemic cells by serum from mice injected with endotoxin. Int J Cancer 25:225-233

Metcalf, D, Nicola, NA (1982) Autoinduction of differentiation in WEHI-3B leukemia cells. Int J Cancer 30:773-780

Morin, MJ, Sartorelli, AC (1984) Inhibition of glycoprotein biosynthesis by the inducers of HL-60 cell differentiation, aclacinomycin A and marcellomycin. Cancer Res 44:2807-2812

Nicola, NA, Begley, CG, Metcalf, D (1985) Identification of the human analogue of a regulator that induces differentiation in murine leukemic cells. Nature 314:625-628

Nicola, NA, Metcalf, D (1984) Binding of differentiation-inducer, granulocyte colony-stimulating factor, to responsive but not unresponsive leukemia cell lines. Proc Natl Acad Sci USA 81:3765-3769

Nicola, NA, Metcalf, D (1985) Binding of ^{125}I-labeled granulocyte colony-stimulating factor to normal murine hemopoietic cells. J Cell Physiol 124:313-321

Park, LS, Waldron, PE, Friend, D, Saasenfeld, HM, Price, V, Anderson, D, Cosman, D, Andrews, RG, Berstein, ID, Urdal, DL (1989) Interleukin-3, GM-CSF, and G-CSF receptor expression on cell lines and primary leukemia cells: receptor heterogeneity and relationship to growth factor responsiveness. Blood 74:56-65

Reiss, M, Gamba-Vitalo, C, Sartorelli, AC (1986) Induction of tumor cell differentiation as a therapeutic approach: preclinical models for hematopoietic and solid neoplasms. Cancer Treat Rep 70:201-218

Sato, S, Sakashita, A, Ishiyama, T, Nakamaki, T, Hino, K-I, Tomoyasu, S, Tsuruoka, N, Honma, Y, Hozumi, M (1992) Possible differentiation treatment with aclacinomycin A in acute myelomonocytic leukemia refractory to conventional chemotherapy. Anticancer Res 12:371-376

Schwartz, EL, Brown, BJ, Nierenburg, M, Marsh, JC, Sartorelli, AC (1983) Evaluation of some anthracycline antibiotics in an in vivo model for studying drug-induced human leukemia cell differentiation. Cancer Res 43:2725-2730

Schwartz, EL, Ishiguro, K, Sartorelli, AC (1983) Induction of leukemia cell differentiation by chemotherapeutic agents. Adv Enz Reg 21:3-20

Schwartz, EL, Sartorelli, AC (1982) Structure-activity relationships for the induction of differentiation of HL-60 human acute promyelocytic leukemia cells by anthracyclines. Cancer Res 42:2651-2655

Schwenk, M, Hegay, E, Lopez del Pino, V (1984) Uridine uptake by isolated intestinal epithelial cells of guinea pig. Biochim Biophys Acta 805:370-374

Toki, H, Matsutomo, S, Okabe, K, Shimokawa, T (1989) Remission in hypoplastic acute myeloid leukemia induced by granulocyte colony-stimulating factor. Lancet 1:1389-1390

Tsuda, H, Neckers, LM, Pluznik, DH (1986) Colony stimulating factor-induced differentiation of murine M1 myeloid leukemic cells is permissive in early G1 phase. Proc Natl Acad Sci USA 83:4317-4321

Uzumaki, H, Okabe, T, Sakaki, N, Hagiwara, K, Takaku, F, Tobita, M, Yasukawa, K, Ito, S, Umezawa, Y (1989) Identification and characterization of receptors for granulocyte colony-stimulating factor on human placenta and trophoblastic cells. Proc Natl Acad Sci USA 86:9323-9326

Warrell, Jr, RP, Frankel, SR, Miller, Jr, WH, Scheinberg, DA, Itri, LM, Hittelman, WN, Vyas, R, Andreeff, M, Tafuri, A, Jakubowski, A, Gabrilove, J, Gordon, M, Dmitrovsky, E, (1991) Differentiation therapy of acute promyelocytic leukemia with tretinoin (all-trans retinoic acid). New Eng J Med 324:1385-1393

Yamasaki, Y, Izumi, Y, Sawada, H, Fujita, K (1991) Probable in vivo induction of differentiation by recombinant human granulocyte colony-stimulating factor (rhG-CSF) in acute promyelocytic leukemia (APL). Brit J Haematol 78:579-580

REGULATION OF MURINE ERYTHROLEUKEMIA CELL DIFFERENTIATION

Asterios S. Tsiftsoglou*, Ioannis S. Vizirianakis and Ioannis S.Pappas

Laboratory of Pharmacology, Department of Pharmaceutical Sciences, Aristotle University of Thessaloniki, Thessaloniki 54006, Macedonia, Greece

Introduction

It has been evident over the past three decades that virus-transformed hematopoietic cells represent progenitor cells unable to differentiate but capable to rapidly proliferate and give rise to leukemias. Such cells have been envisioned as cells deviated from the normal pathway of growth and differentiation and lost their responsiveness to physiological regulators (growth factors) (Sachs, 1980). In most cases leukemic cells appear to be less differentiated or undifferentiated, exhibit karyotypic abnormalities and fail to produce molecules characteristic of their differentiated phenotype.

Thus far conventional chemotherapy is aimed to kill the majority, if not all of the malignant cells with single or more than one antineoplastic agents (combination) and reduce tumor cell load markedly. The realization however, that in most, if not all tumors undergoing treatment, multidrug resistant (MDR) cells emerge (Biedler,1994) led to searching for alternative approaches to counteract rapid tumor growth and malignancy.

The early observations of Sachs (1964), Friend (1971) and Pierce (1974) that some neoplastic cells can be differentiated *in vitro* and/or *in vivo* into cells resembling their normal counterparts *via* treatment with a large variety of agents including growth factors led to a new way of thinking to combat malignacy under relatively less cytotoxic conditions. The so called "Differentiation Therapy of Neoplasms" (Lotan et al.,1990) has been already applied as an effective approach of cancer treatment in the case of human acute promyelocytic leukemia (AML) (Castaigne et al., 1990). Moreover, induction of

---

* To whom correspondence should be addressed.

NATO ASI Series, Vol. H 99
Tumor Biology
Regulation of Cell Growth, Differentiation and Genetics in Cancer
Edited by Asterios S. Tsiftsoglou, Alan C. Sartorelli, David E. Housman and T. Michael Dexter

differentiation has been achieved in a large number of neoplastic cell systems (Reiss et al.,1986). In most cases studied, induction of differentiation led to less or non-malignant cells that fail to proliferate and support malignant growth. Finally, in most of the differentiation systems of neoplastic cells, the same principles and cellular behaviour appear to occur although the precise mechanism(s) have not been fully elucidated (Tsiftsoglou and Robinson,1985).

One of the most widely used model systems of leukemic cell differentiation is the so called Friend or murine erythroleukemia (MEL) cell differentiation system reviewed elsewhere (Marks and Rifkind,1978; Tsiftsoglou and Robinson,1985; Tsiftsoglou and Wong,1985). We have been studying this system over the years to delineate the molecular basis of leukemic cell differentiation on one hand and hemopoietic cell development on the other. The present study will focus on experiments designed to: (a) analyze the developmental program of MEL cells, (b) investigate the central role of commitment in leukemic cell growth and differentiation, (c) examine the role of inducer-binding proteins and (d) identify posttranscriptional modifications of RNA as an integral part of the differentiation process.

Differentiation of MEL cells is a multistep coordinated process

It is now well established that treatment of MEL cells in culture with a variety of structural unrelated agents leads to early events, that is, a set of biochemical events that occur within the first 12 to 24hr before cells irreversibly commit to erythroid maturation and begin to synthesize hemoglobin (Housman et al.,1980). In this early period, that precedes commitment (latent period) the majority of the events appear to be membrane-associated and may be responsible for signal transduction. This period, is also characterized by the ability of cells to express "memory" that is, an ability to remember previous exposure to inducer (Levenson and Housman,1979). These early events are transient, easily reversible and unable to commit cells to maturation upon removal of the inducer. Upon commitment to differentiation, (a point in time required for individual inducer-treated cells to irreversibly differentiate into mature cells), however, cells undergo programmed loss of proliferation and expression of erythroid markers (e.g. hemoglobin) after 24 to 26 hrs. These findings indicate that MEL cell differentiation along the erythrocytic pathway is a multistep process characterized by subsets of coordinating events. The distinction of events of differentiation in early and late and in subsets allowed us and others (Marks and Rifkind,1978; Tsiftsoglou and Robinson,1985; Tsiftsoglou and Wong,1985; Housman et al.,1980) to address a number of fundamental questions such as how do the early process(es) lead to irreversible events? how do cells make decision to differentiate? and what impact do these events have on the expression of several genes involved in the control of cell growth and differentiation?

Expression of "memory" precedes commitment to maturation

In late 70s, it was observed that MEL cells exposed to an inducer of differentiation for several hours (6 to 12), then washed out and grown in inducer-free medium acquire a unique ability to remember the previous exposure and respond rapidly upon rechallenge with the same or other inducer without recapitulating the entire latent period (12hr prior commitment) (Levenson and Housman,1979). This phenomenon so called "memory to previous exposure" suggested for the first time that the inducing agent used evokes a response before an individual cell commits irreversibly to terminal maturation. Apparently, this response is maintained for short or long time (depends on the subclone of MEL cells used; Tsiftsoglou, unpublished observations) and decay thereafter if commitment does not occur. This suggests that "memory" is one of those initiative events which enable the cells to recognize signals promoted by the inducer upon its interaction with cellular components. Although the molecular basis of "memory" is not known, experiments with metabolic inhibitors indicated that "memory" is erased by treatment with several structurally unrelated agents including inhibitors of new protein and RNA synthesis (Housman et al.,1980), inhibitors of posttranscriptional methylation of RNA (Vizirianakis and Tsiftsoglou,1995) and even glycocorticoids (Tsiftsoglou et al.,1983). All these findings suggest that "memory" is carried by macromolecules that accumulate into the cells and reach a threshold level just prior to commitment. Even today, it is not known whether "memory" is a general response of cells to different growth factors or differentiators and how does this occur. It is known however, that both "memory" and commitment are general properties of differentiated MEL cells, since both are initiated by one inducer and can be terminated by another. "Memory" and commitment have also been observed in RD/TE-671 cells, a human neoplastic cell line of neuroectodermal origin that respond to same inducing agents as MEL cells (Pappas et al.,1996).

The central role of commitment in hemopoietic cell differentiation

A) Commitment leads to loss of proliferation. That individual MEL cells exposed to an inducer undergo a binary decision either to differentiate or not led to the discovery of *commitment* (Gusella et al.,1976). As a process, commitment is pivotal and occurs stochastically. Committed cells continue to complete the differentiation program even in the absence of the inducer (Gusella et al.,1976) and divide only a few times before suffer cell death and disintegration. This central event in development (Tsiftsoglou and Robinson,1985), appears also to occur in normal bone marrow enviroment responsive to natural growth and differentiation factors (e.g. erythropoietin, GM-CSF e.t.c.) (Sachs,1994). Initiation of commitment depends like "memory" on new RNA and protein synthesis (Levenson and Housman,1979) as well as on a methionine-sensitive event, like methylation of RNA (Vizirianakis et al.,1992). It appears however, that commitment regulates cell cycle and causes G_1 cellular arrest (Tsiftsoglou and Robinson,1985). On the light of this evidence, one can assume that commitment may regulate the cyclins cascade, although direct evidence is still missing. The observation that several oncogenes like c-myc and p53 are irreversibly repressed in committed cells, indicates that commitment regulates also the expression of proto-oncogenes and perhaps other cell cycle genes (cyclin genes).

B) Commitment sets on discrete patterns of gene expression. Several studies over the past years have established that some RNA transcripts accumulate gradually into the cytoplasm while others disappear in the differentiating MEL cells. The terminally differentiated cells synthesize heme and large amounts of hemoglobin *via* activation of the cascade of enzymes involved in heme biosynthesis (Sassa,1992). By using several DNA probes (cDNA and/or genomic), it was observed that MEL cells undergoing terminal differentiation regulate various RNA transcripts in discrete patterns (see Table I). One set of RNA transcripts like those involved in heme biosynthesis (e.g. e-ALAs gene, transferrin receptor gene, globin genes and their regulators) accumulate gradually and reach high levels in differentiating cells. p53, c-myc and c-myb are genes affected upon exposure to inducer but cease to produce RNA transcripts later on in the differentiated state. In contrast, genes encoding house-keeping proteins continued to be expressed constitutively for long time. On the top of all, there are rapidly responsive genes like c-fos and c-jun (Tsiftsoglou and Robinson,1985) that are also repressed later on in differentiation. All these observations indicate that there are rapidly responsive genes which are activated prior to commitment and others that are regulated by commitment thereafter. The latter group includes genes regulated by commitment and involved in the cell cycle and hemoglobin synthesis. These two sets of genes appear to function independently after commitment has been initiated in MEL cells (Gusella et al.,1982).

Induction of MEL cell differentiation involves receptor-mediated processes

The structural diversity of various inducers (physical, chemical or natural) discovered thus far, didn't lead to precise structure-activity relationships (SAR). This has been a critical point, since SAR could be very valuable in predicting compounds able to promote differentiation and in identifying cellular components selectively recognized by the inducers. Most recently, we developed a series of ureido-derivatives of pyridine (UDPs) as potent inducing agents (Pappas et al.,1992; Pappas et al.,1996). Radiolebeling one of them and studying its accumulation and interaction with cellular components in both MEL and RD/TE-671 revealed that this agent selectively recognizes an inducer-binding protein (inducerphilin) of 40kDa (Pappas and Tsiftsoglou,1995). Additional experiments indicated that this 40kDa protein may mediates induction of differentiation by other inducers of MEL cells, like hexamethylene-bis-acetamide (HMBA) and butyric acid (unpublished observations).

Posttranscriptional modification of RNA as part of the developmental program of MEL cells: the role of RNA methylation

The observations that discrete patterns of gene expression operate in differentiating MEL cells (see Table I) suggest that the steady-state cytoplasmic accumulation of RNA transcripts results either from transcriptional activation, repression, or posttranscriptional processes. In fact, previous studies have considered DNA hypomethylation in differentiating MEL cells responsible for activation of transcription of globin genes (Christman et al.,1977). Although, this observation can explain in part the discrete patterns of gene expression seen in differentiating cells, the possibility that such changes may be due to differential stability of RNA has drawn a lot of attention (Volloch and Housman,1981; Volloch et al.,1987). The latter possibility prompted us to investigate the role of posttranscriptional modifications of RNA in MEL cell differentiation and RNA stability. Biochemical studies with cordycepin (3'-deoxyadenosine) (an inhibitor of polyadenylation

TABLE I

MEL cell differentiation and patterns of gene expression

mRNA	Effect of differentiation on mRNA steady-state levels: Before commitment	After commitment	Reference
β^{major}-globin	—	↑	Curtis et al.,1980
β^{minor}-globin	—	↑	Curtis et al.,1980
α_1-globin	—	↑	Sheffery et al.,1984
Tf-R	—	↑	Klinken et al.,1987
e-ALAS	—	↑	Fujita et al.,1991
NF-E2	NR	↑	Nagai et al.,1995
c-myc	↓↑ (Biphasic)	↓	Lachman and Skoultchi,1984; Tsiftsoglou et al.,1987
c-fos	↑	↓	Ramsay et al.,1986;
c-myb	↓↑ (Biphasic)	↓	Ramsay et al.,1986; Spotts and Hann,1990
p53	↓↑ (Biphasic)	↓	Ramsay et al.,1986; Tsiftsoglou et al.,1987
jun B	↑	↓	Francastel et al.,1994
c-K-ras	—	—	Coppola et al.,1989
rRNAs	↓	↓	Tsiftsoglou et al.,1982
MER5	↑	↓	Yamamoto et al.,1989
β-actin	—	→↓	Lachman and Skoultchi,1984
γ-actin	—	→↓	Richon et al.,1992
GAPDH	—	→↓	Francastel et al.,1994
$H1^0$ histone	↑	↓	Cheng and Skoultchi,1989
H3.1.histone	NR	↓	Brown et al.,1985
H3.2 histone	↓	↓	Brown et al.,1985;1988
H3.3 histone	NR	—	Brown et al.,1985
H2a histones	↓	↓	Brown et al.,1985;1988
H2b histones	NR	↓	Brown et al.,1985
HSP70	↓	↓	Hensold and Housman,1988
pRB	NR	↑	Coppola et al.,1990; Richon et al., 1992
β_2-microglobulin	—	—	Khochbin et al.,1988
PrP	—	↓	Vizirianakis et al.,1994
COX II	—	↓	UD
COX IV	—	→↓	UD
clone 11.1 [a]	—	↓	UD
clone 6.2 [b]	—	↓	UD
eEF-Tu	—	↓	Roth et al., 1987
PU.1 factor	↓	↓	Schuetze et al.,1992
Vimentin	↓	↓	Ngai et al., 1984
ODC	↓	↓	Klinken et al., 1987

↓ = decrease, ↑ = increase, — = not changed, and →↓ = decrease only at late stages of differentiation, NR= not referred, UD= Vizirianakis et al., upublished data.

[a] clone 11.1 is a 702bp cDNA fragment isolated from MEL cells that shares significant structural homology to rat S5 ribosomal protein mRNA.

[b] clone 6.2 is a 208bp cDNA fragment isolated from MEL cells that shares significant structural homology to B22 subunit of NADH-ubiquinone oxidoreductase complex of mitochondria.

Abbreviations: eEF-Tu: eukaryotic elongation factor Tu; ODC: ornithine decarboxylase; e-ALAS: erythroid form of δ-aminolevulinate synthase; NF-E2: erythroid-specific nuclear factor 2; COX: mitochondrial cytochrome c oxidase; pRB: Petinoblastoma protein; GAPDH: glyceraldehyde 3-phosphate dehydrogenase; Pu.1: *ets*-related transcription factor; MER5: gene preferentially expressed in murine eythroleukemia cells; PrP: Prion protein; NF-E2: erythroid-specific nuclear factor 2;

and methylation of RNA) and more recently with N^6-methyladenosine (N^6mAdo) have suggested that initiation of commitment of inducer-treated MEL cells to terminal erythroid maturation may depend on both the synthesis and posttranscriptional modifications of RNA, such as methylation (Kredich, 1980; Vizirianakis et al.,1992). The process of posttranscriptional methylation of RNA, involves transfer of methyl-groups from S-adenosylmethionine (SAM) into specific base residues of RNA *via* RNA methyltransferases. Several RNA species in eukaryotic cells (snRNAs, tRNAs, rRNAs, mRNAs) have found to be methylated at specific base residues located at unique structures like the 5'-cap stucture (Banerjee, 1980; Choi et al.,1982; Kersten, 1984; Maden, 1990).

Most recently, we observed that MEL cell differentiation is associated with hypermethylation of RNA (Vizirianakis and Tsiftsoglou, 1996[b]). In particular, we have shown that treatment of MEL cells with the inducer dimethylsulfoxide (DMSO) led to an increase in methylation of total cytoplasmic and polyA$^+$ RNA as well as a transient increase in methylation of ribosomal RNAs (Vizirianakis and Tsiftsoglou, 1995[a]). By using a reversed-phase HPLC to assess the level of various methylated nucleosides in polyA$^-$ and polyA$^+$ RNA fractions of control and differentiating MEL cells, we observed the following: a) polyA$^-$ RNA from control and differentiating MEL cells contain several methylated nucleosides in different quantities (Fig. 1). The level of 5-methylcytidine, 1-methylguanosine and N^2-methylguanosine was higher in polyA$^-$ RNA isolated from DMSO-treated than from control untreated cells; and b) polyA$^+$ RNA was hypermethylated in DMSO-treated cells mainly in the 5′-cap structure (Vizirianakis and Tsiftsoglou,1996[b]). Furthermore, we observed that the increase in RNA methylation in differentiating MEL cells was quantitatively associated with alterations in the level of both S-adenosylhomocysteine (SAH) and S-adenosylmethionine (SAM) thus resulting in a lower ratio SAH/SAM. These changes in SAH/SAM ratio appears to influence the extent of RNA methylation, since RNA methyltransferases may no longer be inhibited by SAH (Cantoni, 1986). Unfortunately, the precise role of changes in methylation of both polyA$^-$ and polyA$^+$ RNA during MEL cell differentiation are not clear yet. Increased methylation of RNA transcripts may affect either the physicochemical behavior of RNA and facilitate the transport of mRNA from nucleus into cytoplasm as previously reported (Camper et

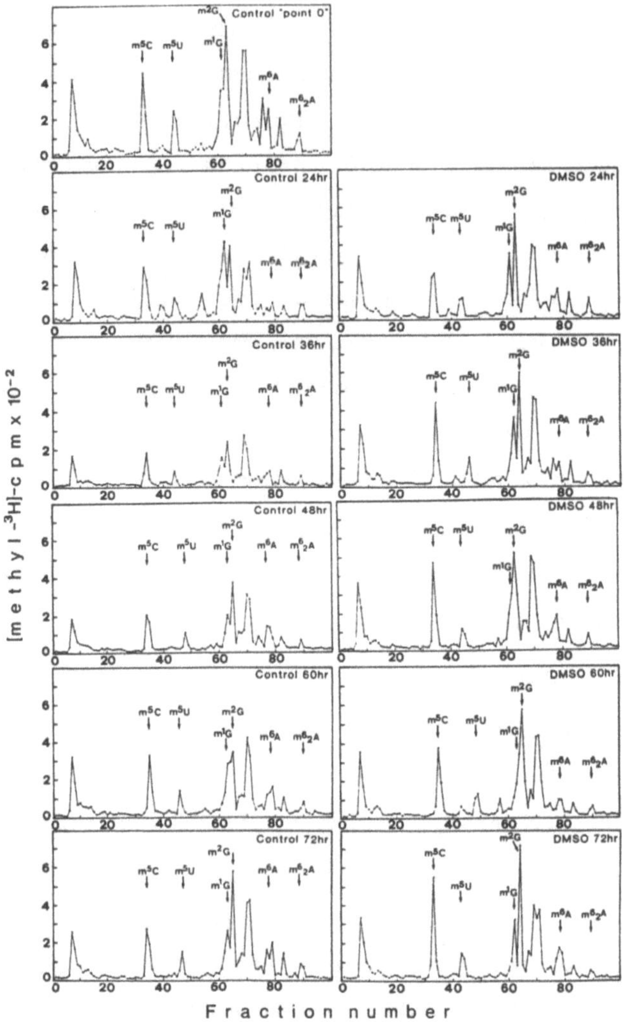

Fig. 1. HPLC separation of products derived by alkaline hydrolysis of [methyl-^{3}H]-polyA$^-$ RNA isolated from control and DMSO-treated MEL cells.

MEL-745PC-4A cells were incubated in the presence or absence of DMSO (1,5% v/v). At times indicated, cells (1-1.5x10^8) were removed from cultures, pulse-labeled with L-[methyl-^{3}H]-methionine (44µCi/ml, sp.act. 80.0Ci/mmol) at 37°C for 3hr and [methyl-^{3}H]-labeled polyA$^-$ RNA was isolated as described elsewhere (Vizirianakis and Tsiftsoglou, 1996^b). 5µg of each sample of [methyl-^{3}H]-polyA$^-$ RNA were hydrolyzed by alkali treatment, then the resulting nucleotides were dephosphorylated by alkaline phosphatase and analyzed by HPLC. Fractions (0.5ml/30sec) were collected and counted for radioactivity. The arrows indicate the peaks of absorbance at 254nm of known nucleosides. Abbreviations: **m^{5}C**: 5-methylcytidine, **m^{1}G**: 1-methylguanosine, **m^{2}G**: N^2-methylguanosine, **m^{6}A**: N^6-methyladenosine and **m$^6{}_2$A**: N^6,N^6-dimethyladenosine.

al.,1984), or alter the configuration of 5′-end of mRNAs (Kim and Sarma, 1978) to make them less susceptible to nucleases that degrade RNA beginning from the 5′-end (Goutts and Brawerman, 1993, and Goutts et al.,1993). Finally, changes in RNA methylation may contribute to tertiary structure and conformation of RNA in a way that enables it to interact with trans-acting factors (Peltz et al.,1991; Belasco and Brawerman, 1993; Klausner et al., 1993, Burd and Dryefuss, 1994). All these possible changes that may result from hypermethylation of RNA at several sites may affect RNA stability, an event that is developmentally regulated (Tsiftsoglou et al.,1992) during MEL cell differentiation (Volloch and Housman,1981; Volloch et al.,1987; Tsiftsoglou et al.,1992).

That RNA methylation is considered an integral part of the differentiation program of MEL cells, was further supported from the results obtained with the use of N^6mAdo (Vizirianakis and Tsiftsoglou, 1995), an agent found to block commitment, to erase "memory", to reduce methylation of total cytoplasmic RNA and to alter the intracellular levels of SAH and SAM. As we observed, this occurs *via* conversion of N^6mAdo intracellularly into S-N^6-methyladenosylhomocysteine (N^6-SAH), an intermediate that inhibits RNA methyltransferases (Pugh et al.,1977; Hoffman,1978). These data are consistent with those obtained with other known inhibitors of RNA methylation, like 3-deazaadenosine, S-5′-isobutylthioadenosine, 5′-methylthioadenosine, neplanocin A, 3-deazaneplanocin A, and cycloleucine (Chiang and Miura, 1986). All these inhibitors of RNA methylation inhibit MEL cell differentiation as well (Fiore et al.,1984; Sherman et al.,1985; Vizirianakis et al.,1992; Vizirianakis and Tsiftsoglou, 1996[b]).

REFERENCES

Banerjee A K (1980) 5'-Terminal cap structure in eukaryotic messenger ribonucleic acids. Microbiol Reviews 44:175-205

Belasco J and Brawerman G (eds) (1993) Control of messenger RNA stability. Academic Press Inc. San Diego

Biedler JL (1994) Drug resistance: genotype *versus* phenotype. Cancer Res 54:666-678

Brown DT, Wellmsan SE and Sittman DB (1985) Changes in the levels of three different classes of histone mRNA during murine erythroleukemia cell differentiation. Mol Cell Biol 5 2879-2886

Brown DT, Yang Y-S and Sittman DB (1988) Histone gene switching in murine erythroleukemia cells is differentiation specific and occurs without loss of cell cycle regulation. Mol Cell Biol 8:4406-4415

Burd CG and Dreyfuss G (1994) Conserved structures and diversity of functions of RNA-binding proteins. Science 265:615-621

Camper SA, Albers RJ, Coward JK, and Rottman FM (1984) Effect of undermethylation on mRNA cytoplasmic appearance and half-life. Mol Cell Biol 4:538-543

Cantoni GL (1986) The centrality of S-adenosylhomocysteinase in the regulation of the biological utilization of S-adenosylmethionine. In: Borchardt RT, Creveling CR and Ueland PM (eds), Biological methylation and drug design, pp. 227-238, Humana Press, Clifton, New Jersey

Castaigne S, Chomienne C, Daniel MT, Ballerini P, Berger R, Fenaux P, and Degos L (1990) All-trans retinoic acid as a differentiation therapy for acute promyelocytic leukemia. I. Clinical results. Blood 76:1704-1709

Cheng G and Skoultchi AI (1989) Rapid induction of polyadenylated H^1 histone mRNAs in mouse erythroleukemia cells is regulated by c-myc. Mol Cell Biol 9:2332-2340

Chiang PK and Miura GA (1986) S-adenosylhomocysteine hydrolase. In: Borchardt RT, Creveling CR and Ueland PM (eds), Biological methylation and drug design pp. 239-251. Humana Press, Clifton, New Jersey

Choi YC, Reddy R and Busch H (1982) Methylation of snRNAs and 18S rRNA of Novikoff hepatoma. In: Usdin E, Borchardt RT and Creveling CR (eds), Biochemistry of S-adenosylmethionine and related compounds, pp. 313-320, Macmillan Press, London

Christman JK, Price P, Pedrinan L and Acs G (1977) Correlation between hypomethylation of DNA and expression of globin genes in Friend erythroleukemia cells.Eur J Biochem 81:53-61

Coppola JA, Lewis BA and Cole MD (1990) Increased retinoblastoma gene expression is associated with late stages of differentiation in many different cell types. Oncogene 5:1731-1733

Coppola JA, Parker JM, Schuler GD and Cole MD (1989) Continued withdrawal from the cell cycle and regulation of cellular genes in mouse erythroleukemia cells blocked in differentiation by c-myc oncogene. Mol Cell Biol 9:1714-1720

Curtis P, Finnigam AC and Rovera G (1980) The β^{major} and β^{minor} globin nuclear transcripts of Friend erythroleukemia cells induced to differentiate in culture. J Biol Chem 225:8971-8974

Fiore PPD, Grieco M, Pinto A, Attadia V, Porcelli M, Cacciapuoti G and Carteni-Farina M (1984) Inhibition of dimethylsulfoxide-induced differentiation of Friend erythroleukemic cells by 5'-methylthioadenosine. Cancer Res 44:4096-4103

Francastel C, Groisman R, Pfarr CM and Robert-Lézénès J (1994) Antisense c-jun overcomes a differentiation block in a murine erythroleukemia cell line. Oncogene 9:1957-1964

Friend C, Scher W, Holland JG and Sato T (1971) Hemoglobin synthesis in murine virus-induced leukemia cells in vitro: stimulation of erythroid differentiation by dimethylsulfoxide. Proc Natl Acad Sci USA 68:378-382

Fujita H, Yamamoto M, Yamagami, T, Hayashi N and Sassa S (1991) Erythroleukemia differentiation: Distinctive responces of the erythroid-specific and the nonspecific ä-aminolevulinate synthase mRNA. J Biol Chem 266:17494-17502

Goutts M and Brawerman G (1993) A 5′ exoribonuclease from cytoplasmic extracts of mouse sarcoma 180 ascites cells Biochem Biophys Acta, 1173:57-62

Goutts M, Krowczynska A and Brawerman G (1993) Protection of mRNA against nucleases in cytoplasmic extracts of mouse sarcoma ascites cells. Biochem Biophys Acta 1173:49-56

Gusella J, Geller R, Clarke B, Weeks V and Housman D (1976) Commitment to erythroid differentiation by Friend erythroleukemia cells: a stochastic analysis. Cell 9:221-229

Gusella JF, Tsiftsoglou AS, Volloch V, Weil SC, Neumann J and Housman DE (1982) Dissociation of hemoglobin accumulation and commitment during murine erythroleukemia cell differentiation by treatment with imidazole. J Cell Physiol 113:179-185

Hensold JO and Housman DE (1988) Decreased expression of the stress protein HSP70 is an early event in murine erythroleukemia cell differentiation. Mol Cell Biol 8:2219-2223

Hoffman JL (1978) Biosynthesis of S-N^6-methyladenosylhomosysteine an inhibitor of RNA methyltransferases. J Biol Chem 253:2905-2907

Housman D, Levenson R, Volloch V, Tsiftsoglou AS, Gusella JF, Parker D, Kernen J, Mitrani A, Weeks V, Witte O and Besmer P (1980) Control of proliferation and differentiation in cells transformed by Friend virus. Cold Spring Harbor Quant Biol 44:1177-1185

Kersten H (1984) On the biological significance of modified nucleosides in tRNA. Prog Nucleic Acid Res Mol Biol 31:59-114

Khochbin S, Principaud E, Chabanas A and Lawrence J-J (1988) Early events in murine erythroleukemia cells induced to differentiate. J Mol Biol 200 55-64

Kim CH and Sarma RH (1978) Spatial configuration of the bizarre 5′ terminus of mammalian mRNA. J Am Chem Soc 100:1571-1590

Klausner RD, Rouault TA and Harford JB (1993) Regulating the fate of mRNA: the control of cellular iron metabolism Cell 72:19-28

Klinken SP, Billelo J, Bauer S, Morse III HC and Thorgeirsson SS (1987) Altered expression of â-globin, transferrin receptor, and ornithine decarboxylase in Friend murine erythroleukemia cells inhibited by α-difluoromethylornithine. Cancer Res 47:2638-2644

Kredich NM (1980) Inhibition of nucleic acid methylation by cordycepin. J Biol Chem 255:7380-7385

Lachman H and Skoultchi AI (1984) Expression of c-myc changes during differentiation of mouse erythroleukemia cells. Nature 310:592-594

Levenson R and Housman D (1979) Momory of MEL cells to a previous exposure to inducer. Cell 17:485-490

Lotan R, Francis GE, Freeman CS and Waxman S (1990) Differentiation therapy. Cancer Res 50:3453-3464.

Maden BEH (1990) The numerous modified nucleotides in eukaryotic ribosomal RNA. Prog Nucleic Acid Res Mol Biol 39:241-303

Marks PA and Rifkind RA (1978) Erythroleukemic differentiation. Ann Rev Biochem 47:419-448

Nagai T, Igarashi K, Fujita H, Hayashi N, Yamamoto M and Sassa S (1995) Effects of DMSO and heme on the regulation of the erythroid-specific transcription factor, NF-E2. Proceedings of the "International Symposium on Porphyrias and Heme Related Disorders: Molecular Basis, Diagnostics, and Clinical Aspects", 28 June-2 July, Helsinki, Finland

Ngai J, Capetanaki YG and Lazarides E (1984) Differentiation of murine erythroleukemia cells results in the rapid repression of vimentin gene expression. J Cell Biol, 99:306-314

Pappas IS and Tsiftsoglou AS (1995) Inducer-receptor mediated processes in murine erythroleukemia cell differentiation. In: Waxman S (edr), Challenges of Modern Medicine: Differentiation Therapy, vol. 10, pp. 269-273, Ares-Serono Symposia Publications, Rome

Pappas IS, Niopas I, and Tsiftsoglou AS (1992) Ureido-derivatives of pyridine: A new class of inducers of murine erythroleukemia cell differentiation. Anticancer Drug Des 7:153-161

Pappas IS, Sophianos D, Tzartos S and Tsiftsoglou AS (1996) Expression of memory, differentiation and repression of c-myc and p53 genes in human RD/TE-671 cells induced by a ureido-derivative of pyridine (UDP-4). Cell Crowth Differ in press

Peltz SW, Brewer G, Bernstein P, Hart A and Ross J (1991) Regulation of mRNA turnover in eukaryotic cells. Critical Reviews in Eukaryotic Gene Expression 1:99-126

Pierce GB (1974) Neoplasms, differentiations and mutations. Am J Pathol 77:103-118

Pugh CSG, Borchardt RT and Stone HO (1977) Inhibition of Newcastle disease virion messenger RNA (guanine-7-)-methyltransferase by analogues of S-adenosyl-homocysteine. Biochemistry 16:3928-3932

Ramsay RG, Ikeda K, Rifkind RA and Marks PA (1986) Changes in gene expression associated with induced differentiation of erythroleukemia: Protoncogenes, globin genes and cell division. Proc Natl Acad Sci USA, 83:6849-6853

Reiss M, Gamba-Vitalo C, Sartorelli AC (1986) Induction of tumor cell differentiation as a therapeutic approach: preclinical models for hematopoietic and solid neoplasms. Cancer Treat Rep 70:201-218

Richon VM, Rifkind RA and Marks PA (1992) Expression and phosphorylation of the retinoblastoma protein during induced differentiation of murine erythroleukemia cells. Cell Growth Differ 3:413-420

Roth WW, Bragg PW, Corrias MV, Reddy NS, Dholakia JN and Wahba AJ (1987) Expression of a gene for mouse eukaryotic elongation factor Tu during murine erythroleukemia cell differentiation. Mol Cell Biol 7:3929-3926

Sachs L (1964) The analysis of regulatory mechanisms in differentiation. In: Sela M (edr), New Perspectives in Biology, pp. 246-260, Elsevier, Amsterdam

Sachs L (1980) Constitutive uncoupling of pathways of gene expression that control growth and differentiation in myeloid leukemia: a model for the origin and progression of malignacy. Proc Natl Acad Sci USA 77:6152-6156

Sachs L (1994) The molecular control of haemopoiesis and leukemia. Eur J Cancer 30A:852-860

Sassa S (1992) Regulation of heme pathway enzyme genes in mammalian cells. In: Murphy, Jr MJ (edr) Concise reviews in clinical and experimental hematology, pp. 331-346, AlphaMed Press, Dayton, Ohio

Schuetze S, Paul R, Gliniak BC and Kabat D (1992) Role of the PU.1 transcription factor in controlling differentiation of Friend erythroleukemia cells. Mol Cell Biol 12:2967-2975

Sheffery M, Marks PA and Rifkind RA (1984) Gene expression in murine erythroleukemia cells: transcriptional control and chromatin structure of the α_1-globin gene. J Mol Biol 172:417-436

Sherman ML, Shafman TD, Spriggs DR and Kufe DW (1985) Inhibition of murine erythroleukemia cell differentiation by 3-deazaadenosine. Cancer Res 45:5830-5834

Spotts GD and Hann SR (1990) Enhanced translation and increased turnover of c-myc proteins occur during differentiation of murine erythroleukemia cells. Mol Cell Biol 10:3952-3964

Tsiftsoglou AS and Robinson SH (1985) Differentiation of leukemic cell lines: A review focusing on murine erythroleukemia and human HL-60 cells. Int J Cell Cloning 3:349-366

Tsiftsoglou AS and Wong W (1985) Molecular and cellular mechanisms of leukemic hemopoietic cell differentiation: An analysis of the Friend system Anticancer Res 5:81-100

Tsiftsoglou AS, Christou IE and Vizirianakis IS (1992) Developmental changes in methylation of polyA$^+$ RNA may regulate its stability posttranscriptionally in murine erythroleukemia (MEL) cells. Proc Amer Assoc Cancer Res 33:48

Tsiftsoglou AS, Hensold J, Robinson SH and Wong W (1987) The regulatory role of commitment in gene expression during induction of leukemic cell differentiation. In: Harrap KR and Connors TA (eds), New Avenues in Developmental Cancer Chemotherapy, pp.205-227, Academic Press, New York

Tsiftsoglou AS, Wong W and Housman D (1983) Dexamethasone-sensitive and insensitive responses during in vitro differentiation of Friend erythroleukemia cells. Biochim Biophys Acta 759:160-169

Tsiftsoglou AS, Wong W, Volloch V, Gusella J and Housman D (1982) Commitment of murine erythroleukemia (MEL) cells to terminal differentiation is associated with coordinated expression of globin and ribosomal genes. Prog Clin Biol Res 102A:69-79

Vizirianakis IS, Tsiftsoglou AS and Sklaviadis T (1994) Developments in slow virus research: Implication of PrP protein involvement in cell growth and differentiation. In: Bradley R and Marchant B (eds), Transmissible Spongiform Encephalopathies, pp. 421-429, Commision of the European Communities, Brucells

Vizirianakis IS and Tsiftsoglou AS (1995) N^6-methyladenosine inhibits murine erythroleukemia cell maturation by blocking methylation of RNA and memory via conversion to S-(N^6-methyl)-adenosylhomocysteine. Biochem Pharmacol 50:1807-1814

Vizirianakis IS and Tsiftsoglou AS (1995[a]) Hypermethylation of RNA as part of the differentiation program of murine erythroleukemia (MEL) cells. In: Waxman S (edr), Challenges of Modern Medicine: Differentiation Therapy, vol. 10, pp. 263-268, Ares-Serono Symposia Publications, Rome

Vizirianakis IS and Tsiftsoglou AS (1996[b]) Induction of murine erythroleukemia cell differentiation is associated with methylation and differential stability of polyA$^+$ RNA transcripts. Biochim Biophys Acta, in press

Vizirianakis IS, Wong W and Tsiftsoglou AS (1992) Analysis of inhibition of commitment of murine erythroleukemia (MEL) cells to terminal maturation by N^6-methyladenosine. Biochem Pharmacol 44:927-936

Volloch V, Schweitzer B and Rits S (1987) Messenger RNA changes during differentiation of murine erythroleukemia cells. Exper Cell Res 173:38-48

Volloch V. and Housman D (1981) Stability of globin mRNA in terminally differentiating murine erythroleukemia cells. Cell 23:509-514

Yamamoto T, Matsui Y, Natori S and Obinata M (1989) Cloning of a housekeeping-type gene (MER5) preferentially expressed in murine erythroleukemia cells. Gene 80:337-343

Expression of Tyrosine Kinases in the Mouse Small Intestine

A. L. Tyner, V. Vasioukhin, M. S. Serfas, E. Y. Siyanova, and X. Llor
Department of Genetics
University of Illinois College of Medicine
Chicago, IL 60607

Abstract

The signal transduction pathways that regulate the regeneration of the mammalian intestinal epithelium are not well understood. To identify tyrosine kinases involved in regulating the continuous proliferation and cell differentiation, we amplified the catalytic domains of tyrosine kinase cDNAs expressed in intestinal crypt cells, using the polymerase chain reaction technique with primers directed against two invariant amino acid sequence motifs found in all tyrosine kinases. These fragments were cloned and a library of kinase catalytic domains was generated. Sequence analysis of unique clones resulted in the identification of the catalytic domains of several characterized tyrosine kinases, including lyn, hck, fgr, tec, JAK2 , itk, and the putative receptor kinase ryk. In addition, we identified two novel catalytic domain sequences. One of these which we named sik (src-related intestinal kinase), is expressed throughout the gastrointestinal tract, with highest levels in the ileum. We isolated cDNA clones encoding sik, a 451 amino acid protein which appears to be the mouse homologue of a recently identified human tyrosine kinase, brk (breast tumor kinase, Mitchell et al., 1994). We have determined that sik expression is developmentally regulated, and confined to epithelia, including the skin and lining of the alimentary canal. The restriction of sik expression to rapidly renewing epithelia, suggests that it may play a role in regeneration, migration, or differentiation.

Identification of Tyrosine Kinases Using PCR

The tyrosine kinase family represents a complex group of proteins that mediate cellular responses to environmental conditions (Hanks and Hunter, 1995). They are key players in signal transduction pathways, and are involved in the regulation of proliferation and cell differentiation in a variety of systems. They include a variety of transmembrane receptors for different growth factors and intracellular kinases that transmit signals from the membrane to the nucleus. Aberrant expression of tyrosine kinases has been demonstrated to be linked to oncogenesis. The important role that tyrosine kinases play during development has

NATO ASI Series, Vol. H 99
Tumor Biology
Regulation of Cell Growth, Differentiation and Genetics in Cancer
Edited by Asterios S. Tsiftsoglou, Alan C. Sartorelli, David E. Housman and T. Michael Dexter

been underscored as several classical developmental mutations in mice, *Drosophila*, and *C. elegans* have been mapped to tyrosine kinase genes. For recent reviews see: van der Geer et al., 1994; Pawson and Hunter, 1994; Marshall, 1995; Perrimon, 1994; Kayne and Sternberg, 1995.

Intestinal crypt epithelial cells represent a dynamic and rapidly proliferating cell population, with some cells dividing every twelve hours (Potten and Loeffler, 1990). A clonal population of stem cells anchored in each crypt gives rise to progenitor cells that differentiate into at least four different cell types. These include enterocytes, goblet cells, enteroendocrine cells, and Paneth cells. Differentiating cells migrate in a bipolar fashion from the crypts, and all of the differentiated cells, except the Paneth cells which remain at the base of the crypts, migrate up the villi and are extruded at the villus tip within 3-5 days (reviewed in Gordon and Hermiston, 1994).

High levels of tyrosine kinase activity found in the proliferating undifferentiated cells of the intestinal crypts suggested that tyrosine kinases may play an important role in regulation in the intestine. The proliferating fetal intestinal epithelium contains high levels of proteins phosphorylated on tyrosine residues, but these levels decrease as maturation proceeds (Maher and Pasquale, 1988). In the adult intestine, cell proliferation is restricted to the crypt cell compartment, which contains the stem cells and the undifferentiated progenitor cells. This region of the intestinal epithelium has been shown to contain fifteen-fold higher levels of tyrosine phosphorylated proteins than villus epithelium, and the majority of the tyrosine kinases appear to be associated with the cytoskeleton (Burgess et al., 1989). Higher levels of cytoskeletal associated pp60 c-src protein and activity are found also found in the crypt epithelial cells (Cartwright et al., 1993). Increased levels of src family kinase activity have been detected in undifferentiated human colon carcinoma cells, as pp60 c-src and p56 lck activity and abundance decrease as these cells are induced to differentiate in vitro (Foss et al., 1989).

In order to identify the tyrosine kinases expressed in the crypt compartment, we isolated crypt epithelial cells using EDTA perfusion as described by Bjerknes and Cheng (Bjerknes and Cheng, 1981). The intestinal epithelium is continuous and covers the villi and extends down into the crypts of Lieberkühn. While perfusion with 30 mM EDTA results in the stripping of the intact epithelium (Fig. 1A), perfusion with 1 mM EDTA followed by vibration permits the sequential isolation of villus (Fig. 1B) and crypt epithelial cell fractions (Fig. 1C). To confirm that we had enriched for crypt cell sequences, cDNA was prepared from isolated crypt cell RNA and hybridized with ^{32}P labeled probes complementary to cryptdin, a gene expressed only in Paneth cells located at the base of the crypts and sucrase isomaltase, which is expressed only in differentiated villus cells. In the cDNA prepared from the crypt cell population, we detected little expression of the villus cell marker sucrase isomaltase, and an approximately 8 fold enrichment for cryptdin in the crypt cell cDNA fractions, indicating significant enrichment for crypt cell sequences.

All tyrosine kinases share homology in their catalytic domains (Hanks and Hunter, 1995). Wilks designed PCR primers PTK I and PTK II based on homologies within conserved motifs VI and IX (Wilks, 1989) . In order to identify intestinal tyrosine kinases, we used the polymerase chain reaction with the primers PTK-1 and PTK-2 and cDNA prepared from the crypt cell mRNA isolated from the crypt cell fractions (Fig. 1C). Amplified fragments were cloned into pBluescript SK- (Stratagene), and following transformation into XL-1 Blue cells, a tyrosine kinase catalytic domain library containing several thousand clones was generated. Over 95% of transformants that contained inserts, of approximately 210 nt, the size expected for tyrosine kinase catalytic domains.

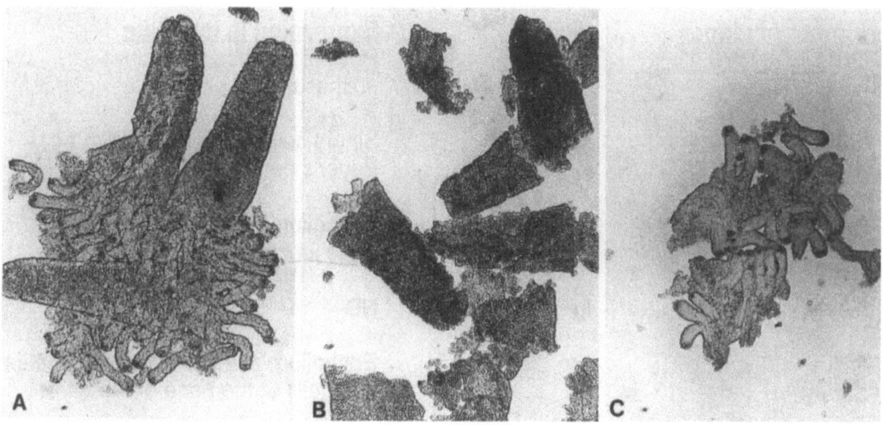

Figure 1. Intestinal epithelial cell fractions were isolated by perfusion of a mouse with EDTA solutions. Total intact epithelium is removed when the animal is perfused with 30 mM EDTA (A). The villus epithelium (B) can be separated from the crypt epithelium (C) when 1 mM EDTA is used for perfusion, followed by vibration. Unstained fractions were visualized using brightfield microscopy (Mag. 150X). Note the differentiated Paneth cells at the base of the crypts.

Random transformants were sequenced by the chain termination method of Sanger, using the enzyme Sequenase and the manufacturer's recommended protocol (United States Biochemical Corporation, Cleveland). We identified a pool of abundant catalytic domain sequences, and used these to further screen the catalytic domain library. Sequences of clones were analyzed using the program GeneWorks

(Intelligenetics, Inc. Mountain View, CA), or databases available from NCBI, using the program BLASTP (Altschul et al., 1990). Several known kinases were identified, including a number of hematopoietic cell kinases, and expression of these tyrosine kinases had not previously been reported in the intestine (see Table 1). These include the src family tyrosine kinases lyn, hck, fgr, and the T-cell specific src related kinase itk (Siliciano et al., 1992). We also isolated the catalytic domains of tec, which has been shown to be expressed predominantly in liver and hematopoietic cells (Mano et al., 1990; Mano et al., 1993), and the Janus kinase JAK2 (Harpur et al., 1992). JAK2 has been shown to play an important role in signal transduction pathways regulated by cytokine receptors (Ihle and Kerr, 1995). The catalytic domain of the tyrosine kinase related molecule ryk was also isolated several times (Hovens et al., 1992; Paul et al., 1992) and this protein appears to represent a novel class of receptors, with greatest homology to met, the receptor for HGF and scatter factor, and to other members of the insulin receptor subfamily.

Number of clones	Identity	Expression in Intestine
~30%	lyn	Peyer's patches, and crypt epithelium
~30%	hck	Single cells in the epithelium near Peyer's patches
~18%	JAK2	Epithelium at the crypt-villus junction, lower villus
~5%	fgr	ND
~4%	ryk	Epithelium at the crypt-villus junction, and cells at the base of the crypt
~1%	itk	ND
1 clone	tec	ND
~3.0%	sik	Epithelium at crypt-villus junction, lower villus
1 clone	ink76	ND

Table 1. Sequenced intestinal catalytic domains. A total of 285 clones were characterized. Approximately 9.0% of the clones appeared to contain only fragments of the expected portions of kinase catalytic domains, or contained portions of catalytic domain sequences that did not hybridize with mouse genomic DNA or RNA. These were not further characterized. Expression patterns were determined by in situ hybridization (Siyanova et al., 1994, and Vasioukhin et al., 1995). ND: Not detected.

Novel catalytic domains that were identified include the **in**testinal **k**inase ink76 (GenBank accession number U13258), and the **s**rc related **i**ntestinal **k**inase sik, which is described in detail below. ink76 is a putative serine threonine kinase and is most closely related to homologues of the yeast SNF1 gene, and shares greatest homology with the Arabidopsis thaliana gene AKin10 (Le Guen et al., 1992). The mammalian serine threonine kinase with greatest homology to ink76 is human SLKA serine threonine kinase (Howard et al., 1992). Homologous sequences were detected in mouse genomic DNA for both sik and ink76. The nearly invariant sequences found in tyrosine kinases and serine threonine kinases are conserved in sik and ink76 respectively (Hanks et al., 1988).

Using PCR we identified 9 kinase catalytic domain sequences expressed in intestinal epithelial cells. Although we screened our catalytic domain library until we no longer could identify additional different tyrosine kinase catalytic domains, we did not find clones encoding several kinases that we presumed we would find, including src (Boulter and Wagner, 1988; Cartwright et al., 1993). Thus, it is possible that our catalogue of tyrosine kinases may not be complete. In addition, most of the kinases that we identified were expressed primarily in villus epithelium, not in the crypt compartment. The two exceptions include lyn and ryk, which we found to be expressed in the crypts at the RNA level using in situ hybridization. Although the approach appears to have had limitations, it allowed us to identify a variety of tyrosine kinases including the novel tyrosine kinase sik, that may play important roles in regulating the processes of proliferation and cell differentiation in the intestine.

After identifying kinase catalytic domains by sequencing cloned PCR products, we examined expression of the RNAs encoding these catalytic domains in the small intestine. We compared levels of expression in multiple tissues using RNase protection assays because of their sensitivity and stringency. Generally, we examined expression in the duodenum of the small intestine, liver, lung, kidney, brain, spleen, testis, and on occasion other tissues such as uterus, ovary, heart and tongue. The hematopoietic cell kinases hck, lyn, fgr, and itk are expressed at highest levels in the spleen . While hck and lyn are ubiquitously expressed, fgr and itk have a much more restricted pattern of expression. The levels of expression in the duodenum vary, with fgr and itk having the lowest levels. We detected expression of tec, JAK2 , and ryk in all tissues examined (Siyanova et al., 1994). Expression of the novel src related kinase sik, was detected only in epithelial tissues, including the small and large intestine, skin and epithelial lining of the tongue (Siyanova et al., 1994; Vasioukhin et al., 1995). Using ^{35}S labeled antisense probes corresponding to the catalytic domains (hck, lyn, and JAK2) or cDNA clones (ryk and sik), we performed in situ hybridization experiments to localize expression of each of the kinases identified by PCR cloning. Patterns of expression detected in the intestine are summarized above in Table 1.

Cloning and characterization of the epithelial cell specific kinase sik

Because it was novel and epithelial cell specific, we focused upon cloning characterizing sik. Using the catalytic domain fragment as a probe, we screened a cDNA library prepared from mouse duodenal pA+ RNA. One cDNA clone with a 1.6 kb insert was isolated. Sequence analysis of this clone indicated that it lacked a possible start site of translation. Two approaches were then utilized to clone the 5' end of the cDNA. First, cDNA synthesis was primed using a sik specific antisense primer, and a sik enriched cDNA library was generated. This library was screened and a number of sik cDNA clones were isolated that extended the sequence to 2 kb. These clones contained the start site of translation. Using the RACE technique (Edwards et al., 1991), additional 5' noncoding sequence was cloned, containing stop codons in all three reading frames.

The nucleotide and predicted amino acid sequence of the sik cDNA have been determined (GenBank accession number U16805). Based on sequence homology, sik is related to the src family of tyrosine kinases and contains src homology 1, 2, and 3 (SH1, SH2, SH3) domains (for reviews see Hanks and Hunter, 1995; Pawson, 1995). The tyrosine kinase catalytic domain (SH1 domain) is conserved among nonreceptor tyrosine kinases. A lysine residue at position 219 of sik corresponds to a conserved lysine at position 295 of c-src that is required for kinase activity (Thomas et al., 1991). The sik protein contains a tyrosine at position 341 that corresponds to the autophosphorylated tyrosine 416 in c-src (Hanks et al., 1988). A possible regulatory tyrosine phosphorylation site is also found at the carboxy end of the sik protein at amino acid 447. This corresponds to the inhibitory tyrosine phosphorylation site at position 527 in src that is phosphorylated by csk (Cooper and Howell, 1993).

SH2 domains are approximately 100 amino acids in length and bind phosphorylated tyrosine residues in specific peptide sequences facilitating protein-protein interactions. SH3 domains are approximately 60 amino acids long and are found in a number of different cytoplasmic proteins including several associated with the plasma membrane or cytoskeleton. This domain has been shown to be capable of binding guanine nucleotide exchange factors leading to the regulation of ras activity, and of directing cellular localization of signaling molecules. For recent reviews see Pawson, 1995 and Schlessinger, 1994). The structure of the sik kinase is diagrammed in Figure 2.

Figure 2. Structure of the sik tyrosine kinase.

The mouse sik cDNA clone was used to screen a normal human small intestine cDNA library that was purchased from Clontech. From a screen of 10^6 recombinant phage, three positive clones were isolated. When sequenced, these clones shared sequence identity with brk (**br**east tumor **k**inase), a novel nonreceptor tyrosine kinase that was isolated from a human metastatic breast tumor (Mitchell et al., 1994). The mouse sik sequence shares 80% identity with human brk. We have concluded that sik and brk are homologues of one another, because the brk and sik probes hybridize with the same fragments in restricted mouse and human DNA in Southern blot experiments.

Sik/brk also share significant homology with RAK/FRK (44% identity) which is also expressed in breast tumor cells (Cance et al., 1994; Lee et al., 1994). In contrast to members of the src family, sik , brk, and RAK/FRK do not contain sites for myristylation at the amino terminus. Sik, brk and RAK/FRK also have the sequence HRDLAARN in their catalytic domains, in contrast to the sequence HRDLRAAN shared by members of the src family. Like the src family tyrosine kinases, these kinases contain both src homology 2 (SH2) and src homology 3 (SH3) domains (Pawson, 1995; Schlessinger, 1994).

By performing western blot analysis using antiphosphotyrosine antibody and lysates of bacterial cells expressing the sik coding sequence, we determined that sik has tyrosine kinase activity. Phosphorylated tyrosines were detected only in bacterial cells expressing the sik protein (Vasioukhin et al., 1995). Northern blot analysis revealed that at least three major sik transcripts, with sizes of approximately 2, 3 and 5 kb in skin, tongue, and throughout the gastrointestinal tract. The major transcripts found in liver were approximately 4.6 kb and 1.8 kb and corresponded to minor transcripts found in the other tissues. The highest level of sik RNA was detected in the ileum and cecum. Hybridization of restricted mouse genomic DNA at the same stringency results in the identification of single bands, indicating that the variety of sik transcripts are encoded by a single gene (Vasioukhin et al., 1995).

The ontogeny of sik expression in the mouse embryo was studied by RNase protection and in situ hybridization. Sik expression was first detected at day 15.5 days of gestation and sik expression increased as gestation proceeded. At embryonic day 15.5, expression was detected primarily in the skin, where sik RNA was localized to the newly differentiating granular layer. At later timepoints, sik was detected in other organs such as the intestine and the epithelial linings of the fetal nasal passages. Striking expression of sik was detected in the d17.5 stomach. At this stage of development, the stomach is already fairly well developed (Rugh, 1991), and sik mRNA was localized to the stratified epithelium of the fundus of the stomach (Vasioukhin et al., 1995).

In the adult gastrointestinal tract, sik expression was detected in differentiating epithelial cells immediately above zones of proliferation. In the intestine, where cell proliferation is restricted to the epithelial cells lining the crypts, we detected highest levels of sik expression in nondividing epithelial cells at the base of the villi. A similar pattern of expression was detected in other epithelial cell linings, such as the skin and tongue where sik expression was localized to the epithelial cells immediately above the proliferating basal cell layer (Vasioukhin et al., 1995).

Currently several approaches are being utilized to determine the function of the sik kinase in the intestine. Three mammalian sik expression constructs have been generated, including: 1) wildtype sik, 2) a dominant negative form of sik in which the catalytic domain has been mutated, and 3) a constitutively activated form of sik in which the regulatory carboxy tyrosine has been altered. We are currently determining if expression of these constructs results in altered growth characteristics, cellular transformation or induction of differentiation. In addition we are examining the interaction of sik with other proteins in the cell using immunoprecipitation and Western blotting techniques. We hope to identify proteins that may activate sik or are phosphorylated by the sik kinase. These experiments should provide further insight about the role of sik in epithelial cells.

Literature References

Altschul, S. F., W. Gish, W. Miller, E. W. Myers and D. J. Lipman (1990). Basic local alignment search tool. J. Mol. Biol. **215**: 403-410.

Bjerknes, M. and H. Cheng (1981). Methods for the isolation of intact epithelium from mouse intestine. Anat. Rec. **199**: 565-574.

Boulter, C. A. and E. F. Wagner (1988). Expression of c-src and c-abl in embryonal carcinoma cells and adult mouse tissues. Exp. Cell. Res. **179**(1): 214-21.

Burgess, D. R., W. Jiang, S. Mamajiwalla and W. Kinsey (1989). Intestinal crypt stem cells possess high levels of cytoskeletal-associated phosphotyrosine containing proteins and tyrosine kinase activity relative to differentiated enterocytes. J. Cell Biol. **109**: 2139-2144.

Cance, W. G., R. J. Craven, M. Bergman, L. Xu, K. Aitalo and E. T. Liu (1994). Rak, a novel nuclear tyrosine kinase expressed in epithelial cells. Cell Growth Diff. **5**: 1347-1355.

Cartwright, C. A., S. Mamajiwalla, S. A. Skolnick, W. Eckhart and D. R. Burgess (1993). Intestinal crypt cells contain higher levels of cytoskeletal-associated pp60$^{c\text{-}src}$ protein kinase activity than do differentiated enterocytes. Oncogene **8**: 1033-1039.

Cooper, J. A. and B. Howell (1993). The when and how of src regulation. Cell **73**: 1051-1054.

Edwards, J. B. D. M., J. Delort and J. Mallet (1991). Oligodeoxyribonucleotide ligation to single-stranded cDNAs: a new tool for cloning 5' ends of mRNAs and for constructing cDNA libraries by in vitro amplification. Nuc Acid Res. **19**: 5227-5232.

Foss, F. M., A. Veillette, O. Sartor, N. Rosen and J. B. Bolen (1989). Alterations in the expression of pp60c-src and p56-lck associated with butyrate-induced differentiation of human colon carcinom cells. Oncogene Res. **5**: 13-23.

Gordon, J. I. and M. L. Hermiston (1994). Differentiation and self-renewal in the mouse gastrointesinal epithelium. Curr. Opin. in Cell Biol. **6**: 795-803.

Hanks, S. K., Quinn. A .M., and T. Hunter (1988). The protein kinase family: conserved features and deduced phylogeny of the catalytic domains. Science **241**: 42-52.

Hanks, S. K. and T. Hunter (1995). The eukaryotic protein kinase superfamily: kinase (catalytic) domain structure and classification. FASEB J. **9**: 576-596.

Harpur, A. G., A. C. Andres, A. Ziemiecki, R. R. Aston and A. F. Wilks (1992). JAK2, a third member of the JAK family of protein tyrosine kinases. Oncogene **7**(7): 1347-53.

Hovens, C. M., S. A. Stacker, A. C. Andres, A. G. Harpur, A. Ziemiecki and A. F. Wilks (1992). RYK, a receptor tyrosine kinase related molecule with unusual kinase domain motifs. Proc. Natl. Acad. Sci. U.S.A. **89**: 11818-11822.

Howard, O. M., S. Cevioros, S. J. O'Brien and W. Farrar (1992). Chromosomal assignment of the human homologue of yeast serine threonine kinase SNF-1. Genbank Reference Unpublished.

Ihle, J. N. and I. M. Kerr (1995). Jaks and STATS in signaling by the cytokine receptor superfamily. TIG **11**: 69-74.

Kayne, P. S. and P. W. Sternberg (1995). Ras pathways in Caenorhabditis elegans. Curr. Opin. Genet. Dev. **5**: 38-43.

Le Guen, L., M. Thomas, M. Bianchi, N. G. Halford and M. Kreis (1992). Structure and expression of a gene from Arabadopsis thaliana encoding a protein related to SNF1 protein kinase. Gene **120**: 249-254.

Lee, J., Z. Wang, S.-M. Luoh, W. L. Wood and D. T. Scadden (1994). Cloning of FRK, a novel human intracellular SRC-like tyrosine kinase-encoding gene. Gene **138**: 247-251.

Maher, P. A. and E. B. Pasquale (1988). Tyrosine phosphorylated proteins in different tissues during chick embryo development. J. Cell Biol. **106**: 1747-1755.

Mano, H., F. Ishikawa, J. Nishida, H. Hirai and F. Takaku (1990). A novel protein-tyrosine kinase, tec, is preferentially expressed in liver. Oncogene **5**: 1781-1786.

Mano, H., K. Mano, B. Tang, M. Koehler, T. Yi, D. J. Gilbert, N. A. Jenkins, N. G. Copeland and J. N. Ihle (1993). Expression of a novel form of Tec kinase in hematopoietic cells and mapping of the gene to chromosome 5 near kit. Oncogene **8**: 417-424.

Marshall, C. J. (1995). Specificity of receptor tyrosine kinase signalling: transient versus sustained extracellular signal-regulated kinase activation. Cell **80**: 179-185.

Mitchell, P. J., K. T. Barker, J. E. Martindale, T. Kamalati, P. N. Lowe, M. J. Page, B. A. Gusterson and M. R. Crompton (1994). Cloning and characterisation of cDNAs encoding a novel non-receptor tyrosine kinase, brk, expressed in human breast tumours. Oncogene **9**: 2383-2390.

Paul, S. R., D. Merberg, H. Finnerty, G. E. Morris, J. C. Morris, S. S. Jones, R. Kriz, K. J. Turner and C. R. Wood (1992). Molecular cloning of the cDNA encoding a receptor tyrosine kinase-related molecule with a catalytic region homologous to c-met. International J. of Cell Cloning **10**: 309-314.

Pawson, T. (1995). Protein modules and signalling networks. Nature **373**: 573-579.

Pawson, T. and T. Hunter (1994). Oncogenes and cell proliferation. Curr. Opin. Gen. Dev. **4**: 1-4.

Perrimon, N. (1994). Signalling pathways initiated by receptor protein tyrosine kinases. Curr. Opin. Cell Biol. **6**: 260-266.

Potten, C. S. and M. Loeffler (1990). Stem cells: attributes, cycles, spirals, pitfalls, and uncertainties. Lessons for and from the crypt. Development **110**: 1001-1020.

Rugh, R. (1991). The mouse: its reproduction and development. New York, Oxford University Press.

Schlessinger, J. (1994). SH2/SH3 signaling proteins. Curr. Opin. Gen. Dev. **4**: 25-30.

Siliciano, J. D., T. A. Morrow and S. V. Desiderio (1992). itk, a T-cell-specific tyrosine kinase gene inducible by interleukin 2. Proc. Natl. Acad. Sci .U S A **89**(23): 11194-8.

Siyanova, E. Y., M. S. Serfas, I. A. Mazo and A. L. Tyner (1994). Tyrosine kinase gene expression in the mouse small intestine. Oncogene **9**: 2053-2057.

Thomas, J. E., P. Soriano and J. S. Brugge (1991). Phosphorylation of c-Src on tyrosine 527 by another protein tyrosine kinase. Science **254**(5031): 568-71.

van der Geer, P., T. Hunter and R. A. Lindberg (1994). Receptor protein tyrosine kinases and their signal transduction pathways. Ann. Rev. Cell Biol. **10**: 251-337.

Vasioukhin, V., M. S. Serfas, E. Y. Siyanova, M. Polonskaia, B. J. Costigan, B. Liu, A. Thomason and A. L. Tyner (1995). A novel intracellular epithelial cell tyrosine kinase is expressed in the skin and gastrointestinal tract. Oncogene **10**: 349-357.

Wilks, A. F. (1989). Two putative protein-tyrosine kinases identified by application of the polymerase chain reaction. Proc. Natl. Acad. Sci. USA **86**: 1603-1607.

The Discovery and Characterization of Neurotrophic and Myotrophic Factors

George D. Yancopoulos
Regeneron Pharmaceuticals, Inc.
777 Old Saw Mill River Road
Tarrytown, New York 10591
USA

Introduction to Neurotrophic Factors, and the Nerve-Muscle Connection

The focus of my group has been to discover and study factors acting on neurons and their innervation targets, with an eye toward eventually using these factors to intervene in neurodegenerative disease. When we first started working in this area, we and others thought that neurotrophic factors were in some way fundamentally different from so-called "conventional" growth factors and cytokines operating elsewhere in the body, mostly because the apparent actions of neurotrophic factors seemed quite unlike the usual proliferative and differentiative effects of "conventional" growth factors and cytokines. For example, neurons are post-mitotic and terminally-differentiated, and neurotrophic factors are known to act on neurons to promote their survival and to elicit dramatically impressive neuritic process outgrowth that appear quite distinct from the effects of "conventional" growth factors and cytokines on non-neuronal cells. It turns out, however, as realized from the work of our group and others, that neurotrophic factors are quite analogous to "conventional" growth factors and cytokines, in that they use very similar receptors and signaling systems (Glass and Yancopoulos, 1993; Ip and Yancopoulos, 1994). Thus some neurotrophic factors use receptors known as receptor tyrosine kinases (RTKs) very similar to those used by "conventional" growthfactors such as fibroblast growth factor (FGF) or epidermal growth factor (EGF), while others use so-called "cytokine" receptors like those used by interleukins. The major difference seems to be that factors considered to be "neurotrophic factors" are thus categorized

NATO ASI Series, Vol. H 99
Tumor Biology
Regulation of Cell Growth, Differentiation and Genetics in Cancer
Edited by Asterios S. Tsiftsoglou, Alan C. Sartorelli, David E. Housman and T. Michael Dexter

because their receptors are highly restricted to neurons in their expression - thus limiting the actions of these factors only to neurons. While using receptors and signaling pathways very similar to those used by "conventional" growth factors and cytokines, responses elicited by neurotrophic factors appear so distinctive simply because the signaling processes initiated at the cell surface are interpreted very differently within the context of a post-mitotic, terminally-differentiated neuron than by, for example, a mitotically-competent fibroblast or hemopoietic precursor (Glass et al., 1991; Glass and Yancopoulos, 1993; Ip and Yancopoulos, 1994).

In this review manuscript, I will focus on one particular class of neurons and their target tissue - that is, the motor neurons in the spinal cord which innervate skeletal muscle. This system is responsible for movement of the body, and a variety of diseases result from problems with any part of this system - thus Amyotropic Lateral Sclerosis (also known as Lou Gehrig's Disease) results from inappropriate death of the motor neurons, muscular dystrophies result from primary defects of the skeletal muscle, while myasthenia gravis results from problems with the specialized connection between nerve and muscle, known as the neuromuscular junction (NMJ). Motor neuron innervation of muscle has been intensely studied for decades, and it has long been known that there are critical reciprocal "trophic" interactions between the motor neuron and the muscle. Excess numbers of motor neurons are born early in development, and the number of surviving neurons depends on competition between these neurons for limiting amounts of neurotrophic factors found in their target, the skeletal muscle; this seems to insure that only the correct numbers of motor neurons survive. The nerve, in turn, provides key signals to the muscle. For example, signals from the nerve induce formation of the NMJ specialization - key features of this specialization is that acetylcholine receptors (AchRs), which sense the acetylcholine released by the nerve and thus mediate synaptic transmission to the muscle, are highly localized on the muscle surface only at the tiny region corresponding to the NMJ where the nerve directly contacts muscle. The nerve also releases substances thought to be trophic for the muscle - that is, involved in maintaining muscle size and mass.

The critical roles of reciprocal trophic interactions between nerve and muscle were revealed decades ago by nerve disruption and target ablation studies. However, the molecular mediators of these interactions still remain largely a mystery. For example, the neurotrophic factor that is provided by muscle to support motor neurons remains unidentified. In addition, while it is thought that one of the critical factors released by the nerve to initiate the clustering of AchRs at the NMJ is a molecule known as "agrin" based on its ability to aggregate AchRs, the mechanism of action of agrin remains a mystery. Similarly, no details are known about any factors that might be provided by nerve that are trophic to muscle. I will talk about some of our recent work that is beginning to help solve these mysteries.

Factors That Are Ruled Out As The Long-Sought Muscle-Derived Motor Neuron Factor

A number of factors are known to promote motor neuron survival when exogenously provided to motor neurons in culture, or to motor neurons in vivo. However, recent gene disruption data has ruled out most of these as playing a critical role during normal motor neuron development in vivo. Examples of such factors include brain derived neurotrophic factor (BDNF) and neurotrophin-4 (NT4), both of which are members of the neurotrophin family defined as being homologous to the prototypical neurotrophic factor, nerve growth factor (NGF) (Glass and Yancopoulos, 1993). The neurotrophins use a family of receptor tyrosine kinases known as the Trks (Glass and Yancopoulos, 1993; Ip et al., 1993a), and both BDNF and NT4 were considered candidates to be the long-sought muscle-derived motor neuron trophic factor based on the finding that the receptor for these factors (TrkB) was specifically expressed on motor neurons, as well as their in vitro and in vivo survival actions on motor neurons. However, recent generation of mice disrupted for both their BDNF and NT4 genes have normal numbers of motor neurons, ruling out an in vivo role for these factors in regulating motor neuron survival (Conover et al., 1995).

Other candidate neurotrophic factor receptor systems include those belonging to the EPH family of receptor tyrosine kinases, members of which we and others cloned based on their specific expression in particular neuronal populations (e.g. Maisonpierre et al., 1993; Valenzuela et al., 1995a). Some EPH family receptors are specifically expressed by developing motor neurons, raising the possibility that ligands for these receptors might correspond to the long-sought motor neuron trophic factor produced by muscle. We and others have recently identified a family of ligands for the EPH receptors (Davis et al., 1994). Strikingly, these ligands are all membrane-bound (due to having transmembrane domains or glycosyl phosphtidylinositol linkages to the membrane), and in fact are the only known ligands for receptor tyrosine kinases that must be membrane-bound in order to activate their receptors (Davis et al., 1994). However, these ligands do not seem to be able to promote survival, and rather may play guidance roles or have other effects on neurons, thus ruling them out as candidates for the long-sought after motor neuron survival factor.

Genetic Evidence Identifies the Long-Sought Muscle-Derived Motor Neuron Factor

Now I want to introduce a neurotrophic factor system which we think is directly involved in regulating motor neuron survival during development. The story starts with ciliary neurotrophic factor (CNTF), though as I will get to, the involved factor is not CNTF but a relative. CNTF was discovered, and named, for its ability to act as a classical survival factor for the neurons which innervate the ciliary body near the eye. When we first started studying CNTF, nothing was known about its receptor system or mechanism of action. We used an expression cloning strategy to clone a CNTF-binding protein (Davis et al., 1991), and were quite surprised to find that it was totally unrelated to the receptor systems used by other classical neurotrophic factors such as NGF. Instead, it was very homologous to a receptor component that had previously been identified for interleukin-6 (IL6). Shortly after our realization that these factors bound related receptors, Fernando Bazan showed that there was indeed a distant structural relationship between the CNTF and IL6 factors themselves, as well as some other cytokines such as leukemia inhibitory factor (LIF) and oncostatin M (OSM), strengthening the idea that these might all be considered related cytokines using related receptor systems. At that time, Tadamitsu Kishimoto's group had found that while the IL6 receptor component homologous to the CNTF receptor component we had cloned could bind IL6 directly, it needed to interact with a second receptor component, known as gp130, to initiate signaling; nothing was really known about how signal initiation occurred upon engagement of gp130. Since the CNTF receptor component we had cloned, which we began referring to as an "alpha" component, had no cytoplasmic domain because it was attached to the membrane via a glycosyl phosphatidylinositol (GPI) linkage, we reasoned it too would need a "beta" component like gp130 if it were to signal (Davis et al., 1991). We went on to try to biochemically identify a "beta" component for CNTF, and found not one but two - the first being gp130 and the second a gp130-related component previously cloned by David Gearing as a binding protein for yet another CNTF/IL6-related cytokine, LIF (Ip et al., 1992; Stahl et al., 1993).

The realization that CNTF used two "beta" components was a critical one - it led us to the notion that both CNTF and IL6 might both use two betas - in the IL6 case two gp130s, and in the CNTF case gp130 with what we began referring to as LIFRbeta - and that beta dimerization might be critical for initiating signaling in both cases (Ip et al., 1992). We in fact proposed "uniform" receptor complexes for all members of the expanding family of cytokines in the CNTF and IL6 family, which now includes LIF, OSM, cardiotrophin-1 (CT1) and interleukin-11 (IL11) - in which all contained "beta" dimers - either homodimers or heterodimers (Ip et al., 1992). Since the "beta" components were very widespread and did not bind to the factors directly on their own, we reasoned that the role of the "alpha" components was to specify which cells could respond to a particular cytokine by binding to the cytokine and allowing it to activate via the "betas" (Ip et al., 1992); thus CNTF was a neurotrophic factor as opposed to a more generally acting hemopoietic cytokine like IL6 only because its alpha component was largely restricted to neurons in its expression (Ip et al., 1993b). These general models have now been largely verified by our group as well as by Kishimoto's group and others. A critical feature of realizing that all these cytokines used "beta" dimers was that it allowed for an explanation of how these ligands initiated signaling. We found that all the receptor components are initially unassembled on the cell surface, and ligand serves to bring them together in step-wise fashion, and in the process activates signaling (Davis et al., 1993). We now know that although the beta components, unlike RTKs, have no intrinsic kinase activity, they are instead constitutively pre-associated with members of the Jak family of kinases in an inactive form (Stahl et al., 1994). Ligand first binds "alpha", then recruits a single "beta" which forms a still inactive intermediate (Davis et al., 1993). What then follows is "beta" dimerization, which seems to activate the associated Jaks in much the same way RTKs seem to be activated by ligand-mediated dimerization, resulting in autophosphorylation of the Jaks as well as tyrosine phosphorylation of the "beta" components themselves (Davis et al., 1993; Stahl et al., 1994).

While the realization that this family of cytokines used members of the Jak family of kinases was very important, which we did in collaboration with Jim

Ihle's group, it raised a major paradox. It turns out that while this family could use some of the very same Jaks as other cytokine families, it could also activate a distinct (albeit overlapping) set of intracellular substrates as compared to these other cytokines - for example, while interferon gamma could activate the same Jaks as the CNTF/IL6 cytokines in particular cells, only the CNTF/IL6 cytokines activated the downstream PTP1D and STAT3 substrates (Boutlon et al., 1994; Boulton et al., 1995). This led us to realize that the Jaks were rather generic kinases, and that they were actually being directed as to what to phosphorylate by tyrosine-phosphorylated motifs in the "beta" components (Stahl et al., 1995). We were able to do a mutational analysis of the receptors which identified the docking sites for various substrates (Stahl et al., 1995). Then, to unequivocably demonstrate that these were true specification sites that could direct the Jaks as to which substrates to choose, we transplanted a site from the CNTF/IL6 "beta" components that seemed to specify STAT3 activation into another cytokine receptor - the erythropoietin receptor - which normally does not activate this STAT, and showed that this engineered erythropoietin receptor did indeed gain the ability to specifically activate STAT3 (Stahl et al., 1995)!

The receptor and signaling systems used by the CNTF/IL6 family of cytokines have provided a good model for other multi-subunit cytokine receptors used by other cytokine families (Stahl and Yancopoulos 1993; Stahl and Yancopoulos, 1994). Furthermore, identifying and understanding the CNTF receptor has led indirectly to the identification of the long-sought after muscle-derived trophic factor for motor neurons. CNTF itself was a promising candidate to be such a factor based on its dramatic survival actions on motor neurons in vitro and in vivo. But, just as with the neurotrophins, disruption of the CNTF gene revealed that it was not the long-sought after motoneurotrophic factor. Not only did mice engineered by Hans Thoenen's group to lack CNTF develop normally with rather insignificant effects on motor neurons, but it was subsequently realized that about 3% of the Japanese population is null for CNTF, and yet these individuals are apparently totally normal neurologically. Thus these studies largely eliminated consideration of CNTF as the critical and long-sought after motor neuron trophic factor. We were still, however, quite struck by the notable expression of the CNTF receptor alpha component in the embryonic nervous system and in particularly in developing motor neurons (Ip et al., 1993b). Thus we decided to disrupt the CNTF receptor alpha gene in mice, reasoning it might be recognizing yet another heretofore unknown member of the CNTF family. As it turned out, mice lacking CNTF receptor alpha were - unlike mice lacking CNTF - drastically abnormal (DeChiara et al., 1995). The mice died on the day of birth, and exhibited dramatic weakness, lack of suckling and inability to feed as revealed here by absence of milk in their stomachs (DeChiara et al., 1995). Underlying these abnormalities appeared to be severe losses of motor neurons in every motor neuron population examined, both in the spinal cord and in cranial motor nuclei (DeChiara et al., 1995). Thus the CNTF receptor alpha mutant mouse phenotype provides very strong genetic evidence for the existence of a second ligand using the CNTF receptor alpha component, and it appears that this "CNTF2" is much more critical developmentally than CNTF, particularly for all populations of motor neurons, and thus appears to correspond to the long-sought after motor neuron trophic factor (DeChiara et al., 1995).

Trophic Factors from Motor Neurons Acting on Muscle

We reasoned that if motor neurons released trophic factors critical for skeletal muscle, receptors for these factors might be specifically expressed on the skeletal muscle, just as receptors for the trophic factors acting on motor neurons are specifically expressed on the motor neurons themselves. We thus set out, and succeeded in cloning, a receptor tyrosine kinase specific for the skeletal muscle lineage (Valenzuela et al., 1995b). This receptor, termed MuSK for Muscle Specific Kinase, is almost absolutely specific to developing muscle during embryogenesis (Valenzuela et al., 1995b). By birth, MuSk is almost exclusively localized to the muscle-side of the neuromuscular junction (NMJ) specialization, making it a promising candidate to be mediating important nerve-muscle interactions. Verification of its importance comes with generation of mice homozygously disrupted for the gene encoding the MuSK receptor: these mice die at birth - apparently from inability to breathe (DeChiara, Yancopoulos et al., in preparation). Underlying this defect, mice lacking MuSK have abnormal NMJs, and notably lack AchR clusters at their NMJs (DeChiara, Yancopoulos et al., in preparation). As mentioned above, "agrin" is a molecule released by nerve that normally promotes AchR clustering on cultured myotubes in vitro. However, in myotubes from mice lacking MuSK, agrin cannot promote AchR clustering, demonstrating that MuSK is required for the organization signal provided by agrin - presumably not only in vitro, but in vivo as well (DeChiara, Yancopoulos et al., in preparation).

In addition to its role in regulating NMJ formation, MuSK may also be required for reception of trophic signals from nerve. Consistent with such a possibility, we find that in all situations in which muscle is compromised and usually atrophies - such as following nerve injury or due to disuse (as when a limb is immobilized with a cast), MuSK expression soars (Valenzuela et al., 1995b). This up-regulation results in MuSK being expressed throughout the muscle fibers, not only at the NMJ as it is normally expressed (Valenzuela et al., 1995b). These situations in which MuSK receptors are up-regulated are situations in which, clinically, it would be very useful to treat with a factor that might prevent the associated muscle atrophy. Ligands that activate the up-regulated MuSK would seem to be ideal candidates to try in these situations - for even if such actions are not normally physiologically relevant for MuSK, MuSK ligands might still be able to act in this manner pharmacologically, in much the same way that BDNF, NT4 and CNTF all appear capable of rescueing motor neurons in clinical relevant situations, although none of these factors may normally be involved in regulating motor neuron survival.

Conclusions

I have reviewed our recent work attempting to understand the factors and mechanisms which mediate the reciprocal trophic interactions between nerve and muscle. Most importantly, gene disruption and other approaches have allowed us to evaluate trophic factors and receptors (e.g. BDNF, NT4, the EPH family ligands, CNTF, the CNTF receptor system, and MuSK) we have isolated as candidates, and to determine which ones are or are not critical mediators of the reciprocal trophic interactions between nerve and muscle.

References

Boulton, T.G., Stahl, N. and Yancopoulos, G.D. (1994) The CNTF/LIF/IL6/OSM family of cytokines induces tyrosine phosphorylation of a common set of proteins overlapping those induced by other cytokines and growth factors. J. Biol. Chem., 269: 11648-11655.

Boulton, T.G., Zhong, Z., Wen, Z., Darnell, J.E., Stahl, N. and Yancopoulos, G.D. (1995) STAT3 Activation by Cytokines Using GP130 and Related Transducers Involves a Secondary Modification Requiring an H7-Sensitive Kinase. Proc. Natl. Acad. Sci., USA, 92: 6915-6919.

Conover, J.C., Erickson, J.T., Katz, D.M., Bianchi, L.M., Pouyemirou, W.T., McClain, J., Pan, L., Helgren, M., Ip, N.Y., Boland, P., Friedman, B., Weigand, S., Vejsada, R., Kato, A., DeChiara, T.M. and Yancopoulos, G.D. (1995) Neuronal Deficits, Not Involving Motor Neurons, In Mice Lacking BDNF and/or NT4. Nature, 375: 235-238.

Davis, S., Aldrich, T.H., Valenzuela, D., Wong, V., Furth, M.E., Squinto, S.P. and Yancopoulos, G.D. (1991) The Receptor for Ciliary Neurotrophic Factor. Science, 253: 59-63.

Davis, S., Aldrich, T.H., Stahl, N., Pan, L., Taga, T., Kishimoto, T., Ip, N.Y. and Yancopoulos, G.D. (1993) LIFRß and gp130 as heterodimerizing signal transducers for the tripartite CNTF receptor. Science, 260: 1805-1808.

Davis, S., Gale, N.W. , Aldrich, T.H., Maisonpierre, P.C., Lhotak, V., Pawson, T. Goldfarb, M. and Yancopoulos, G.D. (1994) Ligands for EPH-Related Receptor Tyrosine Kinases That Require Membrane-Attachment or Clustering for Activity. Science, 266: 816-819.

DeChiara, T.M., Vejsada, R., Pouyemirou, W.T., Acheson, A., McClain, J., Pan, L., Ip, N.Y., Stahl, N., Suri, C., Conover, J.C., Kato, A. and Yancopoulos, G.D. (1995) Mice lacking the CNTF receptor, unlike mice lacking CNTF, exhibit profound motor neuron deficits at birth. Cell, 83: 313-322.

Glass, D.J., Nye, S.H., Hantzopoulos, P., Macchi, M.J., Squinto, S.P., Goldfarb, M. and Yancopoulos, G.D. (1991) TrkB mediates BDNF/NT-3-dependent survival and proliferation in fibroblasts lacking the low-affinity NGF receptor. Cell, 66: 405-413.

Glass, D. and Yancopoulos, G.D. (1993) The neurotrophins and their receptors. Trends in Cell Biology, 3: 262-268.

Ip, N.Y., Nye, S.H., Boulton, T.G., Davis, S., Taga, T., Li, Y., Birren, S.J., Yasukawa, K., Kishimoto, T., Anderson, D.J.,Stahl, N., and Yancopoulos, G.D. (1992) CNTF and LIF Act on Neuronal Cells via Shared Signaling Pathways that Involve the IL-6 Signal Transducing Receptor Component gp130. Cell,69: 1121-1132.

Ip, N.Y., Stitt, T. N., Tapley, P., Klein, R., Glass, D.J., Fandl, J., Greene, L.A., Barbacid, M.,and Yancopoulos, G.D. (1993a) Similarities and differences in the way neurotrophins interact with the Trk receptors in neuronal and non-neuronal cells. Neuron, 10:137-149.

Ip, N.Y., McClain, J., Barrezueta, N.X., Aldrich, T.H., Pan, L., Li, Y., Wiegand, S.J., Friedman, B., Davis, S. and Yancopoulos, G.D. (1993b) The α chain of the CNTF receptor is required for signalling and defines potential CNTF targets in the Adult and During Development. Neuron, 10: 89-102.

Ip, N.Y. and Yancopoulos, G.D. (1994) Neurotrophic Factor Receptors: Just like other growth factor and cytokine receptors? Current Opinions in Neurobiology, 4: 400-405.

Maisonpierre, P., Barrezueta, N.X. and Yancopoulos, G.D. (1993) Ehk-1 and Ehk-2: Two novel members of the Eph receptor-like tyrosine kinase family with distinctive structures and neuronal expression. Oncogene, 8: 3277-3288.

Stahl, N. and Yancopoulos, G.D. (1993) The alphas, betas and kinases of cytokine receptor complexes. Cell, 74: 587-590.

Stahl, N., Davis, S., Wong, V., Taga, T., Kishimoto, T., Ip, N.Y. and Yancopoulos, G.D. (1993) Cross-linking identifies Leukemia Inhibitory Factor-Binding

Protein as a Ciliary Neurotrophic Factor Receptor Component. Journal of Biological Chemistry, 268: 7628-7631.

Stahl, N. and Yancopoulos, G.D. (1994) The tripartite CNTF receptor complex: Activation and signaling involves components shared with other cytokines. J. Neurobiology, 25: 1454-1466.

Stahl, N., Boulton, T.G., Farruggella, T., Ip, N.Y., David, S., Witthuhn, B.A., Quelle, F.W., Silvennoinen, O., Barbieri, G., Pellegrini, S., Ihle, J.N., and Yancopoulos, G.D. (1994) Association and activation of Jak-Tyk kinases by CNTF-LIF-OSM-IL6 ß receptor components. Science, 263: 92-95.

Stahl, N., Farugella, T.J., Boulton, T.G., Zhong, Z., Darnell, J.E. and Yancopoulos, G.D. (1995) Modular Tyrosine-Based Motifs in Cytokine Receptors Specify Choice of STATs and other Substrates. Science, 267: 1349-1353.

Valenzuela, D.M., Rojas, E., Griffiths, J., Compton, D.L., Gisser, M., Ip, N.Y., Goldfarb, M. and Yancopoulos, G.D. (1995a) Identification of Full-Length and Truncated Forms of Ehk-3, A Novel Member of the Eph Receptor Tyrosine Kinase Family. Oncogene, 10, 1573-1580.

Valenzuela, D.M., Stitt, T.N., DiStefano, P.S., Rojas, E., Mattsson, K., Compton, D.L., Nuñez, L., Park, J.S., Stark, J.L., Gies, D.R., Thomas, S., Le Beau, M.M., Fernald, A.A., Copeland, N.G., Jenkins, N.A., Burden, S.J., Glass, D.J., and Yancopoulos, G.D. (1995b) Receptor Tyrosine Kinase Specific for the Skeletal Muscle Lineage: Expression in Embryonic Muscle, at the Neuromuscular Junction, and After Injury. Neuron, 15: 573-584.

SUBJECT INDEX

D

E

F

M

N

O

P

NATO ASI Series H

Vol. 74: Molecular Mechanisms of Membrane Traffic.
Edited by D. J. Morré, K. E. Howell, and J. J. M. Bergeron. 429 pages. 1993.

Vol. 75: Cancer Therapy. Differentiation, Immunomodulation and Angiogenesis.
Edited by N. D'Alessandro, E. Mihich, L. Rausa, H. Tapiero, and T. R. Tritton. 299 pages. 1993.

Vol. 76: Tyrosine Phosphorylation/Dephosphorylation and Downstream Signalling.
Edited by L. M. G. Heilmeyer Jr. 388 pages. 1993.

Vol. 77: Ataxia-Telangiectasia. Edited by R. A. Gatti, R. B. Painter. 306 pages. 1993.

Vol. 78: Toxoplasmosis. Edited by J. E. Smith. 272 pages. 1993.

Vol. 79: Cellular Mechanisms of Sensory Processing. The Somatosensory System.
Edited by L. Urban. 514 pages. 1994.

Vol. 80: Autoimmunity: Experimental Aspects.
Edited by M. Zouali. 318 pages. 1994.

Vol. 81: Plant Molecular Biology. Molecular Genetic Analysis of Plant Development and Metabolism.
Edited by G. Coruzzi, P. Puigdomènech. 579 pages. 1994.

Vol. 82: Biological Membranes: Structure, Biogenesis and Dynamics.
Edited by Jos A. F. Op den Kamp. 367 pages. 1994.

Vol. 83: Molecular Biology of Mitochondrial Transport Systems.
Edited by M. Forte, M. Colombini. 420 pages. 1994.

Vol. 84: Biomechanics of Active Movement and Division of Cells.
Edited by N. Akkaş. 587 pages. 1994.

Vol. 85: Cellular and Molecular Effects of Mineral and Synthetic Dusts and Fibres.
Edited by J. M. G. Davis, M.-C. Jaurand. 448 pages. 1994.

Vol. 86: Biochemical and Cellular Mechanisms of Stress Tolerance in Plants.
Edited by J. H. Cherry. 616 pages. 1994.

Vol. 87: NMR of Biological Macromolecules.
Edited by C. I. Stassinopoulou. 616 pages. 1994.

Vol. 88: Liver Carcinogenesis. The Molecular Pathways.
Edited by G. G. Skouteris. 502 pages. 1994.

Vol. 89: Molecular and Cellular Mechanisms of H^+ Transport.
Edited by B. H. Hirst. 504 pages. 1994.

Vol. 90: Molecular Aspects of Oxidative Drug Metabolizing Enzymes: Their Significance in Environmental Toxicology, Chemical Carcinogenesis and Health.
Edited by E. Arınç, J. B. Schenkman, E. Hodgson. 623 pages. 1994.

NATO ASI Series H

Vol. 91: Trafficking of Intracellular Membranes:
From Molecular Sorting to Membrane Fusion.
Edited by M. C. Pedroso de Lima, N. Düzgüneş, D. Hoekstra.
371 pages. 1995.

Vol. 92: Signalling Mechanisms – from Transcription Factors to Oxidative Stress.
Edited by L. Packer, K. Wirtz. 466 pages. 1995.

Vol. 93: Modulation of Cellular Responses in Toxicity.
Edited by C. L. Galli, A. M. Goldberg, M. Marinovich.
379 pages. 1995.

Vol. 94: Gene Technology.
Edited by A. R. Zander, W. Ostertag, B. V. Afanasiev, F. Grosveld.
556 pages. 1996.

Vol. 95: Flow and Image Cytometry.
Edited by A. Jacquemin-Sablon. 241 pages. 1996.

Vol. 96: Molecular Dynamics of Biomembranes.
Edited by J. A. F. Op den Kamp. 424 pages. 1996.

Vol. 97: Post-transcriptional Control of Gene Expression.
Edited by O. Resnekov, A. von Gabain. 283 pages. 1996.

Vol. 98: Lactic Acid Bacteria: Current Advances in Metabolism, Genetics and Applications.
Edited by T. F. Bozoğlu, B. Ray. 412 pages. 1996.

Vol. 99: Tumor Biology
Regulation of Cell Growth, Differentiation and Genetics in Cancer.
Edited by A. S. Tsiftsoglou, A. C. Sartorelli, D. E. Housman, T. Michael Dexter. 342 pages. 1996.

MIX
Papier aus verantwortungsvollen Quellen
Paper from responsible sources
FSC® C105338

If you have any concerns about our products,
you can contact us on
ProductSafety@springernature.com

In case Publisher is established outside the EU,
the EU authorized representative is:
Springer Nature Customer Service Center GmbH
Europaplatz 3, 69115 Heidelberg, Germany

Printed by Libri Plureos GmbH
in Hamburg, Germany